Lecture Notes in Mathematics　　　2098

For further volumes:
http://www.springer.com/series/304

Habib Ammari • Josselin Garnier • Wenjia Jing •
Hyeonbae Kang • Mikyoung Lim • Knut Sølna •
Han Wang

Mathematical and Statistical Methods for Multistatic Imaging

 Springer

Habib Ammari
Wenjia Jing
Han Wang
Department of Mathematics
   and Applications
École Normale Supérieure
Paris, France

Josselin Garnier
Laboratory of Probability and Random Mod
University Paris VII
Paris, France

Hyeonbae Kang
Department of Mathematics
Inha University
Incheon, Korea

Mikyoung Lim
Department of Mathematical Sciences
Korean Advanced Institute of Science
   and Technology (KASIT)
Daejeon, Korea

Knut Sølna
Department of Mathematics
University of California
Irvine School of Physical Sciences
Irvine, CA, USA

ISBN 978-3-319-02584-1   ISBN 978-3-319-02585-8 (eBook)
DOI 10.1007/978-3-319-02585-8
Springer Cham Heidelberg New York Dordrecht London

Lecture Notes in Mathematics ISSN print edition: 0075-8434
                         ISSN electronic edition: 1617-9692

Library of Congress Control Number: 2013953715

Mathematics Subject Classification (2010): 35R30, 35B30

Printed on acid-free paper

Springer is part of Springer Science+Business Media (www.springer.com)

Introduction

In multistatic imaging one uses waves to probe for information about an unknown medium. These waves can be acoustic, elastic, or electromagnetic. They can be at zero-frequency and consequently modeled by the conductivity equation or at nonzero-frequency and hence modeled by the Helmholtz equation. They are generated by an array of transmitters and recorded by an array of receivers (transducers in acoustics, seismographs in geophysics, or antennas in electromagnetics).

Multistatic imaging usually involves two steps. The first step is experimental. It consists in recording the waves generated by sources on an array of receivers. The second step is numerical. It consists in processing the recorded data in order to estimate some relevant features of the medium (source or reflector locations and shapes).

This book covers recent mathematical, numerical, and statistical approaches for multistatic imaging of targets with waves at single or multiple frequencies. The waves are generated by point sources on a transmitter array and measured on a receiver array. For the sake of simplicity, we consider coincident transmitter and receiver arrays. There are two interesting problems: one is finding small targets and the other is reconstructing shape deformations of an extended target. A target is called small when its characteristic size times the operating frequency is less than one while it is called extended when this factor is much larger than one. In both situations, we are interested in imaging small perturbations with respect to known situations.

Our approach is based on an asymptotic analysis of the measured data in terms of the size of the unknown targets or the order of magnitude of the shape deformation. The asymptotic analysis plays a key role in characterizing all the information about the small target or the shape deformations of an extended target that can be stably reconstructed from the measured data. It provides robust and accurate reconstruction of the location and some geometric features of the small targets as well as small changes of the shape of an extended target, even with moderately noisy data.

When we are dealing with target imaging problems, a fundamental problem is to have a shape representation well suited for solving the inverse problem. Parametric representations do not do a good job. In fact, the data have highly nonlinear dependence with respect to parametric representations. However, high-order polarization tensors can be stably reconstructed from the data by solving a least-squares problem. Therefore, they provide a well-suited representation of the target. Moreover, they capture both the high-frequency information in its shape and its topology. The high-order polarization tensors generalize the (classical) magnetic and electric polarization tensors associated to a small target with a given set of electromagnetic parameters. For an arbitrary shape, we can find an equivalent ellipse (or ellipsoid) with the same first-order polarization tensor. Using high-order polarization tensors we can not only recover finer details of the shape of a given target but also separate its electromagnetic parameters from its volume. In this book, we derive an optimization procedure to reconstruct a target from its high-order polarization tensors. We also present a dictionary matching technique based on new invariants for the generalized polarization tensors. For extended targets, a concept equivalent to the polarization tensor can be introduced and direct algorithms can be designed for reconstructing small shape changes.

The main applications that we have in mind are medical imaging (such as microwave and electrical impedance breast cancer detections), airport security screening, geophysical exploration, and nondestructive testing. For such applications, the general purpose of multistatic imaging is, from imperfect information (rough forward models, limited and noisy data), to estimate parts of the unknown structure that is of interest. In this book we consider, in the presence of noise, the detection and localization of sources, reflectors, and the reconstruction of small inclusions and shape deformations.

An important problem in multistatic imaging is to quantify and understand the trade-offs between data size, computational complexity, signal-to-noise ratio, and resolution. For instance, in geophysics, very large amount of data are collected and the computational complexity of the imaging algorithm is a limiting factor. In this book we carefully address the trade-off between resolution and stability when the data are noisy. We provide imaging algorithms and analyze their resolution and stability with respect to noise in the measurements. Resolution analysis is to estimate the size of the finest detail that can be reconstructed from the data while stability analysis is to quantify the localization error in the presence of noise. The noise models discussed in this book are measurement and medium (or clutter) noises. They affect the stability and resolution of the imaging functionals in very different ways.

The book is organized as follows. Chapter 1 reviews some of the fundamental mathematical and statistical concepts that are key to understanding imaging principles. Chapter 2 collects some preliminary results regarding layer potentials. This chapter offers a comprehensive treatment of the subject of integral equations and provides key identities for solving imaging problems. Chapter 3 covers the method of small volume expansions. It provides the leading-order term in the asymptotic expansion of the solution to the conductivity or Helmholtz equation with respect to the size of a small inclusion. In Chap. 4 we introduce the concept of high-order polarization tensors (also called generalized polarization tensors) associated with a conductivity inclusion and present their main properties. Chapter 5 is devoted to the frequency-dependent polarization tensors associated with an electromagnetic (or acoustic) inclusion. We introduce the notion of scattering coefficients and prove some of their properties. Scattering coefficients can be obtained from far-field measurements and are the Fourier coefficients of the scattering amplitude. Chapter 6 deals with the structure of the multistatic response matrix. It introduces a Hadamard technique for noise reduction and provides statistical distributions of significant singular values of the multistatic response matrices associated with point reflectors and inclusions. In Chap. 7, using multipolar expansions for the solutions to the conductivity and Helmholtz equations, we analyze the structure of the corresponding multistatic response matrices. Based on the method of small volume expansions we derive in Chap. 8 localization techniques for inclusions in the continuum approximation that take advantage of the smallness of the inclusions. Direct algorithms for imaging small conductivity and electromagnetic inclusions are introduced and their stability with respect to medium and measurement noises as well as their resolution is investigated. Chapter 9 outlines detection and localization techniques from noisy multistatic measurements. The results of Chap. 6 on the statistical properties of the multistatic response matrix in the presence of noise are used to design detection tests. The detection test is to decide whether a point reflector is present or not. An extension of Berens' modeling for point reflector detection is given. Chapter 10 deals with the reconstruction of the generalized polarization tensors from multistatic response measurements. A stability analysis for the reconstruction in the presence of measurement noise which quantifies the ill-posedness of the reconstruction problem is provided. Chapter 11 is devoted to target identification from multistatic data using generalized polarization tensors. It provides a fast and efficient procedure for target identification in multistatic imaging based on matching on a dictionary of precomputed generalized polarization tensors. The approach is based on the use of invariants for the generalized polarization tensors. Chapter 11 also applies an extended Kalman filter to track both the location and the orientation of a mobile target from multistatic measurements. Chapters 12–14 discuss multistatic imaging techniques for extended targets. We start with inverse source problems and introduce time reversal techniques. Then we focus on reconstructing shape changes of an extended target. We introduce several algorithms and analyze their resolution and stability for the linearized reconstruction problem. Finally, we describe optimal control approaches for solving the nonlinear problem.

Chapters 15 and 16 present results on electromagnetic invisibility. Electromagnetic invisibility is to make a target invisible for electromagnetic probing. Many schemes are under active current investigations. These include active cloaking, transmission line cloaking, interior cloaking, and exterior cloaking. Chapter 15 focusses on interior cloaking while Chap. 16 is devoted to exterior cloaking.

The main tool to obtain interior cloaking, where the target to be cloaked is inside the cloaking structure, is to use a change of variables scheme (also called transformation optics). The change of variables-based cloaking method uses a singular transformation to boost the material property so that it makes a cloaking region look like a point to outside measurements. However, this transformation induces the singularity of material constants in the transversal direction (also in the tangential direction in two dimensions), which causes difficulty both in the theory and in the applications. To overcome this weakness, the so-called near-cloaking is naturally considered, which is a regularization or an approximation of singular cloaking. Instead of the singular transformation, one can use a regular one to push forward the material constant in the conductivity equation, in which a small ball is blown up to the cloaking region. The aim of Chap. 15 is to discuss recent advances in near-cloaking. We first provide a method of constructing effective near-cloaking structures for the conductivity problem. These new structures are such that their first generalized polarization tensors vanish. We show that this in particular significantly enhances the cloaking effect. Then we extend this method to scattering problems. We construct very effective near-cloaking structures for the scattering problem at a fixed frequency. These new structures are, before using the transformation optics, layered structures and are designed so that their first scattering coefficients approximately vanish. Inside the cloaking region, any target has near-zero scattering cross section for a band of frequencies. As for the conductivity problem, we analytically show that this new construction significantly enhances the cloaking effect for the Helmholtz equation. Chapter 16 aims to give a mathematical justification of exterior cloaking due to anomalous localized resonance. We consider the dielectric problem with a source term in a structure with a layer of plasmonic material. The real part of the permittivity inside the plasmonic layer is negative. In the case of concentric disk structure, we show that for any source supported outside a critical radius cloaking does not take place, and for sources located inside the critical radius satisfying certain conditions cloaking does take place as the loss parameter inside the metamaterial layer goes to zero.

The last part of this book provides the reader with practical implementations and performance evaluations of the described imaging methods and techniques.

Chapter 17 provides MATLAB codes for the main algorithms described in this book. Chapter 18 presents numerical illustrations using these codes in order to highlight the performance and show the limitations of our numerical approaches for multistatic imaging.

The bibliography provides a list of relevant references. It is by no means comprehensive. However, it should provide the reader with some useful guidance in searching for further details on the main ideas and approaches discussed in this book.

The book has grown out of lecture notes for a summer school on mathematical and statistical methods in imaging at the Institute of Computational Mathematics of Chinese Academy of Sciences in Beijing. It was also taught through a series of intensive lectures at the Korean Advanced Institute of Science and Technology. We are very grateful to the organizers of these two events. Some of the material in this book is from our wonderful collaborations with Elie Bretin, Thomas Boulier, Giulio Ciraolo, Pierre Garapon, Vincent Jugnon, Hyundae Lee, Graeme Milton, Abdul Wahab, Sanghyeon Yu, and Habib Zribi. We feel indebted to all of them. We would also like to acknowledge the support of the European Research Council Project MULTIMOD and of Korean ministry of education, science, and technology through grant NRF 2010-0017532. M. Lim was supported by TJ Park Junior Faculty Fellowship.

Paris, France   Habib Ammari, Josselin Garnier
                                                  Wenjia Jing, Han Wang
Incheon, Korea   Hyeonbae Kang
Daejeon, Korea   Mikyoung Lim
Irvine, CA   Knut, Sølna

Contents

Part IV   Localization and Detection Algorithms

Part V   Dictionary Matching and Tracking Algorithms

Part I
Mathematical and Probabilistic Tools

Chapter 1
Preliminaries

This chapter reviews some mathematical and statistical concepts essential for understanding multistatic imaging principles. We first review commonly used special functions, function spaces, and an integral transform: the Fourier transform. We then collect basic facts about the Moore-Penrose generalized inverse, singular value decomposition, and compact operators. The theory of regularization of ill-posed inverse problems is briefly discussed. Then we introduce useful probabilistic tools for imaging in the presence of noise. In particular, we review results on the statistics of the singular values of a random matrix. Such results will be of help to us when dealing with inclusion detection tests. Finally, we examine image characteristics with respect to various data acquisition and processing schemes. We focus specifically on issues related to image resolution, signal-to-noise ratio, and image artifacts.

1.1 Sobolev Spaces

The following Sobolev spaces are needed for the study of mapping properties of layer potentials in Chap. 2.

For ease of notation we will sometimes use ∂ and ∂^2 to denote the gradient and the Hessian, respectively.

Let $D \subset \mathbb{R}^d$ be a bounded smooth domain. We define the Banach spaces $W^{1,p}(D), 1 \leq p < +\infty$, by

$$W^{1,p}(D) = \left\{ u \in L^p(D) : \int_D |u|^p + \int_D |\nabla u|^p < +\infty \right\},$$

where ∇u is interpreted as a distribution, and $L^p(D)$ is defined in the usual way, with the norm

H. Ammari et al., *Mathematical and Statistical Methods for Multistatic Imaging*,
Lecture Notes in Mathematics 2098, DOI 10.1007/978-3-319-02585-8_1,
© Springer International Publishing Switzerland 2013

$$\|u\|_{L^p(D)} = \left(\int_D |u|^p\right)^{1/p}.$$

The space $W^{1,p}(D)$ is equipped with the norm

$$\|u\|_{W^{1,p}(D)} = \left(\int_D |u|^p + \int_D |\nabla u|^p\right)^{1/p}.$$

Another Banach space $W_0^{1,p}(D)$ arises by taking the closure of $\mathcal{C}_0^\infty(D)$, the set of infinitely differentiable functions with compact support in D, in $W^{1,p}(D)$. The spaces $W^{1,p}(D)$ and $W_0^{1,p}(D)$ do not coincide for bounded D. The case $p = 2$ is special, since the spaces $W^{1,2}(D)$ and $W_0^{1,2}(D)$ are Hilbert spaces under the scalar product

$$(u, v) = \int_D u\,v + \int_D \nabla u \cdot \nabla v.$$

We will also need the space $W_{\mathrm{loc}}^{1,2}(\mathbb{R}^d \setminus \overline{D})$ of functions $u \in L_{\mathrm{loc}}^2(\mathbb{R}^d \setminus \overline{D})$, the set of locally square summable functions in $\mathbb{R}^d \setminus \overline{D}$, such that

$$hu \in W^{1,2}(\mathbb{R}^d \setminus \overline{D}), \forall\, h \in \mathcal{C}_0^\infty(\mathbb{R}^d \setminus \overline{D}).$$

Further, we define $W^{2,2}(D)$ as the space of functions $u \in W^{1,2}(D)$ such that $\partial^2 u \in L^2(D)$ and the space $W^{3/2,2}(D)$ as the interpolation space $[W^{1,2}(D), W^{2,2}(D)]_{1/2}$; see, for example, the book by Bergh and Löfström [49].

It is known that the trace operator $u \mapsto u|_{\partial D}$ is a bounded linear surjective operator from $W^{1,2}(D)$ into $W_{1/2}^2(\partial D)$, where $f \in W_{1/2}^2(\partial D)$ if and only if $f \in L^2(\partial D)$ and

$$\int_{\partial D} \int_{\partial D} \frac{|f(x) - f(y)|^2}{|x - y|^d}\, d\sigma(x)\, d\sigma(y) < +\infty.$$

We set $W_{-1/2}^2(\partial D) = (W_{1/2}^2(\partial D))^*$ and let $\langle, \rangle_{1/2, -1/2}$ denote the duality pair between these dual spaces.

Finally, let $\{\tau_1, \dots, \tau_{d-1}\}$ be an orthonormal basis for the tangent plane to ∂D at x and let

$$\partial/\partial \tau = \sum_{p=1}^{d-1} (\partial/\partial \tau_p)\, \tau_p$$

denote the tangential derivative on ∂D. We say that $f \in W_1^2(\partial D)$ if $f \in L^2(\partial D)$ and $\partial f/\partial \tau \in L^2(\partial D)$.

1.2 Fourier Analysis

1.2.1 Fourier Transform

The Fourier transform plays an important role in imaging and in the analysis of waves. In both cases, the notion of frequency content of a signal is important.

For $f \in L^1(\mathbb{R}^d)$, the Fourier transform $\mathcal{F}(f)$ and the inverse Fourier transform $\mathcal{F}^{-1}(f)$ are defined by

$$\mathcal{F}(f)(\xi) \quad = (2\pi)^{-d/2} \int_{\mathbb{R}^d} e^{-ix\cdot\xi} f(x) \, dx \,,$$

$$\mathcal{F}^{-1}(f)(\xi) = (2\pi)^{-d/2} \int_{\mathbb{R}^d} e^{ix\cdot\xi} f(x) \, dx \,.$$

We use both transforms for other classes of functions, such as for functions in $L^2(\mathbb{R}^d)$ and even for the tempered distributions $\mathcal{S}'(\mathbb{R}^d)$, the dual of the Schwartz space of rapidly decreasing functions:

$$\mathcal{S}(\mathbb{R}^d) = \left\{ u \in \mathcal{C}^\infty(\mathbb{R}^d) : x^\beta \partial^\alpha u \in L^\infty(\mathbb{R}^d) \text{ for all } \alpha, \beta \geq 0 \right\},$$

where $x^\beta = x_1^{\beta_1} \ldots x_d^{\beta_d}, \partial^\alpha = \partial_1^{\alpha_1} \ldots \partial_d^{\alpha_d}$, with $\partial_j = \partial/\partial x_j$.

We list a few properties of the Fourier transform. It is easy to verify that $\mathcal{F} : \mathcal{S}(\mathbb{R}^d) \to \mathcal{S}(\mathbb{R}^d)$ and

$$i^{|\alpha|} \xi^\alpha \partial_\xi^\beta \mathcal{F}(f)(\xi) = (-i)^{|\beta|} \mathcal{F}(\partial^\alpha (x^\beta f))(\xi) \,.$$

If $f_r(x) = f(rx), r > 0$, we have

$$\mathcal{F}(f_r)(\xi) = r^{-d} \mathcal{F}(f)(r^{-1}\xi) \,.$$

Likewise, if $f_y(x) = f(x + y)$ for $y \in \mathbb{R}^d$, then

$$\mathcal{F}(f_y)(\xi) = e^{i\xi\cdot y} \mathcal{F}(f)(\xi) \,.$$

We have the inversion formula: $\mathcal{F}\mathcal{F}^{-1} = \mathcal{F}^{-1}\mathcal{F} = I$ on both $\mathcal{S}(\mathbb{R}^d)$ and $\mathcal{S}'(\mathbb{R}^d)$. If $f \in L^2(\mathbb{R}^d)$, then $\mathcal{F}(f) \in L^2(\mathbb{R}^d)$, too. Plancherel's theorem says that $\mathcal{F} : L^2(\mathbb{R}^d) \to L^2(\mathbb{R}^d)$ is unitary, so that $\mathcal{F}^{-1}$ is the adjoint.

In general, if $f, g \in L^2(\mathbb{R}^d)$, then we have Parseval's relation:

$$\int_{\mathbb{R}^d} \mathcal{F}(f)g \, dx = \int_{\mathbb{R}^d} f\mathcal{F}(g) \, dx \,. \tag{1.1}$$

Since $\mathcal{F}^{-1}(\overline{f}) = \overline{\mathcal{F}(f)}$, this relation has its counterpart for $\mathcal{F}^{-1}$. This indeed also gives

$$\int_{\mathbb{R}^d} |f|^2\, dx = \int_{\mathbb{R}^d} |\mathcal{F}(f)|^2\, d\xi\ .$$

We now make some comments on the relation between the Fourier transform and convolutions. For $f \in \mathcal{S}'(\mathbb{R}^d)$, $g \in \mathcal{S}(\mathbb{R}^d)$, the convolution is defined by

$$(f \star g)(x) = \int_{\mathbb{R}^d} f(x-y)g(y)\, dy\ ,$$

and we have

$$\mathcal{F}(f \star g) = (2\pi)^{d/2}\mathcal{F}(f)\mathcal{F}(g), \quad \mathcal{F}(fg) = (2\pi)^{-d/2}\mathcal{F}(f) \star \mathcal{F}(g)\ .$$

Moreover, for a real-valued function f, we have

$$\mathcal{F}(f(-x)) = \overline{\mathcal{F}(f)}\ , \tag{1.2}$$

and

$$\mathcal{F}\Big(\int_{\mathbb{R}^d} f(y)g(x+y)\, dy\Big) = (2\pi)^{d/2}\overline{\mathcal{F}(f)}\mathcal{F}(g)\ . \tag{1.3}$$

These simple formulas have important interpretations in imaging. Identity (1.2) expresses the fact that the time reversal operation in the time domain ($x \in \mathbb{R}$ variable) is equivalent to the complex conjugation in the frequency domain (ξ variable). Identity (1.3) shows that the cross correlation of two signals involves a product of the two Fourier transforms in the frequency domain, one of the transform being complex conjugated.

Fourier transforms of a few special functions will be needed. For h a Gaussian function,

$$h(x) := e^{-|x|^2/2}, \quad x \in \mathbb{R}^d\ ,$$

we have

$$\mathcal{F}(h)(\xi) = e^{-|\xi|^2/2}, \quad \xi \in \mathbb{R}^d\ . \tag{1.4}$$

For δ_0 the Dirac function at the origin, i.e., $\delta_0 \in \mathcal{S}'(\mathbb{R}^d)$ and $\delta_0(f) = f(0)$ for $f \in \mathcal{S}(\mathbb{R}^d)$, we have

$$\mathcal{F}(\delta_0) = (2\pi)^{-d/2}\ . \tag{1.5}$$

Another useful result is the classification of distributions supported at a single point. If $f \in \mathcal{S}'(\mathbb{R}^d)$ is supported at $\{0\}$, then there exist an integer n and real numbers a_α such that

$$f = \sum_{|\alpha| \leq n} a_\alpha \partial^\alpha \delta_0 \ .$$

Let $\mathbb{Z}$ denote the set of all integers. The Shah distribution

$$\text{shah}_K = \sum_{l \in \mathbb{Z}^d} \delta_{Kl} \ ,$$

where $\delta_y(f) = f(y)$, has the Fourier transform

$$\mathcal{F}(\text{shah}_{2\pi/K}) = (2\pi)^{-d/2} K^d \text{shah}_K \ .$$

This is Poisson's formula. More generally, we have for $f \in \mathcal{S}(\mathbb{R}^d)$

$$\sum_{l \in \mathbb{Z}^d} \mathcal{F}(f)(\xi - \frac{2\pi l}{K}) = (2\pi)^{-d/2} K^d \sum_{l \in \mathbb{Z}^d} f(Kl) e^{-iK\xi \cdot l} \ . \qquad (1.6)$$

1.2.2 Shannon's Sampling Theorem

We call a function (or distribution) in $\mathbb{R}^d, d \geq 1$, whose Fourier transform vanishes outside $|\xi| \leq K$ band-limited with bandwidth K. Shannon's sampling theorem for $d = 1$ is the following. The reader is referred to [126, p. 41] for a proof.

Theorem 1.1 (Shannon's Sampling Theorem). *Let $f \in L^2(\mathbb{R})$ be band-limited with bandwidth K, and let $0 < \Delta x \leq \pi/K$. Then f is uniquely determined by the values $f(l\Delta x), l \in \mathbb{Z}$. The smallest detail represented by such a function is then of size $2\pi/K$. We also have the explicit formula*

$$f(x) = \sum_{l \in \mathbb{Z}} f\left(\frac{l\pi}{K}\right) \frac{\sin(Kx - l\pi)}{Kx - l\pi} \ . \qquad (1.7)$$

The sampling interval π/K is often imposed by computation or storage constraints. Moreover, if the support of $\mathcal{F}(f)$ is not included in $[-K, K]$, then the interpolation formula (1.7) does not recover f. We give a filtering procedure to reduce the resulting error, known as the aliasing artifact. To apply Shannon's sampling theorem, f is approximated by the closest function

$\tilde{f}$ whose Fourier transform has a support in $[-K, K]$. Plancherel's theorem gives that

$$||f - \tilde{f}||^2 = \int_{-\infty}^{+\infty} |\mathcal{F}(f)(\xi) - \mathcal{F}(\tilde{f})(\xi)|^2 \, d\xi$$

$$= \int_{|\xi|>K} |\mathcal{F}(f)(\xi)|^2 \, d\xi + \int_{|\xi|<K} |\mathcal{F}(f)(\xi) - \mathcal{F}(\tilde{f})(\xi)|^2 \, d\xi \; .$$

The distance is minimal when the second integral is zero and hence

$$\mathcal{F}(\tilde{f})(\xi) = \mathcal{F}(f)(\xi)\chi([-K, K])(\xi) = \sqrt{2\pi}\mathcal{F}(\tilde{\delta}_K)(\xi)\mathcal{F}(f)(\xi) \; ,$$

where $\chi([-K, K])$ is the characteristic function of the interval $[-K, K]$ and

$$\tilde{\delta}_K(x) = \sin(K|x|)/(\pi K|x|) \; ;$$

see (1.18). This corresponds to $\tilde{f} = f \star \tilde{\delta}_K$. The filtering of $f(x)$ by $\tilde{\delta}_K(x)$ removes any frequency larger than K. Since $\mathcal{F}(\tilde{f})$ has a support in $[-K, K]$, the sampling theorem proves that $\tilde{f}$ can be recovered from the samples $\tilde{f}(l\pi/K)$.

In the two-dimensional case, we use the separable extension principle. This not only simplifies the mathematics but also leads to faster numerical algorithms along the rows and columns of images. If $\mathcal{F}(f)$ has a support included in $[-K_1, K_1] \times [-K_2, K_2]$, then the following two-dimensional sampling formula holds:

$$f(x, y) = \sum_{l=(l_1, l_2)\in\mathbb{Z}^2} f\left(\frac{l_1\pi}{K_1}, \frac{l_2\pi}{K_2}\right) \frac{\sin(K_1 x - l_1\pi)}{K_1 x - l_1\pi} \frac{\sin(K_2 y - l_2\pi)}{K_2 y - l_2\pi} \; . \qquad (1.8)$$

If the support of $\mathcal{F}(f)$ is not included in the low-frequency rectangle $[-K_1, K_1] \times [-K_2, K_2]$, then we have to filter f with the low-pass separable filter $\tilde{\delta}_{K_1}(x) \, \tilde{\delta}_{K_2}(y)$.

1.2.3   Fast Fourier Transform

The fast Fourier transform technique (denoted by FFT) is used for the evaluation of the Fourier transform and is often utilized in imaging algorithms. Assume that f vanishes outside $[-K, K]$ and is sampled with stepsize h. Applying the trapezoidal rule to $\mathcal{F}(f)$ leads to the approximation

$$\mathcal{F}(f)(\xi) \approx \frac{1}{\sqrt{2\pi}} h \sum_{n=-N}^{N-1} e^{-i\xi hn} f(hn) \; ,$$

where $N = K/h$. Since $\mathcal{F}(f)$ is band-limited with bandwidth K, $\mathcal{F}(f)$ needs to be sampled with stepsize $\leq \pi/K$. If we choose the coarsest possible stepsize π/K, we have to evaluate

$$\mathcal{F}(f)(m\pi/K) = \frac{1}{\sqrt{2\pi}}h \sum_{n=-N}^{N-1} e^{-i\pi mn/N} f(hn) \tag{1.9}$$

for $m = -N, \ldots, N-1$. Evaluating (1.9) is a discrete Fourier transform of length $2N$ and requires $O(N^2)$ complex multiplications and additions. Any algorithm of lower complexity, usually $O(N \log_2 N)$, is called a fast Fourier transform. The possibility of doing this arises from observing redundancies and reorganizing the calculations. Standard references are [56,141]. We briefly describe the well-known FFT algorithm of Cooley and Tukey for N a power of 2. The basic idea in the Cooley-Tukey algorithm is to break the sum into one part with n even and the rest with n odd. We have

$$\mathcal{F}(f)(2m\pi/K) = \frac{1}{\sqrt{2\pi}}h \sum_{n=-N/2}^{N/2-1} e^{\frac{-i\pi mn}{N/2}} \left(f(hn) + (-1)^m f(h(n - N/2)) \right.$$

$$\left. + (-1)^m f(h(n + N/2)) \right),$$

and

$$\mathcal{F}(f)((2m+1)\pi/K)$$

$$= \frac{1}{\sqrt{2\pi}}h \sum_{n=-N/2}^{N/2-1} e^{\frac{-i\pi mn}{N/2}} \left(f(hn) + (-1)^m e^{i\pi/2} f(h(n - N/2)) \right.$$

$$\left. + (-1)^m e^{-i\pi/2} f(h(n + N/2)) \right) e^{-i\pi n/N}.$$

A discrete Fourier transform of length $2N$ may thus be calculated with two discrete Fourier transforms of size N plus $O(N)$ operations. If this is done in a recursive way, we arrive at $C(N) = O(N \log_2 N)$, where $C(N)$ is the number of elementary operations needed to compute a discrete Fourier transform with the FFT. In fact, we have $C(N) = 2C(N/2) + O(N)$. With the change of variable $l = \log_2 N$ and the change of function $T(l) = C(N)/N$, we derive that $T(l) = T(l-1) + O(1)$. Since $C(1) = 0$, so $T(1) = 0$ and we obtain $T(l) = O(l)$ and in turn $C(N) = O(N \log_2 N)$.

1.3   Special Functions

1.3.1   Bessel Functions

Bessel functions of the first kind of real order ν, denoted by $J_\nu(x)$, are useful for describing some imaging effects. One definition of $J_\nu(x)$ is given in terms of the series representation

$$J_\nu(x) = (\frac{x}{2})^\nu \sum_{l=0}^{+\infty} \frac{(-x^2/4)^l}{l!\Gamma(\nu+l+1)} , \tag{1.10}$$

where the gamma function Γ is defined by

$$\Gamma(z) = \int_0^{+\infty} e^{-t}t^{z-1}\, dt \quad \text{for } \Re e(z) > 0 .$$

Another formula, valid for $\Re e\,\nu > -\frac{1}{2}$, is

$$J_\nu(x) = [\Gamma(\frac{1}{2})\Gamma(\nu+\frac{1}{2})]^{-1}(\frac{x}{2})^\nu \int_{-1}^1 (1-t^2)^{\nu-\frac{1}{2}}e^{ixt}\, dt . \tag{1.11}$$

Some useful identities for Bessel functions are summarized below. For further details, we refer the reader to [153, pp. 225–233].

We have the recurrence relation

$$(\frac{d}{dx} + \frac{\nu}{x})J_\nu(x) = J_{\nu-1}(x) . \tag{1.12}$$

For $n \in \mathbb{Z}$, we have the integral representation

$$J_n(x) = \frac{1}{2\pi} \int_{-\pi}^{\pi} e^{ix\sin\phi - in\phi}d\phi , \tag{1.13}$$

i.e., the functions $J_n(x)$ are the Fourier coefficients of $e^{ix\sin\phi}$. Therefore, the Jacobi-Anger expansion holds:

$$e^{ix\sin\phi} = \sum_{n\in\mathbb{Z}} J_n(x)e^{in\phi} . \tag{1.14}$$

Formula (1.14) can be used in two dimensions to expand a plane wave as a sum of cylindrical waves. We have

$$e^{i\xi\cdot x} = \sum_{n\in\mathbb{Z}} e^{in(\frac{\pi}{2}-\theta_\xi)} J_n(|\xi||x|)e^{in\theta_x} , \tag{1.15}$$

Fig. 1.1 Plots of Bessel functions $J_n(x), n = 0, \ldots, 5$

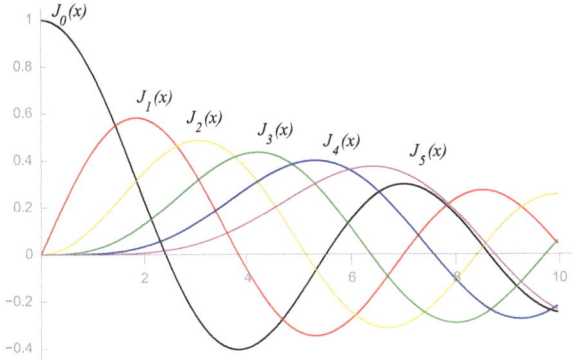

where $x = (|x|, \theta_x)$ and $\xi = (|\xi|, \theta_\xi)$ in the polar coordinates. The function $x \mapsto J_n(|\xi||x|)e^{in\theta_x}$ is called a cylindrical wave (Fig. 1.1).

The following identities will be useful in Chap. 13. For $n, l \in \mathbb{Z}$, we have

$$\int_0^{\pi/2} J_{2n}(2x \sin \phi) \, d\phi = \frac{\pi}{2} J_n^2(x) \,, \tag{1.16}$$

and

$$\int_0^{\pi} J_l(2x \cos \phi) \cos((2n - l)\phi) \, d\phi = 2\pi J_n(x) J_{l-n}(x) \,. \tag{1.17}$$

Formula (1.17) is known as Neumann's formula.

Bessel functions appear in an approximation of the Dirac function δ_0 at the origin. In view of (1.5), an approximation $\tilde{\delta}_K$ to δ_0 can be defined by

$$\mathcal{F}(\tilde{\delta}_K)(\xi) = \begin{cases} (2\pi)^{-d/2}, & |\xi| < K \,, \\ 0, & |\xi| \geq K \,, \end{cases} \tag{1.18}$$

or equivalently by

$$\tilde{\delta}_K(x) = (2\pi)^{-d/2} \frac{J_{d/2}(K|x|)}{(K|x|)^{d/2}} \,, \tag{1.19}$$

where $J_{d/2}$ is the Bessel function of the first kind of order $d/2$. As will be seen later, (1.19) gives a typical convolution kernel function in multistatic imaging.

For arguments $x < \nu$, the Bessel functions look qualitatively like simple powers law, with the asymptotic form for $0 < x \ll \nu$

$$J_\nu(x) \approx \frac{1}{\Gamma(\nu + 1)} \left(\frac{x}{2}\right)^\nu \approx \frac{1}{\sqrt{2\pi\nu}} \left(\frac{ex}{2\nu}\right)^\nu \,. \tag{1.20}$$

For $x > \nu$, the Bessel functions look qualitatively like cosine waves whose amplitude decay as $x^{-1/2}$. The asymptotic form for $x \gg \nu$ is

$$J_\nu(x) \approx \sqrt{\frac{2}{\pi x}} \cos\left(x - \frac{\nu\pi}{2} - \frac{\pi}{4}\right) . \tag{1.21}$$

In the transition region where $x \approx \nu$, the typical amplitude of the Bessel functions is

$$J_\nu(\nu) \approx \frac{2^{1/3}}{3^{2/3}\Gamma\left(\frac{2}{3}\right)} \frac{1}{\nu^{1/3}} \approx \frac{0.4473}{\nu^{1/3}} ,$$

which holds asymptotically for large ν.

The Bessel function J_ν solves the ODE, known as Bessel's equation,

$$\left(\frac{d^2}{dx^2} + \frac{1}{x}\frac{d}{dx} + \left(1 - \frac{\nu^2}{x^2}\right)\right) J_\nu(x) = 0 , \tag{1.22}$$

or equivalently,

$$\left(\frac{d}{dx} - \frac{\nu-1}{x}\right)\left(\frac{d}{dx} + \frac{\nu}{x}\right) J_\nu(x) = -J_\nu(x) . \tag{1.23}$$

Note that adding and subtracting (1.12) and (1.23) produce the identities

$$2J_\nu'(x) = J_{\nu-1}(x) - J_{\nu+1}(x) ,$$

$$\frac{2\nu}{x} J_\nu(x) = J_{\nu-1}(x) + J_{\nu+1}(x) .$$

Equation (1.22), for each ν, has a two-dimensional solution space. Note that $J_{-\nu}$ is also a solution. From the expression (1.10) it is clear that J_ν and $J_{-\nu}$ are linearly independent provided ν is not an integer. On the other hand, comparison of power series shows

$$J_{-n}(x) = (-1)^n J_n(x), \quad n \in \mathbb{N} .$$

A calculation of the Wronskian shows that

$$W(J_\nu, J_{-\nu})(x) = -2\frac{\sin \pi\nu}{\pi x} .$$

Therefore, J_ν and $J_{-\nu}$ are linearly independent, and consequently they form a basis of solutions to (1.22), if and only if ν is not an integer. To construct a basis of solutions uniformly good for all ν, it is natural to set

$$Y_\nu(x) = \frac{J_\nu(x)\cos \pi\nu - J_{-\nu}(x)}{\sin \pi\nu} \tag{1.24}$$

when ν is not an integer, and define for integer n

$$Y_n(x) = \lim_{\nu \to n} Y_\nu(x) .$$

We have

$$W(J_\nu, Y_\nu)(x) = \frac{2}{\pi x} ,$$

for all ν.

Another important pair of solutions to Bessel's equation is that of Hankel functions

$$H_\nu^{(1)}(x) = J_\nu(x) + iY_\nu(x), \quad H_\nu^{(2)}(x) = J_\nu(x) - iY_\nu(x) . \tag{1.25}$$

We will need the following behavior of $H_\nu^{(1)}$ for fixed ν and $x \to 0$. We have

$$H_\nu^{(1)}(x) \approx -\frac{i2^\nu \Gamma(\nu)}{\pi} x^{-\nu} . \tag{1.26}$$

For n an integer, it is also known that, as $x \to 0$,

$$J_n(x) = \frac{x^n}{2^n} \left(\frac{1}{\Gamma(n+1)} - \frac{\frac{1}{4}x^2}{\Gamma(n+2)} + \frac{(\frac{1}{4}x^2)^2}{2!\Gamma(n+3)} - \frac{(\frac{1}{4}x^2)^3}{3!\Gamma(n+4)} + \cdots \right) , \tag{1.27}$$

$$Y_n(x) = -\frac{(\frac{1}{2}x)^{-n}}{\pi} \sum_{l=0}^{n-1} \frac{(n-l-1)!}{l!} (\frac{1}{4}x^2)^l + \frac{2}{\pi} \log(\frac{1}{2}x) J_n(x)$$

$$- \frac{(\frac{1}{2}x)^n}{\pi} \sum_{l=0}^{\infty} (\psi(l+1) + \psi(n+l+1)) \frac{(-\frac{1}{4}x^2)^l}{l!(n+l)!} , \tag{1.28}$$

where $\psi(1) = -\gamma$ and

$$\psi(n) = -\gamma + \sum_{l=1}^{n-1} \frac{1}{l} \quad \text{for } n \geq 2$$

with γ being the Euler constant. In particular, if $n = 0$, we have

$$J_0(x) = 1 - \frac{1}{4}x^2 + \frac{1}{64}x^4 + O(x^6) ,$$

$$Y_0(x) = \frac{2}{\pi} \log x + \frac{2}{\pi}(\gamma - \log 2) - \frac{1}{2\pi}x^2 \log x + \left(\frac{1}{2\pi} - \frac{1}{2\pi}(\gamma - \log 2)\right)x^2$$

$$+ O(x^4 \log x) .$$

It is worth pointing out that the Bessel functions $J_{n+1/2}(x)$, for n an integer, are elementary functions. For $\nu = n + 1/2$, the integrand in (1.11) involves $(1 - t^2)^n$, so the integral can be evaluated explicitly. We have, in particular,

$$J_{1/2}(x) = (\frac{2}{\pi x})^{1/2} \sin x \ .$$

Then (1.12) gives

$$J_{-1/2}(x) = (\frac{2}{\pi x})^{1/2} \cos x \ ,$$

which by (1.24) is equal to $-Y_{1/2}(x)$. Applying (1.23) and (1.12) repeatedly gives

$$J_{n+1/2}(x) = (-1)^n \prod_{l=1}^{n} (\frac{d}{dx} - \frac{l - \frac{1}{2}}{x}) \frac{\sin x}{\sqrt{2\pi x}}$$

and the same sort of formula for $J_{-n-1/2}(x)$, with the $(-1)^n$ removed, and $\sin x$ replaced by $\cos x$.

The functions

$$j_n(x) := \sqrt{\frac{\pi}{2}} \frac{J_{n+\frac{1}{2}}(x)}{\sqrt{x}} \ , \tag{1.29}$$

and

$$y_n(x) := \sqrt{\frac{\pi}{2}} \frac{Y_{n+\frac{1}{2}}(x)}{\sqrt{x}} \tag{1.30}$$

are known as the spherical Bessel functions and form a basis for the solution space of the spherical Bessel equation

$$\left(\frac{d^2}{dx^2} + \frac{2}{x} \frac{d}{dx} + (1 - \frac{n(n+1)}{x^2}) \right) f(x) = 0 \ .$$

Analogously to (1.25), we define $h_n^{(1)}$ and $h_n^{(2)}$ by

$$h_n^{(1)}(x) = j_n(x) + i y_n(x), \quad h_n^{(2)}(x) = j_n(x) - i y_n(x) \ . \tag{1.31}$$

In three dimensions, there is an analogue to (1.15). The following plane wave expansion, also known as the Rayleigh equation, holds:

$$e^{i\xi \cdot x} = 4\pi \sum_{l=0}^{+\infty} \sum_{m=-l}^{l} i^l j_l(|\xi||x|) Y_{lm}(\theta_x, \phi_x) \overline{Y}_{lm}(\theta_\xi, \phi_\xi) \ , \tag{1.32}$$

where Y_{lm} are the spherical harmonic functions and $\xi = (|\xi|, \theta_\xi, \phi_\xi)$, $x = (|x|, \theta_x, \phi_x)$ in the spherical coordinates.

We will need the closure relation

$$\int_0^{+\infty} x J_\nu(tx) J_\nu(sx)\, dx = \frac{1}{t}\delta_0(t-s) \qquad (1.33)$$

for $\nu > -1/2$, which is equivalent to

$$\int_0^{+\infty} x^2 j_\nu(tx) j_\nu(sx)\, dx = \frac{\pi}{2t^2}\delta_0(t-s)\ .$$

The cylindrical waves form a complete set. We have the completeness relation

$$\frac{\delta_0(r-r_0)\delta_0(\theta-\theta_0)}{r} = \sum_{m\in\mathbb{Z}} \frac{1}{2\pi} \int_0^{+\infty} t J_m(tr) J_m(tr_0)\, dt\, e^{im(\theta-\theta_0)}\ , \qquad (1.34)$$

which is the analogue of the completeness relation for plane waves

$$\delta_0(x-x_0)\delta_0(y-y_0) = \left(\frac{1}{2\pi}\right)^2 \int_\mathbb{R}\int_\mathbb{R} e^{i(\xi_x x+\xi_y y)} e^{-i(\xi_x x_0+\xi_y y_0)}\, d\xi_x d\xi_y$$
$$= \frac{1}{2\pi}\mathcal{F}^{-1}(1)(x-x_0, y-y_0)\ .$$

The connecting link between these two relations is the plane wave expansion (1.14).

Finally, we will need the relation

$$\int_0^{+\infty} x^2 j_0(tx)\, dx = 2\pi^2 \delta_0(t)\ , \qquad (1.35)$$

which can be obtained by integrating the spherical plane wave representation (1.32).

1.3.2 Prolate Spheroidal Functions

The following results on prolate spherical functions will be used in Chap. 14 for a resolution analysis.

Let $C > 0$. The prolate spheroidal functions $\psi^{(l)}(x)$ are the eigenfunctions of the sinc kernel:

$$\int_{-1}^1 \frac{\sin[C(x-y)]}{\pi(x-y)}\psi^{(l)}(y)\, dy = \sigma^{(l)}\psi^{(l)}(x)\ . \qquad (1.36)$$

The symmetric sinc kernel $\frac{\sin C(x-y)}{\pi(x-y)}$ is positive definite. Its spectrum $(\sigma^{(l)})_{l\geq 1}$ is discrete and positive, $\sigma^{(1)} > \sigma^{(2)} > \cdots > 0$ and $\sigma^{(l)} \to 0$ as $l \to \infty$. The real-valued eigenfunctions $\psi^{(l)}$ are orthonormal on $]-1,1[$:

$$\int_{-1}^{1} \psi^{(l)}(x)\psi^{(j)}(x)\,dx = \delta_{jl} \,, \tag{1.37}$$

where δ_{jl} denotes the Kronecker symbol.

By the spectral representation of the sinc kernel, we have

$$\sum_{l=1}^{\infty} \sigma^{(l)}\psi^{(l)}(x)\psi^{(l)}(y) = \frac{\sin[C(x-y)]}{\pi(x-y)} \quad \text{for } x,y \in]-1,1[\,, \tag{1.38}$$

$$\sum_{l=1}^{\infty} \psi^{(l)}(x)\psi^{(l)}(y) = \delta(x-y) \quad \text{for } x,y \in]-1,1[\,. \tag{1.39}$$

When C is large, the eigenvalues $\sigma^{(l)}$ stay close to one for small l and then they plunge to 0 near the threshold value $[2C/\pi]$:

$$\sigma^{(l)} \overset{C\to\infty}{\longrightarrow} \begin{cases} 1 & \text{if } l = \left[\dfrac{2C}{\pi}(1-\varepsilon)\right], \quad \varepsilon > 0\,, \\[2mm] \dfrac{1}{1+e^{\pi b}} & \text{if } l = \left[\dfrac{2C}{\pi} + \dfrac{b}{\pi}\log C\right], \quad b \in \mathbb{R}\,, \\[2mm] 0 & \text{if } l = \left[\dfrac{2C}{\pi}(1+\varepsilon)\right], \quad \varepsilon > 0\,. \end{cases} \tag{1.40}$$

Here, $[x]$ denotes the smallest integer not smaller than x. Finally, we have for any $x \in \mathbb{R}$ and $l \geq 1$:

$$\int_{-1}^{1} e^{-iCxy}\psi^{(l)}(y)\,dy = i^{l+1}\sqrt{\frac{2\pi\sigma^{(l)}}{C}}\psi^{(l)}(x)\,. \tag{1.41}$$

1.3.3 Airy Function

For real arguments x the Airy function is defined by the integral

$$\mathrm{Ai}(x) = \frac{1}{2\pi}\int_{\mathbb{R}} e^{i\frac{t^3}{3}+ixt}\,dt\,. \tag{1.42}$$

For arbitrary complex arguments it is a similar integral with a modified contour. The representation (1.42) brings out that the Airy function is a generalization of the δ-function and involves the integration variable not only in a linear way but also as the third power. The Airy function satisfies the differential equation

$$\mathrm{Ai}''(x) - x\mathrm{Ai}(x) = 0 \, .$$

1.4 The Moore-Penrose Generalized Inverse

Let A be a bounded operator from a Hilbert space H into a Hilbert space K. Let A^* denote the adjoint of A (see for example [153, p. 487]). The Moore-Penrose generalized solution f^+ to $Af = g$ is defined as follows: f^+ is the element with the smallest norm in the set of the minimizers of $\|Af - g\|$ (if this set is nonempty, i.e., if $g \in \mathrm{Range}(A) + \mathrm{Range}(A)^\perp$). It can be shown that f^+ is the unique solution to the normal equation

$$A^* A f = A^* g$$

in $\overline{\mathrm{Range}(A^*)}$. The linear operator A^+ defined by

$$f^+ = A^+ g \quad \text{for } g \in \mathrm{Range}(A) + \mathrm{Range}(A)^\perp$$

is called the Moore-Penrose generalized inverse and f^+ is called the least-squares solution.

1.5 Singular Value Decomposition

Let A be a bounded linear operator from a Hilbert space H into a Hilbert space K. By the singular value decomposition (SVD) we mean a representation of A in the form

$$Af = \sum_l \sigma_l \, (f, f_l) \, g_l \, ,$$

where $(f_l), (g_l)$ are orthonormal systems in H, K, respectively, and σ_l are positive numbers, the singular values of A. The sum may be finite or infinite. The adjoint of A is given by

$$A^* g = \sum_l \sigma_l \, (g, g_l) \, f_l \, ,$$

and the operators

$$A^* A f = \sum_l \sigma_l^2 \, (f, f_l) \, f_l \, ,$$

$$A A^* g = \sum_l \sigma_l^2 \, (g, g_l) \, g_l \, ,$$

are self-adjoint operators in H, K, respectively. The spectrum of $A^* A, AA^*$ consists of the eigenvalues σ_l^2 and possibly the eigenvalue 0, whose multiplicity may be infinite.

The Moore-Penrose generalized inverse is given by

$$A^+ g = \sum_l \sigma_l^{-1} \, (g, g_l) \, f_l \, .$$

Indeed this is the least-squares solution to $Af = g$ of minimum norm.

Let us now review the basic concepts of singular value decomposition of a matrix. Let $M_{m,n}(\mathbb{C})$ denote the set of all m-by-n matrices over $\mathbb{C}$. The set $M_{n,n}(\mathbb{C})$ is abbreviated to $M_n(\mathbb{C})$. The spectral theorem applied to the positive semi-definite matrices AA^* and $A^* A$ gives the following singular value decomposition of a matrix $A \in M_{m,n}(\mathbb{C})$. Here $A^* := \overline{A}^T$, where T denotes the transpose.

Theorem 1.2 (Spectral Theorem). *Let $A \in M_{m,n}(\mathbb{C})$ be given, and let $q = \min\{m, n\}$. There is a matrix $\Sigma = (\Sigma_{ij}) \in M_{m,n}(\mathbb{R})$ with $\Sigma_{ij} = 0$ for all $i \neq j$ and $\Sigma_{11} \geq \Sigma_{22} \geq \ldots \geq \Sigma_{qq} \geq 0$, and there are two unitary matrices $V \in M_m(\mathbb{C})$ and $W \in M_n(\mathbb{C})$ such that $A = V \Sigma W^*$. The numbers $\{\Sigma_{ii}\}$ are the nonnegative square roots of the eigenvalues of AA^*, and hence are uniquely determined. The columns of V are eigenvectors of AA^* and the columns of W are eigenvectors of $A^* A$ (arranged in the same order as the corresponding eigenvalues Σ_{ii}^2).*

The diagonal entries Σ_{ii}, $i = 1, \ldots, q = \min\{m, n\}$ of Σ are called the singular values of A, and the columns of V and the columns of W are the (respectively, left and right) singular vectors of A.

The SVD has the following desirable computational properties:

(i) The rank of A can be easily determined from its SVD. Specifically, rank(A) equals to the number of nonzero singular values of A.

(ii) The Frobenius norm of A, $||A||_F := \sqrt{\mathrm{Tr}(A\overline{A}^T)}$ with Tr being the trace, is given by $||A||_F = \sqrt{\sum_{m=1}^q \Sigma_{mm}^2}$.

(iii) SVD is an effective computational tool for finding lower-rank approximations to a given matrix. Specifically, let $p < \mathrm{rank}(A)$. Then the rank p matrix A_p minimizing $||A - A_p||_F$ is given by $A_p = V \Sigma_p W^*$, where the matrix Σ_p is obtained from Σ after the singular values $\Sigma_{nn}, p + 1 \leq n \leq q$, are set to zero.

1.6   Compact Operators

Let H be a Banach space. A bounded linear operator A on H is compact if whenever $\{x_j\}$ is a bounded sequence in H, the sequence $\{Ax_j\}$ has a convergent subsequence. The operator A is said to be of finite rank if Range(A) is finite-dimensional. Clearly every operator of finite rank is compact.

We recall some basic results on compact operators.

(i) The set of compact operators on H is a closed two-sided ideal in the algebra of bounded operators on H with the norm topology.

(ii) If A is a linear bounded operator on the Banach space H and there is a sequence $\{A_N\}_{N\in\mathbb{N}}$ of linear operators of finite rank such that $\|A_N - A\| \to 0$, then A is compact.

(iii) The operator A is compact on the Banach space H if and only if the dual operator A^* is compact on the dual space H^*.

We also recall the main structure theorem for compact operators. Let A be a compact operator on the Hilbert space H (which we identify with its dual). For each $\lambda \in \mathbb{C}$, let $V_\lambda = \{x \in H : Ax = \lambda x\}$ and $V_{\bar{\lambda}} = \{x \in H : A^*x = \bar{\lambda}x\}$. Then

(i) The set of $\lambda \in \mathbb{C}$ for which $V_\lambda \neq \{0\}$ is finite or countable, and in the latter case its only accumulation point is zero. Moreover, $\dim(V_\lambda) < +\infty$ for all $\lambda \neq 0$.

(ii) If $\lambda \neq 0, \dim(V_\lambda) = \dim(V_{\bar{\lambda}})$.

(iii) If $\lambda \neq 0$, the range of $\lambda I - A$ is closed.

Suppose $\lambda \neq 0$. Then

(i) The equation $(\lambda I - A)x = y$ has a solution if and only if $y \perp V_{\bar{\lambda}}$.

(ii) $(\lambda I - A)$ is surjective if and only if it is injective.

We recall the concept of a Fredholm operator acting between Banach spaces H and K. We say that a bounded linear operator $A : H \to K$ is Fredholm if the subspace Range(A) is closed in K and the subspace Ker(A) and the quotient space $K/\text{Range}(A)$ are finite-dimensional. In this case, the index of A is the integer defined by

$$\text{index } (A) = \dim \text{Ker}(A) - \dim(K/\text{Range}(A)) .$$

In the sequel, we encapsulate the main conclusion of Fredholm's original theory. If $A = I + B$, where $B : H \to H$ is compact, then $A : H \to H$ is Fredholm with index zero. If $A : H \to K$ is Fredholm and $B : H \to K$ is compact, then their sum $A + B : H \to K$ is Fredholm, and index$(A + B) =$ index(A). This shows that the index is stable under compact perturbations.

Finally, we recall that a compact operator A on H is a Hilbert-Schmidt operator if the sequence of its singular values is in

$$l_2 = \{(\mu_n)_{n \in \mathbb{Z}} : \sum_{n \in \mathbb{Z}} |\mu_n|^2 < \infty\}.$$

An equivalent characterization is $\sum_n ||A\varphi_n||^2 < \infty$ for any orthonormal basis (φ_n) of H.

1.7 Regularization of Ill-Posed Problems

In this section we review some of the most commonly used methods for solving ill-posed inverse problems. These methods are called regularization methods. Although the emphasis in this book is not on classical regularization techniques, it is quite important to understand the philosophy behind them and how they work in practice.

1.7.1 Stability

Problems in image reconstruction are usually not well-posed in the sense of Hadamard. This means that they suffer from one of the following deficiencies:

(i) They may not be solvable (in the strict sense) at all.
(ii) The solution, if exists, may not be unique.
(iii) The solution may not depend continuously on the data.

A classical ill-posed inverse problem is the deconvolution problem. Define the compact operator $A : L^2(\mathbb{R}) \to L^2(\mathbb{R})$ by

$$(Af)(x) := \int_{-\infty}^{+\infty} h(x-y)f(y)\, dy \ ,$$

where h is a Gaussian convolution kernel,

$$h(x) := \frac{1}{\sqrt{2\pi}} e^{-x^2/2} \ .$$

The operator A is injective, which can be seen by applying the Fourier transform on Af, yielding

$$\mathcal{F}(Af) = \mathcal{F}(h \star f) = \mathcal{F}(h)\mathcal{F}(f) \ ,$$

with $\mathcal{F}(h)$ given by (1.4). Therefore, if $Af = 0$, we have $\mathcal{F}(f) = 0$, hence $f = 0$. Formally, the solution to the equation $Af = g$ is

$$f(x) = \mathcal{F}^{-1}\left(\frac{\mathcal{F}(g)}{\mathcal{F}(h)}\right)(x), \quad x \in \mathbb{R}. \tag{1.43}$$

However, the above formula is not well defined for general $g \in L^2(\mathbb{R})$ (or even in $\mathcal{S}'(\mathbb{R})$) since $1/\mathcal{F}(h)$ grows as $e^{\xi^2/2}$.

To explain the basic ideas of regularization, let A be a bounded linear operator from a Hilbert space H into a Hilbert space K. Consider the problem of solving

$$Af = g \tag{1.44}$$

for f. Item (i) means that g may not be in the range of A, (ii) means that A may not be injective, and (iii) means that A^{-1} may not be continuous.

One could do away with (i) and (ii) by using the generalized inverse A^+. But A^+ does not have to be continuous. Thus, small error in g may cause errors of arbitrary size in f. To restore continuity, we introduce the notion of a regularization of A^+. This is a family $(T_\gamma)_{\gamma>0}$ of linear continuous operators $T_\gamma : K \to H$, which are defined on all of K and for which

$$\lim_{\gamma \to 0} T_\gamma g = A^+ g$$

on the domain of A^+. Obviously, $\|T_\gamma\| \to +\infty$ as $\gamma \to 0$ if A^+ is unbounded. With the help of regularization, we can solve (1.44) in the following way. Let $g^\epsilon \in K$ be an approximation to g such that $\|g - g^\epsilon\| \le \epsilon$. Let $\gamma(\epsilon)$ be such that, as $\epsilon \to 0$,

$$\gamma(\epsilon) \to 0, \quad \|T_{\gamma(\epsilon)}\|\epsilon \to 0.$$

Then, as $\epsilon \to 0$,

$$\|T_{\gamma(\epsilon)}g^\epsilon - A^+g\| \le \|T_{\gamma(\epsilon)}(g^\epsilon - g)\| + \|T_{\gamma(\epsilon)}g - A^+g\|$$
$$\le \|T_{\gamma(\epsilon)}\|\epsilon + \|T_{\gamma(\epsilon)}g - A^+g\|$$
$$\to 0.$$

Hence $T_{\gamma(\epsilon)}g^\epsilon$ is close to A^+g if g^ϵ is close to g.

The number γ is called a regularization parameter. Determining a good regularization parameter is a major issue in the theory of ill-posed problems.

Measurement errors of arbitrarily small L^2-norm in g may cause g to be not in Range(A) and the inversion formula (1.43) practically useless. Therefore, instead of trying to solve (1.44) exactly, one seeks to find a nearby problem that is uniquely solvable and that is robust in the sense that small errors in the data do not corrupt excessively this approximate solution.

We briefly discuss three families of classical regularization methods: (1) regularization by singular value truncation, (2) the Tikhonov-Phillips regularization and (3) regularization by truncated iterative methods.

1.7.2 The Truncated SVD

Let

$$Af = \sum_l \sigma_l \, (f, f_l) \, g_l$$

be the SVD of A. Then

$$T_\gamma g = \sum_{\sigma_l \geq \gamma} \sigma_l^{-1} \, (g, g_l) \, f_l \tag{1.45}$$

is a regularization with $\|T_\gamma\| \leq 1/\gamma$.

A good measure for the degree of ill-posedness of (1.44) is the rate of decay of the singular value σ_l. It is clear from (1.45) that the ill-posedness is more pronounced as the rate of decay increases. A polynomial decay is usually considered manageable, while an exponential decay indicates that only very poor approximations to f in (1.44) can be computed. The SVD gives us all the information we need about an ill-posed problem.

There is a rule for choosing the truncation level, that is often referred to as the discrepancy principle. This principle states that we cannot expect the approximate solution f_γ to yield a smaller residual error, $Af_\gamma - g$, than the noise level ϵ, since otherwise we would be fitting the solution to the noise. This leads to the following selection criterion for γ: choose the largest γ that satisfies $\|g - \sum_{\sigma_l \geq \gamma} (g, g_l) g_l\| \leq \epsilon$.

1.7.3 Tikhonov-Phillips Regularization

Linear Problems

The discussion in the previous subsection demonstrates that when solving (1.44) for a compact operator A, serious problems occur when the singular values of A tend to zero rapidly, causing the norm of the approximate solution to go to infinity as the regularization parameter γ goes to zero. The idea in the basic Tikhonov-Phillips regularization scheme is to control simultaneously the norm of the residual, $Af_\gamma - g$, and the norm of the approximate solution f_γ.

To do so, we set

$$T_\gamma = (A^*A + \gamma I)^{-1}A^* .$$

Equivalently, $f_\gamma = T_\gamma g$ can be defined by minimizing $||Af - g||^2 + \gamma||f||^2$. Here the regularization parameter γ plays essentially the role of a Lagrange multiplier. In terms of the SVD of A presented in Sect. 1.5, we have

$$T_\gamma g = \sum_l F_\gamma(\sigma_l)\sigma_l^{-1}(g, g_l)\, f_l ,$$

where $F_\gamma(\sigma) = \sigma^2/(\sigma^2 + \gamma)$.

The choice of the value of the regularization parameter γ based on the noise level of the measurement g is a central issue in the literature discussing Tikhonov-Phillips regularization. Several methods for choosing γ have been proposed. The most common one is known as the Morozov discrepancy principle. This principle is essentially the same as the discrepancy principle discussed in connection with the singular value truncation principle. It is rather straightforward to implement the principle numerically.

Let ϵ be the measurement error. Let

$$\varphi : \mathbb{R}^+ \to \mathbb{R}^+, \quad \varphi(\gamma) = ||Af_\gamma - g||$$

be the discrepancy related to the regularization parameter γ. The Morozov discrepancy principle says that γ should be chosen from the condition

$$\varphi(\gamma) = \epsilon , \tag{1.46}$$

if possible, i.e., the regularized solution should not try to satisfy the data more accurately than up to the noise level ϵ. Equation (1.46) has a unique solution $\gamma = \gamma(\epsilon)$ if and only if (1) any component in the data g that is orthogonal to Range(A) must be due to noise and (2) the error level ϵ should not exceed the signal level $||g||$.

Nonlinear Problems

Tikhonov-Phillips regularization method is sometimes applicable also when non-linear problems are considered. Let H and K be (real) Hilbert spaces. Let $A : H \to K$ be a nonlinear mapping. We want to find $f \in H$ satisfying

$$A(f) = g + \epsilon , \tag{1.47}$$

where ϵ is observation noise. If A is such that large changes in f may produce small changes in $A(f)$, the problem of finding f a solution to (1.47) is ill-posed and numerical methods, typically, iterative ones, may fail to find a satisfactory estimate of f.

The nonlinear Tikhonov-Phillips regularization scheme amounts to searching for f that minimizes the functional

$$||A(f) - g||^2 + \gamma G(f) , \tag{1.48}$$

where $G : H \rightarrow \mathbb{R}$ is a nonnegative functional. The most common penalty term is $G(f) = ||f||^2$ although a lot of work has been recently devoted to the analysis of L^1-type penalization methods; see, for instance, [148]. We restrict ourselves to this choice and suppose that A is Fréchet differentiable. In this case, the most common method to search for a minimizer of (1.48) is to use an iterative scheme based on successive linearizations of A. The linearization of A around a given point f_0 leads that the minimizer of (1.48) (around f_0) is

$$f = (R_{f_0}^* R_{f_0} + \gamma I)^{-1} R_{f_0}^* \left(g - A(f_0) + R_{f_0} f_0 \right) ,$$

where R_{f_0} is the Fréchet derivative of A at f_0. We recall that A is Fréchet differentiable at f_0 if it allows an expansion of the form

$$A(f_0 + h) = A(f_0) + R_{f_0} h + o(||h||) ,$$

where R_{f_0} is a continuous linear operator.

1.7.4   Regularization by Truncated Iterative Methods

The most common iterative methods to solve (1.44) are Landweber iteration, Kaczmarz iteration, and Krylov subspace methods. The best known of the Krylov iterative methods when the matrix A is symmetric and positive definite is the conjugate gradient method. In this section, we only discuss regularizing properties of Landweber and Kaczmarz iterations. We refer to [103] and the references therein concerning the Krylov subspace methods.

Landweber Iteration

The drawback of the Thikhonov-Phillips regularization is that it requires to invert the regularization of the normal operator $A^*A + \gamma I$. This inversion may be costly in practice. The Landweber iteration method is an iterative technique in which no inversion is necessary. It is defined to solve the equation $Af = g$ as follows:

$$f^0 = 0, \quad f^{k+1} = (I - rA^*A)f^k + rA^*g, \quad k \geq 0 ,$$

for some $r > 0$. By induction, we verify that $f^k = T_\gamma g$, with $\gamma = 1/k, k \geq 1$, and

$$T_\gamma g = r \sum_{l=0}^{1/\gamma - 1} (I - rA^*A)^l A^* g .$$

Let $q \leq +\infty$ be the number of singular values of A. Let σ_l be the singular values arranged in a decreasing sequence and g_l be the associated singular vectors. Since

$$r \sum_{l=0}^{1/\gamma - 1} (I - rA^*A)^l A^* g = \sum_{l=1}^{q} \frac{1}{\sigma_l}(1 - (1 - r\sigma_l^2)^{1/\gamma})(g, g_l) f_l ,$$

where $f_l = (f, g_l)$, a good choice of r is thus $r \approx \sigma_1^{-2}$.

Kaczmarz Iteration

Kaczmarz's method (also known as the algebraic reconstruction technique) is an iterative method for solving linear systems of equations. Let $H, H_j, j = 1, \ldots, p$, be (real) Hilbert spaces, and let

$$A_j : H \rightarrow H_j, \quad j = 1, \ldots, p ,$$

be bounded linear maps from H onto H_j with $\mathrm{Range}(A_j) = H_j$. Let $g_j \in H_j$ be given. We want to compute $f \in H$ such that

$$A_j f = g_j, \quad j = 1, \ldots, p . \tag{1.49}$$

Kaczmarz's method for the solution of (1.49) reads:

Algorithm 1.1 Kaczmarz's method

---

1. $f_0 = f^k$,
2. $f_j = f_{j-1} + \gamma A_j^*(A_j A_j^*)^{-1}(g_j - A_j f_{j-1}), \quad j = 1, \ldots, p$,
3. $f^{k+1} = f_p$, with $f^0 \in H$ arbitrary.

Here γ is a regularization parameter. Under certain assumptions, f^k converges to a solution of (1.49) if (1.49) has a solution and to a generalized solution if not.

1.8   Introduction to Probability Theory

The noise models discussed in this book are measurement and medium (or
cluttered) noises. They affect the stability and resolution of the imaging
functionals in very different ways.

Imaging involves measurement and processing of activated signals ema-
nating from an object. Any practical measurement always contains an
undesirable component that is uncorrelated with (i.e., independent of) the
desired signal. This component is referred to as measurement noise. On the
other hand, medium noise models the uncertainty in the physical parameters
of the background medium. In many practical situations, the physical
parameters of the background medium fluctuate spatially around a known
background. Of great concern in imaging is the question of how measurement
and medium noises are modeled and how the imaging process handles them-
that is, whether they are suppressed or amplified. We give in this section an
introduction to probability theory that provides the basic tools for modeling
imaging schemes with waves in the presence of noise.

1.8.1   Random Variables

A characteristic of noise is that it does not have fixed values in repeated
measurements. Such a quantity is described by a random variable. The
statistical distribution of a random variable can be characterized by its
probability density function (PDF). The PDF of a (real-valued) random
variable ξ is often denoted by $p_\xi(x)$, which represents the probability density
of obtaining a specific value x for ξ in a particular measurement:

$$\mathbb{P}(\xi \in [a, b]) = \int_a^b p_\xi(x)\,dx \ .$$

Note that p_ξ is a nonnegative function whose total integral is equal to one.
Given the PDF it is possible to compute the expectation of a nice function
(bounded or positive) of the random variable $\phi(\xi)$, which is the weighted
average of ϕ with respect to the PDF p_ξ:

$$\mathbb{E}[\phi(\xi)] = \int_{\mathbb{R}} \phi(x)p_\xi(x)\,dx \ .$$

The most important expectations are the first- and second-order moments.
The mean of the random variable ξ is defined as

$$\mathbb{E}[\xi] = \int xp_\xi(x)\,dx \ .$$

It is the first-order statistical moment. The variance is defined as

$$\mathrm{Var}[\xi] = \mathbb{E}[|\xi - \mathbb{E}[\xi]|^2] = \mathbb{E}[\xi^2] - \mathbb{E}[\xi]^2 \,,$$

which is a second-order statistical moment. The variable $\sigma_\xi := \sqrt{\mathrm{Var}[\xi]}$ is called the standard deviation, which is a measure of the average deviation from the mean.

The PDF of measurement noise is not always known in practical situations. We often use parameters such as mean and variance to describe it. It is then usual to assume that the noise has Gaussian PDF. This can be justified by the maximum of entropy principle, which assumes that the PDF maximizes the entropy $-\int p_\xi(x) \log p_\xi(x) \; dx$ with the constraints $\int p_\xi(x)dx = 1$, $\int x p_\xi(x) \; dx = x_0$, and $\int (x - x_0)^2 p_\xi(x)dx = \sigma^2$. This PDF is nothing else than the Gaussian PDF

$$p_\xi(x) = \frac{1}{\sqrt{2\pi}\sigma} \exp\left(-\frac{(x - x_0)^2}{2\sigma^2} \right) , \tag{1.50}$$

with mean x_0 and variance σ^2. Moreover, the measurement error often results from the cumulative effect of many uncorrelated sources of uncertainty. As a consequence, based on the central limit theorem, most measurement noise can be treated as Gaussian noise. Recall here the central limit theorem: When a random variable ξ is the sum of n independent and identically distributed random variables (with finite variance), then the distribution of ξ is a Gaussian distribution with the appropriate mean and variance in the limit $n \to +\infty$. In terms of PDFs, this means that, if a function $h(x)$ is convolved with itself n times, in the limit $n \to +\infty$, the convolution product is a Gaussian function with a variance that is n times the variance of $h(x)$, provided the area, mean, and variance of $h(x)$ are finite.

The following theorem, which is a consequence of Slutsky's theorem, will be useful.

Theorem 1.3. *Let (ξ_n) and (ζ_n) be sequences of random variables. If ξ_n converges in distribution to a random variable ξ and ζ_n converges in probability to a non zero constant c, then $\zeta_n^{-1}\xi_n$ converges in distribution to $c^{-1}\xi$.*

Throughout this book, if ξ has the PDF (1.50), then we write $\xi \sim \mathcal{N}(x_0, \sigma^2)$ with $\mathcal{N}(x_0, \sigma^2)$ being the normal distribution of mean x_0 and variance σ^2.

1.8.2 Random Vectors

A d-dimensional random vector ξ is collection of d (real-valued) random variables $(\xi_1, \ldots, \xi_d)^T$. The distribution of a random vector is characterized by the PDF p_ξ:

$$\mathbb{P}(\xi \in [a_1, b_1] \times \cdots \times [a_d, b_d]) = \int_{[a_1,b_1] \times \cdots [a_d,b_d]} p_\xi(x)\, dx, \qquad \forall a_j \le b_j \ .$$

The vector $\xi = (\xi_1, \ldots, \xi_d)^T$ is independent if its PDF can be written as a product of the one-dimensional PDFs of the components of the vector:

$$p_\xi(x) = \prod_{j=1}^{d} p_{\xi_j}(x_j) \text{ for all } x = (x_1, \ldots, x_d) \in \mathbb{R}^d \ ,$$

or equivalently,

$$\mathbb{E}\big[\phi_1(\xi_1) \cdots \phi_d(\xi_d)\big] = \mathbb{E}\big[\phi_1(\xi_1)\big] \cdots \mathbb{E}\big[\phi_d(\xi_d)\big], \qquad \forall \phi_1, \ldots, \phi_d \in \mathcal{C}_b(\mathbb{R}, \mathbb{R}) \ .$$

Here, $\mathcal{C}_b(\mathbb{R}, \mathbb{R})$ denotes the space of all bounded continuous real-valued functions.

Example: a d-dimensional normalized Gaussian random vector ξ has the Gaussian PDF

$$p_\xi(x) = \frac{1}{\sqrt{(2\pi)^d}} \exp\left(-\frac{|x|^2}{2}\right) \ .$$

This PDF can be factorized into the product of one-dimensional Gaussian PDFs, which shows that ξ is a vector of independent random normalized Gaussian variables $(\xi_1, \ldots, \xi_d)^T$.

In the general case, two formulas are of interest. The marginal formula gives the PDF of a subvector extracted from a random vector: If $\begin{pmatrix} \xi_1 \\ \xi_2 \end{pmatrix}$ is a random vector with PDF $p_{\xi_1, \xi_2}(x_1, x_2)$, then ξ_2 is a random vector with PDF

$$p_{\xi_2}(x_2) = \int p_{\xi_1, \xi_2}(x_1, x_2)\, dx_1 \ ,$$

since for any test function ϕ we have

$$\mathbb{E}[\phi(\xi_2)] = \iint \phi(x_2) p_{\xi_1, \xi_2}(x_1, x_2)\, dx_1\, dx_2 = \int \phi(x_2) p_{\xi_2}(x_2)\, dx_2 \ .$$

The conditioning formula gives the PDF of a subvector extracted from a random vector given the observation of the complementary subvector: If $\begin{pmatrix} \xi_1 \\ \xi_2 \end{pmatrix}$ is a random vector with PDF $p_{\xi_1, \xi_2}(x_1, x_2)$, then, given $\xi_2 = x_2$, ξ_1 is a random vector with PDF

$$p_{\xi_1}(x_1 | \xi_2 = x_2) = \frac{p_{\xi_1, \xi_2}(x_1, x_2)}{p_{\xi_2}(x_2)} \ . \tag{1.51}$$

Indeed, this can be seen by taking the limit $\delta x_1 \to 0$ and $\delta x_2 \to 0$ in

$$\mathbb{P}(\xi_1 \in [x_1, x_1 + \delta x_1)|\xi_2 \in [x_2, x_2 + \delta x_2))$$

$$= \frac{\mathbb{P}(\xi_1 \in [x_1, x_1 + \delta x_1), \xi_2 \in [x_2, x_2 + \delta x_2))}{\mathbb{P}(\xi_2 \in [x_2, x_2 + \delta x_2))} \approx \frac{p_{\xi_1,\xi_2}(x_1, x_2)\delta x_1 \delta x_2}{p_{\xi_2}(x_2)\delta x_2} .$$

It is worth emphasizing that formula (1.51) holds if $p_{\xi_2}(x_2) > 0$, otherwise one defines $p_{\xi_1}(x_1|\xi_2 = x_2) = p_0(x_1)$, where p_0 is an arbitrary PDF which plays no role.

Of course, if the vectors ξ_1 and ξ_2 are independent, then $p_{\xi_1}(x_1|\xi_2 = x_2) = p_{\xi_1}(x_1)$ since the knowledge of ξ_2 does not bring any information about ξ_1.

As in the case of random variables, we may not always require or may not be able to give a complete statistical description of a random vector. In such cases, we work only with the first and second statistical moments. Let $\xi = (\xi_i)_{i=1,...,d}$ be a random vector. The mean of ξ is the vector $\mu = (\mu_j)_{j=1,...,d}$:

$$\mu_j = \mathbb{E}[\xi_j] .$$

The covariance matrix of ξ is the matrix $C = (C_{jl})_{j,l=1,...,d}$:

$$C_{jl} = \mathbb{E}\big[(\xi_j - \mathbb{E}[\xi_j])(\xi_l - \mathbb{E}[\xi_l])\big] .$$

These statistical moments are enough to characterize the first two moments of any linear combination of the components of ξ. Indeed, if $\beta = (\beta_j)_{j=1,...,d} \in \mathbb{R}^d$, then the random variable $Z = \beta \cdot \xi = \sum_{j=1}^d \beta_j \xi_j$ has mean:

$$\mathbb{E}[Z] = \beta \cdot \mu = \sum_{j=1}^d \beta_j \mathbb{E}[\xi_j] ,$$

and variance:

$$\mathrm{Var}(Z) = \beta \cdot C\beta = \sum_{j,l=1}^d C_{jl}\beta_j\beta_l .$$

As a byproduct of this result, we can see that the matrix C is nonnegative.

If the variables are independent, then the covariance matrix is diagonal. In particular:

$$\mathrm{Var}\left(\sum_{j=1}^d \xi_j\right) = \sum_{j=1}^d \mathrm{Var}(\xi_j) .$$

The reciprocal is false in general (i.e., the fact that the covariance matrix is diagonal does not ensure that the vector is independent).

In Chap. 14, we will need Bayes' theorem. Let $\begin{pmatrix} \xi_1 \\ \xi_2 \end{pmatrix}$ be a random vector with PDF $p(\xi_1, \xi_2)$ (with ξ_1 of size d_1 and ξ_2 of size d_2). Bayes' theorem says that

$$p_{\xi_1}(x_1|\xi_2 = x_2) = \frac{p_{\xi_2}(x_2|\xi_1 = x_1)p_{\xi_1}(x_1)}{p_{\xi_2}(x_2)} \ . \tag{1.52}$$

In the situation in which ξ_1 is the model parameters and ξ_2 is the data set, the likelihood $p_{\xi_2}(x_2|\xi_1 = x_1)$ is the distribution of the data given the distribution of the model parameters, and the posteriori distribution $p_{\xi_1}(x_1|\xi_2 = x_2)$ describes the distribution of the model parameters given the measured data. Bayes' theorem says that the posteriori distribution $p_{\xi_1}(x_1|\xi_2 = x_2)$ is proportional to the likelihood function $p_{\xi_2}(x_2|\xi_1 = x_1)$ and the prior $p_{\xi_1}(x_1)$. Given the likelihood function, an important problem is how to define the prior distribution. If prior information about the parameter ξ_1 is known, then it should be incorporated in the prior density. If we have no prior information, then we want a prior with minimal influence on the posteriori distribution. We call such a prior a noninformative prior. Jeffreys' prior is based on the Fisher information matrix

$$I(\xi_1) = \int \partial_{\xi_1} \log p_{\xi_2}(x_2|\xi_1 = x_1) \, \partial_{\xi_1} \log p_{\xi_2}(x_2|\xi_1 = x_1)^T \, p_{\xi_2}(x_2|\xi_1 = x_1) \, dx_2 \ ,$$

and is given by $p_{\xi_1}(x_1)$ proportional to $\sqrt{\det I(\xi_1)}$, where det denotes the determinant. Fisher information is an indicator of the amount of information brought by the observation $\xi_1 = x_1$. Jeffreys' prior is motivated by the fact that to favor the values for ξ_1 of which $I(\xi_1)$ is large is equivalent to minimize the influence of the prior.

1.8.3 Gaussian Random Vectors

A Gaussian random vector $\xi = (\xi_1, \dots, \xi_d)^T$ with mean μ and covariance matrix R (we write $\xi \sim \mathcal{N}(\mu, R)$ with $\mathcal{N}(\mu, R)$ being the normal distribution of mean μ and covariance matrix R) has the PDF

$$p(x) = \frac{1}{(2\pi)^{d/2}\sqrt{\det R}} \exp\left(-\frac{(x - \mu) \cdot R^{-1}(x - \mu)}{2} \right) ,$$

provided R is symmetric and positive definite. As mentioned in the case of random variables, the Gaussian statistics is the one that is obtained from the maximum entropy principle (given that the first- and second-order moments of the random vector are specified) and also from the central limit theorem. This distribution is characterized by

$$\mathbb{E}[e^{i\lambda \cdot \xi}] = \int_{\mathbb{R}^d} e^{i\lambda \cdot x} p(x)\,dx = \exp\left(i\lambda \cdot \mu - \frac{\lambda \cdot R\lambda}{2}\right), \qquad \lambda \in \mathbb{R}^d, \quad (1.53)$$

which also shows that, if $\lambda \in \mathbb{R}^d$, then the linear combination $\lambda \cdot \xi$ is a Gaussian random variable $\mathcal{N}(\lambda \cdot \mu, \lambda \cdot R\lambda)$.

The Gaussian property is robust: it is stable with respect to any linear transform, and it is also stable with respect to conditioning. Indeed, if $\mathcal{L}$ denotes the distribution and $\begin{pmatrix} \xi_1 \\ \xi_2 \end{pmatrix}$ is a Gaussian random vector (with ξ_1 of size d_1 and ξ_2 of size d_2):

$$\mathcal{L}\left(\begin{pmatrix} \xi_1 \\ \xi_2 \end{pmatrix}\right) = \mathcal{N}\left(\begin{pmatrix} \mu_1 \\ \mu_2 \end{pmatrix}, \begin{pmatrix} R_{11} & R_{12} \\ R_{21} & R_{22} \end{pmatrix}\right),$$

with the means μ_1 and μ_2 of sizes d_1 and d_2, the covariance matrices R_{11} of size $d_1 \times d_1$, R_{12} of size $d_1 \times d_2$, $R_{21} = R_{12}^T$ of size $d_2 \times d_1$, and R_{22} of size $d_2 \times d_2$, then the distribution of ξ_1 conditionally on $\xi_2 = x_2$ is Gaussian:

$$\mathcal{L}(\xi_1 | \xi_2 = x_2) = \mathcal{N}\left(\mu_1 + R_{12} R_{22}^{-1}(x_2 - \mu_2), R_{11} - R_{12} R_{22}^{-1} R_{21}\right).$$

This result is obtained from the application of the general conditioning formula (1.51) and from the use of the block inversion formula

$$\begin{pmatrix} R_{11} & R_{12} \\ R_{21} & R_{22} \end{pmatrix}^{-1} = \begin{pmatrix} Q^{-1} & -Q^{-1} R_{12} R_{22}^{-1} \\ -R_{22}^{-1} R_{21} Q^{-1} & R_{22}^{-1} + R_{22}^{-1} R_{21} Q^{-1} R_{12} R_{22}^{-1} \end{pmatrix},$$

where $Q = R_{11} - R_{12} R_{22}^{-1} R_{21}$ is the Schur complement.

The extension to complex-valued random vectors is straightforward: a complex-valued random vector $\xi = (\xi_1, \dots, \xi_d)^T$ has Gaussian statistics if $(\Re e\,\xi_1, \dots, \Re e\,\xi_d, \Im m\,\xi_1, \dots, \Im m\,\xi_d)^T$ is a real-valued Gaussian random vector.

Let $\xi = (\xi_1, \dots, \xi_d)^T$ be a complex Gaussian random vector. We say that ξ is circularly symmetric if $e^{i\phi}\xi$ has the same probability distribution as ξ for all real ϕ. For $d = 1$, i.e., for the case where ξ is a complex Gaussian random variable, circular symmetry holds if and only if $\Re e\,\xi$ and $\Im m\,\xi$ are statistically independent and identically distributed Gaussian statistics with mean zero and equal variance.

1.8.4 Random Processes

Random signals measured in an imaging experiment are conveniently modeled
as random functions of time, which are known as random (or stochastic)
processes.

Remember that a random variable is a random number, in the sense that
a realization of the random variable is a real number and that the statistical
distribution of the random variable is characterized by its PDF. In the
same way, a random process $(\xi(t))_{t \in \mathbb{R}^d}$ is a random function, in the sense
that a realization of the random process is a function from $\mathbb{R}^d$ to $\mathbb{R}$, and
that the distribution of $(\xi(t))_{t \in \mathbb{R}^d}$ is characterized by the finite-dimensional
distributions $(\xi(t_1), \ldots, \xi(t_n))$, for any n, $t_1, \ldots, t_n \in \mathbb{R}^d$ (the fact that the
finite-dimensional distributions completely characterize the distribution of
the random process is not completely trivial and it follows from Kolmogorov's
extension theorem).

As in the case of real random variables, we may not always require a
complete statistical description of a random process, or we may not be able
to obtain it even if desired. In such cases, we work with the first and second
statistical moments. The most important ones are

(i) Mean: $\mathbb{E}[\xi(t)]$;
(ii) Variance: $\mathrm{Var}[\xi(t)] = \mathbb{E}[(\xi(t) - \mathbb{E}[\xi(t)])^2]$;
(iii) Covariance function: $R(t, t+\tau) = \mathbb{E}[(\xi(t) - \mathbb{E}[\xi(t)])(\xi(t+\tau) - \mathbb{E}[\xi(t+\tau)])]$.

We say that a random process $(\xi(t))_{t \in \mathbb{R}^d}$ is a real-valued Gaussian if any
linear combination $\xi_\lambda = \sum_{i=1}^{n} \lambda_i \xi(t_i)$ has Gaussian distribution. In this case ξ_λ
has Gaussian distribution with PDF

$$p_{\xi_\lambda}(x) = \frac{1}{\sqrt{2\pi}\sigma_\lambda} \exp\left(-\frac{(x - m_\lambda)^2}{2\sigma_\lambda^2}\right), \qquad x \in \mathbb{R},$$

where the mean and variance are given by

$$m_\lambda = \sum_{i=1}^{n} \lambda_i \mathbb{E}[\xi(t_i)], \qquad \sigma_\lambda^2 = \sum_{i,j=1}^{n} \lambda_i \lambda_j \mathbb{E}[\xi(t_i)\xi(t_j)] - m_\lambda^2.$$

The first two moments

$$m(t_j) = \mathbb{E}[\xi(t_j)] \quad \text{and} \quad R(t_j, t_l) = \mathbb{E}[(\xi(t_j) - \mathbb{E}[\xi(t_j)])(\xi(t_l) - \mathbb{E}[\xi(t_l)])]$$

characterize the finite-dimensional distribution of the process $(\xi(t))_{t \in \mathbb{R}^d}$.
Indeed, the finite-dimensional distribution of $(\xi(t_1), \ldots, \xi(t_n))$ has PDF
$p(x_1, \ldots, x_n)$ that can be obtained by applying an inverse Fourier transform to

$$\int e^{i\sum_{j=1}^{n}\lambda_j x_j} p(x_1, \ldots, x_n) dx_1 \cdots dx_n$$

$$= \mathbb{E}[e^{i\sum_{j=1}^{n}\lambda_j \xi(t_j)}] = \mathbb{E}[e^{i\xi_\lambda}] = \int e^{ix} p_{\xi_\lambda}(x) dx = \exp\left(im_\lambda - \frac{\sigma_\lambda^2}{2}\right)$$

$$= \exp\left(i\sum_{j=1}^{n}\lambda_j m(t_j) - \frac{1}{2}\sum_{j,l=1}^{n}\lambda_j \lambda_l R(t_j, t_l)\right),$$

which shows with (1.53) that $(\xi(t_1), \ldots, \xi(t_n))$ has a Gaussian PDF with mean $(m(t_j))_{j=1,\ldots,n}$ and covariance matrix $(R(t_j, t_l))_{j,l=1,\ldots,n}$. As a consequence the distribution of a Gaussian process is characterized by the mean function $(m(t_1))_{t_1 \in \mathbb{R}^d}$ and the covariance function $(R(t_1, t_2))_{t_1, t_2 \in \mathbb{R}^d}$.

It is rather easy to simulate a realization of a Gaussian process $(\mu(t))_{t \in \mathbb{R}^d}$ whose mean $m(t)$ and covariance function $R(t, t')$ are given. If $(t_1, \ldots, t_n)$ is a grid of points, then the following algorithm is a random generator of $(\mu(t_1), \ldots, \mu(t_n))$:

Algorithm 1.2 Random generator

---

1. Compute the mean vector $M_i = \mathbb{E}[\xi(t_i)]$ and the covariance matrix $C_{ij} = \mathbb{E}[\zeta(t_i)\xi(t_j)] - \mathbb{E}[\xi(t_i)]\mathbb{E}[\xi(t_j)]$.
2. Generate a random vector $Z = (Z_1, \ldots, Z_n)$ of n independent Gaussian random variables with mean 0 and variance 1 (use **randn** in MATLAB, or use the Box-Müller algorithm).
3. Compute $Y = M + C^{1/2}Z$.
Output: The vector Y has the distribution of $(\xi(t_1), \ldots, \xi(t_n))$ because it has Gaussian distribution and it has the correct mean and covariance.

Note that the computation of the square root is expensive from the computational point of view, and one usually chooses a Cholesky algorithm to compute it. Then the realization is given by $Y = M + GC$ for $C = GG^T$ with G being lower-triangular.

1.8.5 Stationary Random Processes

We say that $(\xi(t))_{t \in \mathbb{R}^d}$ is a stationary stochastic process if the statistics of the process is invariant to a shift in the origin: for any $t_0 \in \mathbb{R}^d$,

$$(\xi(t_0 + t))_t \stackrel{\text{distribution}}{=} (\xi(t))_t.$$

It is a statistical steady state. A necessary and sufficient condition for a stochastic process to be stationary is that, for any integer n, for any

$t_0, t_1, \ldots, t_n \in \mathbb{R}^d$, for any bounded continuous function $\phi \in \mathcal{C}_b(\mathbb{R}^n, \mathbb{R})$, we have

$$\mathbb{E}\left[\phi(\xi(t_0 + t_1), \ldots, \xi(t_0 + t_n))\right] = \mathbb{E}\left[\phi(\xi(t_1), \ldots, \xi(t_n))\right] .$$

Let us consider a stationary process such that $\mathbb{E}[|\xi(t)|] < \infty$. We set $\bar{\xi} = \mathbb{E}[\xi(t)]$. The ergodic theorem claims that the time (or spatial) average can be replaced by the statistical average under the so-called ergodic hypothesis.

Theorem 1.4. *If $(\xi(t))_{t \in \mathbb{R}^d}$ is a stationary process that satisfies the ergodic hypothesis, then*

$$\frac{1}{N^d} \int_{[0,N]^d} \xi(t)\, dt \overset{N \to \infty}{\longrightarrow} \bar{\xi} \quad \text{with probability one.}$$

The ergodic hypothesis requires that the orbit $(\xi(t))_{t \in \mathbb{R}^d}$ visits all of phase space. It is not easy to state and to understand it (see Remark 1.1 below), although it appears as an intuitive notion. The following presents an example of a non-ergodic process.

As an example, let ξ_1 and ξ_2 be two ergodic processes (satisfying the conclusion of Theorem 1.4), and denote $\bar{\xi}_j = \mathbb{E}[\xi_j(t)]$, $j = 1, 2$. Assume $\bar{\xi}_1 \neq \bar{\xi}_2$. Now flip a coin independently of ξ_1 and ξ_2, whose result is $\chi = 1$ with probability $1/2$ and 0 with probability $1/2$. Let $\xi(t) = \chi \xi_1(t) + (1 - \chi)\xi_2(t)$, which is a stationary process with mean $\bar{\xi} = \frac{1}{2}(\bar{\xi}_1 + \bar{\xi}_2)$.

The time- (or spatially-)averaged process satisfies

$$\frac{1}{N^d} \int_{[0,N]^d} \xi(t)\, dt \;=\; \chi \left(\frac{1}{N^d} \int_{[0,N]^d} \xi_1(t)\, dt \right) + (1 - \chi) \left(\frac{1}{N^d} \int_{[0,N]^d} \xi_2(t)\, dt \right)$$

$$\overset{N \to \infty}{\longrightarrow} \chi \bar{\xi}_1 + (1 - \chi)\bar{\xi}_2 ,$$

which is a random limit different from $\bar{\xi}$. The time-averaged limit depends on χ because ξ has been trapped in a part of phase space. The process $(\xi(t))_{t \in \mathbb{R}^d}$ is not ergodic.

Remark 1.1 (Complement on the Ergodic Theory in Dimension One). Here we give a rigorous statement of an ergodic theorem (it is not necessary for the sequel). Let $(\Omega, \mathcal{A}, \mathbb{P})$ be a probability space, that is:

(i) Ω is a non-empty set (the set of all possible realizations);
(ii) $\mathcal{A}$ is a σ-algebra on Ω (the set of events);
(iii) $\mathbb{P} : \mathcal{A} \to [0, 1]$ is a probability, i.e.,

$$\mathbb{P}(\Omega) = 1 \quad \text{and} \quad \mathbb{P}(\cup_j A_j) = \sum_j \mathbb{P}(A_j)$$

for any countable family of disjoint sets $A_j \in \mathcal{A}$.

Let $\theta_t : \Omega \to \Omega$, $t \geq 0$, be a measurable semi-group of shift operators (i.e., $\theta_t^{-1}(A) \in \mathcal{A}$ for any $A \in \mathcal{A}$ and $t \geq 0$, $\theta_0 = I_d$ and $\theta_{t+s} = \theta_t \circ \theta_s$ for any $t, s \geq 0$) that preserves the probability $\mathbb{P}$ (i.e., $\mathbb{P}(\theta_t^{-1}(A)) = \mathbb{P}(A)$ for any $A \in \mathcal{A}$ and $t \geq 0$). The semi-group $(\theta_t)_{t \geq 0}$ is said to be ergodic if the invariant sets are negligible or of negligible complementary, i.e.,

$$\theta_t^{-1}(A) = A \quad \forall t \geq 0 \implies \mathbb{P}(A) = 0 \text{ or } 1 .$$

We then have the following result.

Proposition 1.5. *Let* $f : (\Omega, \mathcal{A}, \mathbb{P}) \to \mathbb{R}$, θ *be a semi-group, and* $\xi(t, \omega) = f(\theta_t(\omega))$. *Then,*

(i) ξ *is a stationary random process.*
(ii) If $f \in L^1(\mathbb{P})$ *and* $(\theta_t)_{t \geq 0}$ *is ergodic, then*

$$\frac{1}{T} \int_0^T \xi(t, \omega) \, dt \stackrel{T \to \infty}{\longrightarrow} \mathbb{E}[f] = \int_\Omega f d\mathbb{P} \quad \text{with probability one.}$$

We now introduce a weaker form of the ergodic theorem that holds true under a simple and explicit condition. Let $(\xi(t))_{t \in \mathbb{R}^d}$ be a stationary process, $\mathbb{E}[\xi^2(0)] < \infty$. In this case, the covariance function depends only on $t_2 - t_1$, so we introduce the autocovariance function

$$c(\tau) = \mathbb{E}\left[(\xi(t) - \bar{\xi})(\xi(t + \tau) - \bar{\xi})\right] ,$$

which does not depend on t since $\xi(t)$ is stationary. By stationarity also, we can see that c is an even function

$$c(-\tau) = \mathbb{E}\left[(\xi(t) - \bar{\xi})(\xi(t - \tau) - \bar{\xi})\right] = \mathbb{E}\left[(\xi(t' + \tau) - \bar{\xi})(\xi(t') - \bar{\xi})\right] = c(\tau) .$$

By Cauchy-Schwarz inequality, we can show that c reaches its maximum at 0:

$$|c(\tau)| \leq \mathbb{E}\left[(\xi(t) - \bar{\xi})^2\right]^{1/2} \mathbb{E}\left[(\xi(t + \tau) - \bar{\xi})^2\right]^{1/2} = c(0) = \text{Var}(\xi(0)) .$$

Proposition 1.6. *Assume that* $\int_0^\infty |c(\tau)| d\tau < \infty$. *Let* $S(N) = \dfrac{1}{N^d} \int_{[0,N]^d} \xi(t) \, dt$. *Then*

$$\mathbb{E}\left[(S(N) - \bar{\xi})^2\right] \stackrel{N \to \infty}{\longrightarrow} 0 ,$$

or more precisely,

$$N^d \mathbb{E}\left[(S(N) - \bar{\xi})^2\right] \stackrel{N \to \infty}{\longrightarrow} \int_{\mathbb{R}^d} c(\tau) d\tau .$$

One should interpret the condition $\int_0^\infty |c(\tau)| d\tau < \infty$ as "the autocovariance function $c(\tau)$ decays to 0 sufficiently fast as $|\tau| \to \infty$." This hypothesis is a mean square version of mixing: $\xi(t)$ and $\xi(t + \tau)$ are approximatively independent for large lags τ.

Proof. The proof consists in a straightforward calculation. We write it in dimension one for simplicity. We have

$$\mathbb{E}\left[(S(N) - \bar{\xi})^2\right] \quad = \quad \mathbb{E}\left[\frac{1}{N^2} \int_0^N dt_1 \int_0^N dt_2 (\xi(t_1) - \bar{\xi})(\xi(t_2) - \bar{\xi})\right]$$

$$\overset{\text{stationarity}}{=} \frac{2}{N^2} \int_0^N dt_1 \int_0^{t_1} dt_2 c(t_1 - t_2)$$

$$\overset{\substack{\tau = t_1 - t_2 \\ h = t_2}}{=} \frac{2}{N^2} \int_0^N d\tau \int_0^{N-\tau} dh c(\tau)$$

$$= \quad \frac{2}{N^2} \int_0^N d\tau (N - \tau) c(\tau) = \frac{2}{N} \int_0^\infty d\tau c_N(\tau) \, ,$$

where $c_N(\tau) = c(\tau)(1 - \tau/N)\chi([0, N])(\tau)$ with $\chi([0, N])$ being the characteristic function of $[0, N]$. By Lebesgue's convergence theorem:

$$N\mathbb{E}\left[(S(N) - \bar{\xi})^2\right] \overset{N \to \infty}{\longrightarrow} 2 \int_0^\infty c(\tau) d\tau \, .$$

This completes the proof.   □

Note that the $L^2(\mathbb{P})$ convergence implies convergence in probability as the limit is deterministic. Indeed, by Chebychev inequality, for any $\delta > 0$,

$$\mathbb{P}\left(|S(N) - \bar{\xi}| \geq \delta\right) \leq \frac{\mathbb{E}\left[(S(N) - \bar{\xi})^2\right]}{\delta^2} \overset{N \to \infty}{\longrightarrow} 0 \, .$$

Note also that we can obtain by the same method that, for any $k \in \mathbb{R}^d$,

$$N^d \mathbb{E}\left[\left|\int_{[0,N]^d} (\xi(t) - \bar{\xi}) e^{ik \cdot t} dt\right|^2\right] \overset{N \to \infty}{\longrightarrow} \int_{\mathbb{R}^d} c(\tau) e^{ik \cdot \tau} d\tau \, ,$$

which shows that the Fourier transform of the autocovariance function of a stationary process is nonnegative. This is a preliminary form of Bochner's theorem which claims that a function $c(\tau)$ is an autocovariance function of a stationary process if and only if its Fourier transform is nonnegative. As a consequence of this fact, for stationary and ergodic or mixing processes, the statistical moments are measurable from any large enough sample.

Define the power spectral density function to be the Fourier transform of the covariance function. If the power spectral density function is a constant over the measurement frequency range, the noise is referred to as white noise in practice.

1.8.6 Stationary Gaussian Random Processes

We finally focus our attention on stationary Gaussian processes. Let us consider a stationary Gaussian process $(\mu(t))_{t\in\mathbb{R}^d}$ with mean zero and autocovariance function $c(\tau) = \mathbb{E}[\mu(t)\mu(t+\tau)]$. The spectral representation of the real-valued stationary Gaussian process $(\mu(t))_{t\in\mathbb{R}^d}$ is

$$\mu(t) = \frac{1}{(2\pi)^{d/2}} \int_{\mathbb{R}^d} e^{ik\cdot t} \sqrt{\mathcal{F}(c)(k)}\hat{n}_k dk \,,$$

with $\hat{n}_k$ is a complex white noise, i.e., $\hat{n}_k$ is complex-valued, Gaussian, $\hat{n}_{-k} = \overline{\hat{n}_k}$, $\mathbb{E}[\hat{n}_k] = 0$, $\mathbb{E}[\hat{n}_k\hat{n}_{k'}] = 0$, and $\mathbb{E}[\hat{n}_k\overline{\hat{n}_{k'}}] = \delta(k-k')$ (the representation is formal, one should in fact use stochastic integrals $d\hat{W}_k = \hat{n}_k dk$). A complex white noise is actually the Fourier transform of a real white noise: we have $\hat{n}_k = (2\pi)^{-d/2}\int e^{ik\cdot t}n(l)dl$ where $n(t)$ is a real white noise, i.e., $n(t)$ real-valued, Gaussian, $\mathbb{E}[n(t)] = 0$, and $\mathbb{E}[n(t)n(t')] = \delta(t-t')$.

It is straightforward to simulate a realization of a stationary Gaussian process (with mean zero and covariance c) using its spectral representation and Fast Fourier Transforms. In dimension $d = 1$, if we fix a grid of points $t_i = (i-1)\Delta t$, $i = 1,\ldots,n$, then one can simulate the vector $(\mu(t_1),\ldots,\mu(t_n))$ by the following algorithm:

Algorithm 1.3 Realization of a stationary Gaussian process

---

1. Evaluate the covariance vector $c = (c(t_1),\ldots,c(t_n))$.
2. Generate a random vector $Z = (Z_1,\ldots,Z_n)$ of n independent Gaussian random variables with mean 0 and variance 1.
3. Filter with the element-wise square root of the (discrete) Fourier transform of c:

$$Y = \text{IFT}\left(\sqrt{\text{DFT}(c)} \times \text{DFT}(Z)\right) .$$

Output: The vector Y is a realization of $(\mu(t_1),\ldots,\mu(t_n))$.

In practice one uses FFT and IFFT instead of DFT and IFT, and one obtains a periodized version of the random vector $(\mu(t_1),\ldots,\mu(t_n))$, due to the FFT. This algorithm is much more efficient than the Cholesky's method.

In imaging problems, a commonly used covariance function is of the form $c(\tau) = e^{-|\tau|^2/l^2}$. Here l is said to be the correlation length of the random process. In this book, to model clutter (or medium noise), we use such a choice for the covariance function. To have a significant interaction between the wave and the medium, the correlation length is chosen to be of order the wavelength.

1.9   Random Matrix Theory

Our main goal in this section is to describe the distribution of the singular values of random matrices. We will first analyze the eigenvalues of simple random diagonalizable matrices and then extend the results to the singular values of the random matrices of interest in the context of imaging.

We will look for different types of results. On the one hand we look for the description of the global distribution of the singular values, and on the other hand we also look for a detailed description of the maximal singular value. There are different types of approaches. Some of them are based on the asymptotic expansions of explicit expressions (for some special models) and they give the more detailed results. Other approaches are based on tools of complex analysis (using in particular Stieltjes transforms of measures) and allow one to obtain asymptotic results for a large class of random matrices. As will be shown later, the distribution of the singular values depends on the structure of the matrix (symmetric, Hermitian, etc.), on the correlation between the random coefficients of the matrix, but not much on the marginal distribution of the coefficients (in the limit of large matrices). A remarkable point is that there are universal results, corresponding to a kind of "central limit theorem", but with unusual scaling and limit distribution.

1.9.1   *Asymptotic Distribution of Eigenvalues of the Invariant Ensembles*

We consider first a simple model of random matrices in

$$\mathcal{S}_n = \{\text{ symmetric real matrices of size } n \times n\} \, .$$

Consider the random matrix M such that

(i) $(M_{jl})_{1 \leq j < l \leq n}$ are independent and identically distributed $\mathcal{N}(0, 1)$;
(ii) $(M_{jj})_{1 \leq j \leq n}$ are independent and identically distributed $\mathcal{N}(0, 2)$;
(iii) $M_{jl} = M_{lj}$ for $1 \leq j < l \leq n$.

We then say that M is a GOE(n)-matrix (GOE stands for Gaussian Orthogonal Ensemble). This model is invariant by any rotation: If P is an orthogonal matrix (i.e., $P^T P = P P^T = I$) and M is a GOE(n)-matrix, then the random matrix $\tilde{M} = PMP^T$ is a GOE(n)-matrix. Indeed, the matrix $\tilde{M}$ is still symmetric, with Gaussian entries, and one can check that the means and covariances of the entries are the ones of the GOE(n)-matrix distribution.

By independence of the coefficients the distribution of M has a PDF with respect of the measure $dM = \prod_{1 \leq j < l \leq n} dM_{jl} \prod_{1 \leq j \leq n} dM_{jj}$:

$$p_S(M) = \prod_{1 \leq j < l \leq n} \frac{1}{\sqrt{2\pi}} \exp\left(-\frac{M_{jl}^2}{2}\right) \prod_{1 \leq j \leq n} \frac{1}{\sqrt{4\pi}} \exp\left(-\frac{M_{jj}^2}{4}\right).$$

Using the symmetry $M_{jl} = M_{lj}$ we can write the PDF as

$$p_S(M) = c_n \exp\left(-\frac{\sum_{1 \leq i,j \leq n} M_{ij}^2}{4}\right),$$

with $c_n = (2\pi)^{-n(n-1)/2}(4\pi)^{-n/2}$. We recognize the trace of M^2 in the exponent which allows us to write

$$p_S(M) = c_n \exp\left(-\frac{\mathrm{Tr}(M^2)}{4}\right). \tag{1.54}$$

The following theorem holds.

Theorem 1.7. *Let M be a $GOE(n)$-matrix. Let $(\lambda_1, \ldots, \lambda_n)$ be the eigenvalues of M sorted in decreasing order. Then the distribution of the random vector $(\lambda_1, \ldots, \lambda_n)$ has the PDF*

$$p(\lambda_1, \ldots, \lambda_n) = \frac{1}{Z_n} \exp\left(-\frac{1}{4} \sum_{i=1}^{n} \lambda_i^2\right) \prod_{1 \leq i < j \leq n} (\lambda_i - \lambda_j),$$

for some normalizing constant Z_n.

We can notice that the eigenvalues are not independent because the PDF cannot be written as a product of one-dimensional PDFs. Compared to the independent case, the presence of the product $\prod_{1 \leq i < j \leq n}(\lambda_i - \lambda_j)$ is a manifestation of level repulsion: the PDF becomes small when two eigenvalues are close to each other. Note finally that this theorem is valid for any n and gives the complete answer to the question: what is the distribution of the eigenvalues of a $GOE(n)$ matrix? However the answer takes a form that is not easy to understand when n is large, and is desirable to determine what is the asymptotic behavior of the eigenvalues in this asymptotic framework.

Proof (of Theorem 1.7). We give the essential steps of the proof. For any symmetric and real matrix M we can associate the eigenvalues $(\lambda_j(M))_{j=1,\ldots,n}$. This map is well-defined provided we require the eigenvalues to be sorted in decreasing order. From the spectral representation of symmetric matrices, the matrix M can be written as

$$M = P\Lambda P^T,$$

where P is an orthogonal matrix and Λ is the diagonal matrix with diagonal entries $\lambda_l(M)$, $l = 1, \ldots, n$. For each $l = 1, \ldots, n$, the vector $(P_{jl})_{j=1,\ldots,n}$ is an eigenvector of M with eigenvalue λ_l, and we can choose the first nonzero coefficient of the vector $(P_{jl})_{j=1,\ldots,n}$ to be positive. Note that the decomposition is then unique provided the eigenvalues are simple, but it is not unique if there are multiple eigenvalues.

If we denote by dP the Haar measure on the set $\mathcal{O}_n$ of orthogonal matrices (which is the uniform measure on $\mathcal{O}_n$) and by $d\Lambda$ the measure $d\lambda_1 \cdots d\lambda_n$, then for any test function f

$$
\begin{aligned}
\mathbb{E}[f(\lambda_1, \ldots, \lambda_n)] &= \int f(\lambda_1(M), \ldots, \lambda_n(M)) p_S(M) dM \\
&= \iint f(\lambda_1, \ldots, \lambda_n) p_S(P\Lambda P^T) \left| \operatorname{Jac} \frac{\partial M}{\partial(\Lambda, P)} \right| dP d\Lambda \\
&= \iint f(\lambda_1, \ldots, \lambda_n) c_n e^{-\frac{1}{4} \sum_{j=1}^{n} \lambda_j^2} \left| \operatorname{Jac} \frac{\partial M}{\partial(\Lambda, P)} \right| dP d\Lambda \,,
\end{aligned}
$$

where we have used the fact that

$$
\operatorname{Tr}((P\Lambda P^T)^2) = \operatorname{Tr}(P\Lambda P^T P\Lambda P^T) = \operatorname{Tr}(\Lambda^2) = \sum_{j=1}^{n} \lambda_j^2 \,.
$$

The set $\mathcal{O}_n$ of orthogonal matrices is a set with $n(n-1)/2$ parameters (it is a compact Lie group of dimension $n(n-1)/2$). Therefore the Jacobian matrix can be written as

$$
\operatorname{Jac} \frac{\partial M}{\partial(\Lambda, P)} = \begin{pmatrix} \frac{\partial M}{\partial \Lambda} \text{ does not depend on } \lambda_j & n \text{ lines} \\ \frac{\partial M}{\partial P} \text{ linear in } \lambda_j & n(n-1)/2 \text{ lines} \end{pmatrix} \,,
$$

since the decomposition $M = P\Lambda P^T$ is linear in Λ. The determinant of the Jacobian matrix is therefore a polynomial in $(\lambda_1, \ldots, \lambda_n)$ of degree $n(n-1)/2$. Moreover, this polynomial cancels when $\lambda_j = \lambda_l$ for some pair of indices (j, l), therefore is must contain the factor $\prod_{j<l}(\lambda_j - \lambda_l)$. Since the degree is $n(n-1)/2$, we must have

$$
\left| \operatorname{Jac} \frac{\partial M}{\partial(\Lambda, P)} \right| = \prod_{j<l}(\lambda_j - \lambda_l) F(P)
$$

for some function F that we do not need to compute. We have

$$
\mathbb{E}[f(\lambda_1, \ldots, \lambda_n)] = \int f(\lambda_1, \ldots, \lambda_n) e^{-\frac{1}{4} \sum_{j=1}^{n} \lambda_j^2} \prod_{j<l}(\lambda_j - \lambda_l) d\Lambda \times \int c_n F(P) dP \,,
$$

which gives the desired result. $\square$

The result can be extended to the Gaussian Unitary Ensemble (GUE). This is a model of random matrices in

$$\mathcal{U}_n = \{ \text{ Hermitian complex matrices of size } n \times n \} .$$

Consider the random matrix M such that

(i) $(M_{jl} = M_{jl}^R + iM_{jl}^I)_{1 \le j < l \le n}$ where M_{jl}^R and M_{jl}^I are independent and identically distributed $\mathcal{N}(0, 1/2)$;

(ii) $(M_{jj})_{1 \le j \le n}$ are independent and identically distributed $\mathcal{N}(0, 1)$;

(iii) $M_{jl} = \overline{M_{lj}}$ for $1 \le j < l \le n$.

We then say that M is a GUE(n)-matrix. This model is invariant by any unitary transform: If P is a unitary matrix and M is a GUE(n)-matrix, then the random matrix $\tilde{M} = PMP^*$ is a GUE(n)-matrix, where * stands for the conjugate transpose. Indeed, the matrix $\tilde{M}$ is still Hermitian, with Gaussian entries, and one can check that the means and covariances of the entries are the ones of the GUE(n)-matrix distribution. By independence of the coefficients the distribution of M has a PDF with respect to the measure $dM = \prod_{1 \le j < l \le n} dM_{jl}^R dM_{jl}^I \prod_{1 \le j \le n} dM_{jj}$ which is of the form

$$p_H(M) = c_n \exp \left(- \frac{\text{Tr}(M^2)}{2} \right)$$

for some constant c_n. Let $(\lambda_1, \dots, \lambda_n)$ be the eigenvalues of M sorted in decreasing order. Then the distribution of the random vector $(\lambda_1, \dots, \lambda_n)$ has the PDF

$$p(\lambda_1, \dots, \lambda_n) = \frac{1}{Z_n} \exp \left(- \frac{1}{2} \sum_{i=1}^n \lambda_i^2 \right) \prod_{1 \le i < j \le n} (\lambda_i - \lambda_j)^2$$

for some normalizing constant Z_n.

Given the complete expressions of the joint PDF of the eigenvalues $(\lambda_1, \dots, \lambda_n)$ of a GOE(n)-matrix or of a GUE(n)-matrix, it is possible to study their asymptotic distributions as $n \to +\infty$ [130]. We first need to take an appropriate scaling. We will analyze the asymptotic behavior of the eigenvalues $(\lambda_1^{(n)}, \dots, \lambda_n^{(n)})$ of the matrix $M/\sqrt{n}$, where M is a GOE(n)-matrix or of a GUE(n)-matrix. Let us first consider the density of states, that is, the random counting measure $\rho^{(n)}$ such that

$$\rho^{(n)}([a, b]) = \frac{1}{n} \text{Card} \left\{ j = 1, \dots, n, \ \lambda_j^{(n)} \in [a, b] \right\} \quad \text{for any } a < b .$$

The random quantity $\rho^{(n)}([a, b])$ is the proportion of eigenvalues of the random matrix $M/\sqrt{n}$ that belong to the interval $[a, b]$. It can be shown that the density of states converges with probability one as $n \to \infty$ to a deterministic continuous measure ρ_{sc}

$$\rho^{(n)}([a,b]) \overset{n \to \infty}{\longrightarrow} \int_a^b \rho_{sc}(\lambda)d\lambda \,, \tag{1.55}$$

where

$$\rho_{sc}(\lambda) = \frac{1}{2\pi} \sqrt{4 - \lambda^2} \chi([-2,2])(\lambda) \,,$$

where $\chi([-2,2])$ is the characteristic function of $[-2,2]$. The distribution with density $\lambda \mapsto (1/2\pi)\sqrt{4 - \lambda^2}\chi([-2,2])(\lambda)$ is called the semi-circle distribution as its graph has the form of a semi-circle. Here, $\rho_{sc}(\lambda)d\lambda$ gives the proportion of eigenvalues in the interval $[\lambda, \lambda + d\lambda]$. We find that the asymptotic density of states is the PDF of the semi-circle distribution $\rho_{sc}(\lambda)$.

Let us now consider the largest eigenvalue

$$\lambda_{\max}^{(n)} = \max\left(\lambda_1^{(n)}, \ldots, \lambda_n^{(n)}\right) \,.$$

From the result (1.55) on the integrated density of states, we can anticipate that $\lambda_{\max}^{(n)}$ converges to 2 as $n \to \infty$, and this is indeed the case. The fluctuations of $\lambda_{\max}^{(n)}$ around its asymptotic limit 2 are interesting because they exhibit an anomalous scaling behavior (which is not the one corresponding to a central limit theorem) and the distribution of the fluctuations is original in the sense that they are not Gaussian but they follow the so-called Tracy-Widom distribution. More exactly, if $\lambda_{\max}^{(n)}$ is the maximal eigenvalue of $M/\sqrt{n}$ where M is a GOE(n)-matrix (resp. a GUE(n)-matrix), then

$$n^{2/3}(\lambda_{\max}^{(n)} - 2)$$

converges in distribution as $n \to +\infty$ to a Tracy-Widom distribution of type 1 (resp. type 2). The type-1 Tracy-Widom distribution has the PDF p_{TW1} such that:

$$\int_{-\infty}^y p_{\text{TW1}}(x)\,dx = \exp\left(-\frac{1}{2}\int_y^\infty (\varphi(x) + (x-y)\varphi^2(x))\,dx\right) \,,$$

where φ is the solution of the Painlevé equation

$$\varphi''(x) = x\varphi(x) + 2\varphi(x)^3, \qquad \varphi(x) \overset{x \to +\infty}{\approx} \text{Ai}(x) \,, \tag{1.56}$$

with Ai being the Airy function defined by (1.42). The expectation is $\int x p_{\text{TW1}}(x)\,dx \approx -1.21$ and the variance is approximately 1.61.

The type-2 Tracy-Widom distribution has the PDF $p_{TW2}(x)$ such that

$$\int_{-\infty}^{y} p_{TW2}(x)\,dx = \exp\left(-\int_{y}^{\infty}(x-y)\varphi^2(x)\,dx\right).$$

The expectation is $\int x p_{TW2}(x)\,dx \approx -1.77$ and the variance is approximately 0.81.

As a warm-up towards the next section devoted to the statistical analysis of singular values of random matrices (with no special symmetry), we consider the Wishart model. Let A be an $m \times n$ real random matrix whose coefficients are independent and identically distributed with the Gaussian distribution with mean zero and variance one. Let us consider the symmetric $n \times n$ real random matrix

$$W = \frac{1}{m}A^T A = \left(\frac{1}{m}\sum_{k=1}^{m} A_{ki}A_{kj}\right)_{i,j=1,\ldots,n}$$

and let us denote by $(\lambda_1^{(n)},\ldots,\lambda_n^{(n)})$ the eigenvalues of W. Define the density of state $\rho^{(n)}$ of W by

$$\rho^{(n)}([a,b]) := \frac{1}{n}\mathrm{Card}\left\{j=1,\ldots,n,\,\lambda_j^{(n)}\in[a,b]\right\}.$$

$\rho^{(n)}$ is a counting measure which consists of a sum of Dirac masses at $\lambda_j^{(n)}$. If $m/n \to c > 0$ as $n \to \infty$, then $\rho^{(n)}$ converges to the measure

$$\rho(\lambda) = (1-c)^+\delta(\lambda) + \frac{1}{2\pi\lambda}\sqrt{[(\lambda-\lambda_-)(\lambda_+ - \lambda)]^+}\,,$$

with $x^+ = \max(x,0)$, $\lambda_\pm = (1+c)\pm 2\sqrt{c}$. In the case $m = n$ we find

$$\rho(\lambda) = \frac{1}{2\pi\sqrt{\lambda}}\sqrt{4-\lambda}\chi([0,4])(\lambda)\,. \tag{1.57}$$

Taking into account that the eigenvalues of W are the squares of the singular values of A, this result shows that the density of states of the singular values of A converges to the quarter-circle law (given by the right-hand side in (1.57)).

1.9.2 SVD of a Gaussian Random Matrix

We consider here an $n \times m$ matrix A with $n \geq m$. We assume that A consists of independent Gaussian noise entries with mean zero and variance $\sigma_{\mathrm{noise}}^2/m$.

We denote by $\sigma_1^{(m)} \geq \sigma_2^{(m)} \geq \sigma_3^{(m)} \geq \cdots \geq \sigma_m^{(m)}$ the singular values of the matrix A sorted in decreasing order and by $\Lambda^{(m)}$ the corresponding density of states defined by

$$\Lambda^{(m)}([a,b]) := \frac{1}{m}\text{Card}\left\{l = 1,\ldots,m\,,\, \sigma_l^{(m)} \in [a,b]\right\} \quad \text{for any } a < b\,.$$

The density of states $\Lambda^{(m)}$ is a counting measure which consists of a sum of Dirac masses:

$$\Lambda^{(m)} = \frac{1}{m}\sum_{j=1}^{m}\delta_{\sigma_j^{(m)}}\,.$$

For large n and m with $n/m = \gamma \geq 1$ fixed we have the following results.

Proposition 1.8. *Let A be an $n \times m$ Gaussian real matrix.*

(i) The random measure $\Lambda^{(m)}$ almost surely converges as $m \to +\infty$ to the deterministic absolutely continuous measure Λ with compact support:

$$\Lambda([\sigma_u,\sigma_v]) = \int_{\sigma_u}^{\sigma_v}\frac{1}{\sigma_{\text{noise}}}\rho_\gamma\left(\frac{\sigma}{\sigma_{\text{noise}}}\right)d\sigma, \qquad 0 \leq \sigma_u \leq \sigma_v\,, \quad (1.58)$$

where ρ_γ is the deformed quarter-circle law given by

$$\rho_\gamma(\sigma) = \begin{cases} \frac{1}{\pi\sigma}\sqrt{\left((\gamma^{1/2}+1)^2 - \sigma^2\right)\left(\sigma^2 - (\gamma^{1/2}-1)^2\right)} \\ \qquad\qquad\qquad \text{if } \gamma^{1/2} - 1 < \sigma \leq \gamma^{1/2} + 1\,, \\ 0 \qquad\quad \text{otherwise.} \end{cases} \quad (1.59)$$

(ii) The normalized l^2-norm of the singular values satisfies

$$m\left[\frac{1}{m}\sum_{j=1}^{m}(\sigma_j^{(m)})^2 - \gamma\sigma_{\text{noise}}^2\right] \xrightarrow{m\to\infty} \sqrt{2\gamma}\sigma_{\text{noise}}^2 Z \text{ in distribution}\,, \quad (1.60)$$

where Z follows a Gaussian distribution with mean zero and variance one.

(iii) The maximal singular value satisfies

$$\sigma_1^{(m)} \approx \sigma_{\text{noise}}\left[\gamma^{1/2} + 1 + \frac{1}{2m^{2/3}}\left(1 + \gamma^{-1/2}\right)^{1/3}Z_1 + o(\frac{1}{m^{2/3}})\right] \text{ in distribution,}$$

$$(1.61)$$

where Z_1 follows a type-1 Tracy Widom distribution.

Proof. The type-1 Tracy-Widom distribution has the PDF p_{TW1} defined in the previous subsection. (i) is due to Marcenko-Pastur [127]. (ii) follows from the expression of the normalized l^2-norm of the singular values in terms of the entries of the matrix:

$$\frac{1}{m}\sum_{j=1}^{m}(\sigma_j^{(m)})^2 = \frac{1}{m}\text{Tr}(A^T A) = \frac{1}{m}\sum_{j=1}^{n}\sum_{l=1}^{m}A_{jl}^2 \,,$$

and from the application of the central limit theorem in the regime $m \gg 1$. (iii) follows from [102]. $\qquad\qquad\qquad\qquad\qquad\qquad\qquad\qquad\qquad\square$

A similar result can be obtained when the matrix A is an $n \times m$ complex-valued matrix with $n \geq m$ and the coefficients of A are independent complex Gaussian random variables with mean zero and variance $\sigma_{\text{noise}}^2/m$ (which means that the real and imaginary parts are independent and identically distributed Gaussian random variables with mean zero and variance $\sigma_{\text{noise}}^2/(2m)$). Then the previous proposition holds, except that the limit of the normalized l^2-norm is $\sqrt{\gamma}\sigma_{\text{noise}}^2 Z$ (instead of $\sqrt{2\gamma}\sigma_{\text{noise}}^2 Z$) and the fluctuations of the maximal singular value are described in terms of a type-2 Tracy-Widom distribution (instead of a type-1 Tracy-Widom distribution).

Proposition 1.9. *Let A be an $n \times m$ Gaussian complex matrix.*

(i) *The random measure $\Lambda^{(m)}$ almost surely converges to the deterministic absolutely continuous measure Λ defined by (1.58).*

(ii) *The normalized l^2-norm of the singular values satisfies*

$$m\left[\frac{1}{m}\sum_{j=1}^{m}(\sigma_j^{(m)})^2 - \gamma\sigma_{\text{noise}}^2\right] \overset{m\to\infty}{\longrightarrow} \sqrt{\gamma}\sigma_{\text{noise}}^2 Z \text{ in distribution,} \quad (1.62)$$

where Z follows a Gaussian distribution with mean zero and variance one.

(iii) *The maximal singular value satisfies*

$$\sigma_1^{(m)} \approx \sigma_{\text{noise}}\left[\gamma^{1/2}+1+\frac{1}{2m^{2/3}}\left(1+\gamma^{-1/2}\right)^{1/3}Z_2+o(\frac{1}{m^{2/3}})\right] \text{ in distribution,}$$
$$(1.63)$$

where Z_2 follows a type-2 Tracy Widom distribution.

1.10 Kalman and Extended Kalman Filters

1.10.1 Kalman Filter

The Kalman filter is a recursive method that uses a stream of noisy observations to produce an optimal estimator of the underlying system state. Consider the following time-discrete dynamical system $(t \geq 1)$:

$$X_t = F_t X_{t-1} + W_t , \tag{1.64}$$

$$Y_t = H_t X_t + V_t . \tag{1.65}$$

where

- X_t is the vector of *system state* at time t;
- Y_t is the vector of *observation* at time t;
- F_t is the state transition matrix which is applied to the previous state X_{t-1};
- H_t is the observation matrix which yields the (noise free) observation from a system state X_t;
- $W_t \sim \mathcal{N}(0, Q_t)$ is the process noise and $V_t \sim \mathcal{N}(0, R_t)$ is the observation noise, with respectively Q_t and R_t the covariance matrix. These two noise components are independent, further, W_t of different time instants are also mutually independent (the same for V_t).

Suppose that X_0 is Gaussian. Then it follows that the process $(X_t, Y_t)_{t \geq 0}$ is Gaussian. The objective is to estimate the system state X_t from the accumulated observations $Y_{1:t} := [Y_1 \dots Y_t]$.

The optimal estimator (in the least-squares sense) of the system state X_t given the observations $Y_{1:t}$ is the conditional expectation

$$\hat{x}_{t|t} = \mathbb{E}[X_t | Y_{1:t}] . \tag{1.66}$$

Since the joint vector $(X_t, Y_{1:t})$ is Gaussian, the conditional expectation $\hat{x}_{t|t}$ is a linear combination of $Y_{1:t}$, which can be written in terms of $\hat{x}_{t-1|t-1}$ and Y_t only. The purpose of the KF is to calculate $\hat{x}_{t|t}$ from $\hat{x}_{t-1|t-1}$ and Y_t.

We summarize the algorithm in the following.

Algorithm 1.4 Kalman filter

Initialization:

$$\hat{x}_{0|0} = \mathbb{E}[X_0], \ P_{0|0} = \mathrm{cov}(X_0) .$$

Prediction:

$$\hat{x}_{t|t-1} = F_t \hat{x}_{t-1|t-1} ,$$

$$\tilde{Y}_t = Y_t - H_t \hat{x}_{t|t-1} ,$$

$$P_{t|t-1} = F_t P_{t-1|t-1} F_t^T + Q_t .$$

Update:

$$S_t = H_t P_{t|t-1} H_t^T + R_t ,$$

$$K_t = P_{t|t-1} H_t^T S_t^{-1} ,$$

$$\hat{x}_{t|t} = \hat{x}_{t|t-1} + K_t \tilde{Y}_t ,$$

$$P_{t|t} = (I - K_t H_t) P_{t-1|t-1} .$$

Note that, to apply the KF algorithm, the covariance matrices Q_t, R_t must be known. Otherwise, they must be estimated from data.

1.10.2 Extended Kalman Filter

Consider now a nonlinear dynamical system:

$$X_t = f_t(X_{t-1}, W_t) , \tag{1.67}$$

$$Y_t = h_t(X_t, V_t) , \tag{1.68}$$

where X_t, Y_t, W_t, V_t are the same as in the Kalman filter, while the functions f_t, h_t are nonlinear and differentiable. Nothing can be said in general on the conditional distribution $X_t | Y_{1:t}$ due to the nonlinearity. The extended Kalman filter (EKF) calculates an approximation of the conditional expectation (1.66) by an appropriate linearization of the state transition and observation models, which makes the general scheme of KF still applicable. However, the resulting algorithm is now not optimal in the least-squares sense due to the approximation.

Let $F_X = \partial_X f(\hat{x}_{t-1|t-1}, 0), F_W = \partial_W f(\hat{x}_{t-1|t-1}, 0)$, the partial derivatives of f (with respect to the system state and the process noise) evaluated at $(\hat{x}_{t-1|t-1}, 0)$, and let $H_X = \partial_X h(\hat{x}_{t|t-1}, 0), H_V = \partial_V h(\hat{x}_{t|t-1}, 0)$ be the partial derivatives of h (with respect to the system state and the observation noise) evaluated at $(\hat{x}_{t|t-1}, 0)$. The EKF algorithm is summarized below.

Algorithm 1.5 Extended Kalman filter

Initialization:

$$\hat{x}_{0|0} = \mathbb{E}[X_0], \ P_{0|0} = \text{cov}(X_0) .$$

Prediction:

$$\hat{x}_{t|t-1} = f(\hat{x}_{t-1|t-1}, 0) ,$$

$$\tilde{Y}_t = Y_t - h(\hat{x}_{t|t-1}, 0) ,$$

$$P_{t|t-1} = F_X P_{t-1|t-1} F_X^T + F_W Q_t F_W^T .$$

Update:

$$S_t = H_X P_{t|t-1} H_X^T + H_V R_t H_V^T ,$$

$$K_t = P_{t|t-1} H_X^T S_t^{-1} ,$$

$$\hat{x}_{t|t} = \hat{x}_{t|t-1} + K_t \tilde{Y}_t ,$$

$$P_{t|t} = (I - K_t H_X) P_{t-1|t-1} .$$

1.11   General Image Characteristics

Irrespectively of the methods used to acquire images, there are a number of criteria by which the image characteristics can be evaluated and compared. The most important of these criteria are spatial resolution and the signal-to-noise ratio. This section covers a number of general concepts applicable to multistatic imaging.

1.11.1   Spatial Resolution

There are a number of measures used to describe the spatial resolution of an imaging modality. We focus on describing a point spread function (PSF) concept and show how to use it to analyze resolution limitation in several practical imaging schemes.

Point Spread Function

Consider an idealized object consisting of a single point. It is likely that the image we obtain from it is a blurred point. Nevertheless, we are still able to identify it as a point. Now, we add another point to the object. If the two points are farther apart, we will see two blurred points. However, as the two points are moving closer to each other, the image looks less like two points. In fact, the two points will merge together to become a single blob when their separation is below a certain threshold. We call this threshold value the resolution limit of the imaging system. Formally stated, the spatial resolution of an imaging system is the smallest separation of two point sources necessary for them to remain resolvable in the resultant image.

In order to arrive at a more quantitative definition of the resolution, we next introduce the point spread function concept. The relationship between an arbitrary object function $I(x)$ and its image $\hat{I}$ can typically be described by $\hat{I}(x) = (I * h)(x)$, where the convolution kernel function $h(x)$ is known as the point spread function since $\hat{I}(x) = h(x)$ for $I(x) = \delta_0(x)$. In a perfect imaging system, the PSF $h(x)$ would be a delta function, and in this case the image would be a perfect representation of the object. If $h(x)$ deviates from a δ-function, $\hat{I}(x)$ will be a blurred version of $I(x)$. The amount of blurring introduced in $\hat{I}(x)$ by an imperfect $h(x)$ can be quantified by the width of $h(x)$. The spatial resolution, W_h, is clearly related to the PSF. It is defined as the full width of $h(x)$ at its half maximum; alternatively the half width of $h(x)$ at its first zero.

If the PSF is a sinc function,

$$h(x) = \frac{\sin kx}{kx} (= j_0(kx)) ,$$

then this definition of resolution coincides with the Rayleigh criterion which states that the two point sources can be resolved if the peak intensity of the *sinc* PSF from one source coincides with the first zero-crossing point of the PSF of the other, i.e., if the two source points are separated by (at least) one-half the wavelength $\lambda := 2\pi/k$. If the PSF is given by

$$h(x) = \frac{J_1(kx)}{kx} \, ,$$

J_1 being the Bessel function of the first-order, then the resolution is given by $W_h \approx 0.61\lambda$ since the first positive zero of J_1 is approximately 3.83.

If the PSF is a Gaussian function,

$$h(x) = e^{-\frac{x^2}{2\sigma^2}} \, ,$$

where σ is the standard deviation of the distribution, then the resolution, defined as the full width at the half maximum, is given by $2\sqrt{2\log 2}\sigma \approx 2.35\,\sigma$.

Consider now the problem of reconstructing a one-dimensional image from its truncated Fourier series. The image reconstructed based on the truncated Fourier series is given by

$$\hat{I}(x) = \frac{1}{\sqrt{2\pi}}\Delta k \sum_{n=-N/2}^{N/2-1} S(n\Delta k)e^{in\Delta k\,x} \, ,$$

where $S(n\Delta k) = \frac{1}{\sqrt{2\pi}} \int_{\mathbb{R}} I(x)e^{-in\Delta k\,x}\, dx$. The underlying PSF is given by

$$h(x) = \Delta k\, \frac{\sin(\pi N\Delta k\,x)}{\sin(\pi\Delta k\,x)} \, ,$$

with Δk being the fundamental frequency and N the number of Fourier samples. Then the resolution (defined as the half width of h at its first zero) is $W_h = 1/(N\Delta k)$. Therefore, we cannot improve image resolution and reduce the number of measured data points at the same time. This assertion is often referred to as the uncertainty relation of Fourier imaging, and in practice, one chooses N as large as signal-to-noise ratio as long as imaging time permits.

1.11.2 Signal-To-Noise Ratio

In imaging it is useful to measure the relative strength of a signal or information versus noise level. For doing so, we define the concept of signal-to-noise ratio.

Let $\hat{I} = I + \xi$ be a measured quantity containing the true signal I and the noise component ξ with zero mean and standard deviation σ_ξ. The signal-to-noise ratio (SNR) for $\hat{I}$ from a single measurement is defined by

$$(\text{SNR})_{\hat{I}} = \frac{|I|}{\sigma_\xi} \ .$$

We remark that the signal-to-noise ratio is sometimes defined by $(|I|/\sigma_\xi)^2$ and that the signal-to-noise ratio in logarithmic decibel scale (dB) is $20 \log(|I|/\sigma_\xi)$.

If N measurements are taken such that $\hat{I}_n = I + \xi_n$ are obtained to produce

$$\frac{1}{N} \sum_{n=1}^{N} \hat{I}_n = I + \frac{1}{N} \sum_{n=1}^{N} \xi_n \ ,$$

then the signal-to-noise ratio for $(1/N) \sum_{n=1}^{N} \hat{I}_n$ is

$$\frac{|I|}{\sqrt{\text{Var}[\frac{1}{N} \sum_{n=1}^{N} \xi_n]}} = \sqrt{N} \frac{|I|}{\sigma_\xi} = \sqrt{N}(\text{SNR})_{\hat{I}} \ ,$$

assuming that the noise for different measurements is uncorrelated. Thus N signal averaging yields an improvement by a factor of $\sqrt{N}$ in the signal-to-noise ratio. Recall that two signals, ξ_1 and ξ_2, are said to be uncorrelated if

$$\mathbb{E}[(\xi_1 - \mathbb{E}[\xi_1])(\xi_2 - \mathbb{E}[\xi_2])] = 0 \ .$$

Bibliography and Discussion

The results on the special functions are from [1, 75, 161]. For a complete account of the mathematical theory of regularization of inverse problems, the reader is referred to the book [74]. See also [103] where regularization methods are analyzed from the point of view of statistics. A convergence proof of the Kaczmarz's method can be found in the book [134]. Landweber iteration method has been introduced in [121] and its convergence analyzed in [92, 96]. Numerical solutions to least-squares problem are discussed in [123]. The results on the prolate spheroidal functions are from [150]. They will be used in Chap. 15 for a resolution analysis. We refer to the book of Breiman [55] for a more complete introduction to probabilistic tools at the level used in this book. A good reference on Random Matrix Theory is the book by Mehta [130]. The Kalman filter has been introduced by Kalman in his famous paper [105]. Since that time, the Kalman filter has been the subject of extensive research and application. For an introduction to the basic discrete Kalman filter and to the extended Kalman filter, we refer to [160].

Chapter 2
Layer Potential Techniques

The inclusion detection, localization, and reconstruction algorithms described in this book rely on asymptotic expansions of the fields when the medium contains inclusions of small volume. Such asymptotics will be investigated in the cases of the conductivity and the Helmholtz equations. As it will be shown in the subsequent chapters, a remarkable feature of these imaging techniques is that they allow a stable and accurate reconstruction of the location and of the geometric features of the inclusions, even for moderately noisy data. The amount of reconstructed information is a function of the signal-to-noise ratio in the data.

We prepare the way in this chapter by reviewing a number of basic facts on the layer potentials for these equations which are very useful for inclusion detection. The most important results in this chapter are on one hand what we call decomposition theorems for transmission problems and on the other hand, the Helmholtz-Kirchhoff identities. For the transmission problems, we prove that the solution is the sum of two functions, one solving the homogeneous problem, the other inheriting geometric properties of the inclusion. These results have many applications. They have been used to prove global uniqueness results for inclusion detection problems [108, 110]. In this book, we will use them to provide asymptotic expansions of the solution perturbations due to presence of small volume inclusions. As will be shown later, the Helmholtz-Kirchhoff identities play a key role in the analysis of resolution in wave imaging.

We begin with the conductivity case by proving a decomposition formula of the steady-state voltage potential into a harmonic part and a refraction part. We then discuss the transmission problem for the Helmholtz equation, and proceed to establish a decomposition formula for the solution to this problem. Compared to the conductivity equation (which models the quasi-static limit for electromagnetic waves), the only new difficulty in establishing a decomposition theorem for the Helmholtz equation is that the equations inside and outside the inclusion are not the same. We should then consider

H. Ammari et al., *Mathematical and Statistical Methods for Multistatic Imaging*, Lecture Notes in Mathematics 2098, DOI 10.1007/978-3-319-02585-8_2, © Springer International Publishing Switzerland 2013

two unknowns and solve a system of equations on the boundary of the inclusion innstead of just one equation. We also note that when dealing with exterior problems for the Helmholtz equation, one should introduce a radiation condition, known as the Sommerfeld radiation condition, to select the physical solution to the problem. Finally, we derive the Helmholtz-Kirchhoff identity, which plays a key role in the resolution analysis, and briefly outline basic results in geometric optics and the implementation of a clutter noise. As will be shown later, the effect of a clutter noise on wave propagation can be modeled by a random travel time model. Random travel time modeling can be also used when the medium is homogeneous but the positions of the sensors are poorly characterized.

2.1   The Laplace Equation

This section deals with the Laplace operator (or Laplacian) in $\mathbb{R}^d$, denoted by Δ. The Laplacian constitutes the simplest example of an elliptic partial differential equation and models the quasi-static approximation for electromagnetic wave propagation. After deriving the fundamental solution for the Laplacian, we shall introduce the single- and double-layer potentials as well as the Neumann-Poincaré operator. We then provide the jump relations and mapping properties of these surface potentials. We review the spectral properties of the Neumann-Poincaré operator. We recall a Calderón identity (also known as Plemelj's symmetrization principle) and apply the symmetrization principle to the Neumann-Poincaré operator. Finally, we introduce the Neumann function and investigate the transmission problem.

2.1.1   Fundamental Solution

To give a fundamental solution to the Laplacian in the general case of the dimension d, we denote by ω_d the area of the unit sphere in $\mathbb{R}^d$. Even though the following result is elementary we give its proof for the reader's convenience.

Lemma 2.1. *A fundamental solution to the Laplacian is given by*

$$\Gamma(x) = \begin{cases} \dfrac{1}{2\pi} \log |x| \,, & d = 2 \,, \\ \dfrac{1}{(2-d)\omega_d} |x|^{2-d} \,, & d \geq 3 \,. \end{cases} \tag{2.1}$$

It satisfies in the sense of distributions $\Delta\Gamma = \delta_0$.

Proof. The Laplacian is radially symmetric, so it is natural to seek Γ in the form $\Gamma(x) = w(r)$ where $r = |x|$. Since

$$\Delta w = \frac{d^2 w}{d^2 r} + \frac{(d-1)}{r}\frac{dw}{dr} = \frac{1}{r^{d-1}}\frac{d}{dr}\left(r^{d-1}\frac{dw}{dr}\right),$$

$\Delta \Gamma = 0$ in $\mathbb{R}^d \setminus \{0\}$ forces that w must satisfy

$$\frac{1}{r^{d-1}}\frac{d}{dr}\left(r^{d-1}\frac{dw}{dr}\right) = 0 \quad \text{for } r > 0,$$

and hence

$$w(r) = \begin{cases} \dfrac{a_d}{(2-d)}\dfrac{1}{r^{d-2}} + b_d & \text{when } d \geq 3, \\ a_2 \log r + b_2 & \text{when } d = 2, \end{cases}$$

for some constants a_d and b_d. The choice of b_d is arbitrary, but a_d is fixed by the requirement that $\Delta \Gamma = \delta_0$ in $\mathbb{R}^d$, where δ_0 is the Dirac function at 0, or in other words

$$\int_{\mathbb{R}^d} \Gamma \Delta \phi = \phi(0) \quad \text{for } \phi \in C_0^\infty(\mathbb{R}^d). \tag{2.2}$$

Any test function $\phi \in C_0^\infty(\mathbb{R}^d)$ has compact support, so we can apply Green's formula over the unbounded domain $\{x : |x| > \epsilon\}$ to arrive at

$$\int_{|x|>\epsilon} \Gamma(x)\Delta\phi(x)\,dx = \int_{|x|=\epsilon} \phi(x)\frac{\partial\Gamma}{\partial\nu}(x)\,d\sigma(x)$$
$$- \int_{|x|=\epsilon} \Gamma(x)\frac{\partial\phi}{\partial\nu}(x)\,d\sigma(x), \tag{2.3}$$

where $\nu = x/|x|$ on $\{|x| = \epsilon\}$. Since

$$\nabla\Gamma(x) = \frac{dw}{dr}\frac{x}{|x|} = \frac{a_d x}{|x|^d} \quad \text{for } d \geq 2,$$

we have

$$\frac{\partial\Gamma}{\partial\nu}(x) = a_d \epsilon^{1-d} \quad \text{for } |x| = \epsilon.$$

Thus by the continuity of ϕ,

$$\int_{|x|=\epsilon} \phi(x)\frac{\partial\Gamma}{\partial\nu}(x)\,d\sigma(x) = \frac{a_d}{\epsilon^{d-1}}\int_{|x|=\epsilon} \phi(x)\,d\sigma(x) \rightarrow a_d \omega_d \phi(0)$$

as $\epsilon \to 0$, whereas

$$\int_{|x|=\epsilon} \Gamma(x)\frac{\partial \phi}{\partial \nu}(x)\, d\sigma(x) = \begin{cases} O(\epsilon) & \text{if } d \geq 3 , \\ O(\epsilon|\log \epsilon|) & \text{if } d = 2 . \end{cases}$$

Thus, if $a_d = 1/\omega_d$, then (2.1) follows from (2.3) after sending $\epsilon \to 0$. $\qquad\square$

Let $a \in \mathbb{R}^d$ and $q \in \mathbb{R}$. Let $\Gamma(x,z)$ be the fundamental solution for a source point at z. The function $q\Gamma(x,z)$ is called the potential due to charges q at the source point z. The function $a \cdot \nabla_z \Gamma(x,z)$ is called the dipole of moment $|a|$ and direction $a/|a|$ at the source point z. It is known that using point charges one can obtain a dipole only approximately (two large charges a small distance apart). See [145].

From now on, we denote by $\Gamma(x,y) := \Gamma(x - y)$ for $x \neq y$. We next prove Green's identity.

Lemma 2.2. *Assume that D is a bounded C^2-domain in $\mathbb{R}^d, d \geq 2$, and let $u \in W^{1,2}(D)$ be a harmonic function. Then for any $x \in D$,*

$$u(x) = \int_{\partial D} \left(u(y)\frac{\partial \Gamma}{\partial \nu_y}(x,y) - \frac{\partial u}{\partial \nu_y}(y)\Gamma(x,y) \right) d\sigma(y) . \qquad (2.4)$$

Proof. For $x \in D$ let $B_\epsilon(x)$ be the ball of center x and radius ϵ. We apply Green's formula to u and $\Gamma(x, \cdot)$ in the domain $D \setminus \overline{B}_\epsilon$ for small ϵ and get

$$\int_{D \setminus B_\epsilon(x)} \left(\Gamma \Delta u - u \Delta \Gamma \right) dy = \int_{\partial D} \left(\Gamma \frac{\partial u}{\partial \nu} - u \frac{\partial \Gamma}{\partial \nu} \right) d\sigma(y)$$
$$- \int_{\partial B_\epsilon(x)} \left(\Gamma \frac{\partial u}{\partial \nu} - u \frac{\partial \Gamma}{\partial \nu} \right) d\sigma(y) .$$

Since $\Delta \Gamma = 0$ in $D \setminus B_\epsilon(x)$, we have

$$\int_{\partial D} \left(\Gamma \frac{\partial u}{\partial \nu} - u \frac{\partial \Gamma}{\partial \nu} \right) d\sigma(y) = \int_{\partial B_\epsilon(x)} \left(\Gamma \frac{\partial u}{\partial \nu} - u \frac{\partial \Gamma}{\partial \nu} \right) d\sigma(y) .$$

For $d \geq 3$, we get by definition of Γ

$$\int_{\partial B_\epsilon(x)} \Gamma \frac{\partial u}{\partial \nu}\, d\sigma(y) = \frac{1}{(2-d)\omega_d}\epsilon^{2-d}\int_{\partial B_\epsilon(x)} \frac{\partial u}{\partial \nu}\, d\sigma(y) = 0$$

and

$$\int_{\partial B_\epsilon(x)} u\frac{\partial \Gamma}{\partial \nu}\, d\sigma(y) = \frac{1}{\omega_d \epsilon^{d-1}}\int_{\partial B_\epsilon(x)} u\, d\sigma(y) = u(x) ,$$

by the mean value property. Proceeding in the same way, we arrive at the same conclusion for $d = 2$.   □

Particularly useful solutions to the Laplace equation in $\mathbb{R}^2$ are homogeneous harmonic polynomials $r^n e^{\pm in\theta}$ with (r, θ) being the polar coordinates.

2.1.2 Layer Potentials

In this subsection we show how important the fundamental solution is to potential theory. It gives rise to integral operators that invert the Laplacian. We need these integral operators (also called layer potentials) in the derivation of the decomposition theorem for solutions to the transmission problem.

Given a bounded $\mathcal{C}^2$-domain D in $\mathbb{R}^d, d \geq 2$, we denote respectively the single- and double-layer potentials of a function $\phi \in L^2(\partial D)$ as $\mathcal{S}_D[\phi]$ and $\mathcal{D}_D[\phi]$, where

$$\mathcal{S}_D[\phi](x) := \int_{\partial D} \Gamma(x, y)\phi(y) \, d\sigma(y) , \quad x \in \mathbb{R}^d , \tag{2.5}$$

$$\mathcal{D}_D[\phi](x) := \int_{\partial D} \frac{\partial}{\partial \nu_y} \Gamma(x, y)\phi(y) \, d\sigma(y) , \quad x \in \mathbb{R}^d \setminus \partial D . \tag{2.6}$$

We begin with the study of their basic properties. We note that for $x \in \mathbb{R}^d \setminus \partial D$ and $y \in \partial D$, $\partial\Gamma/\partial\nu_y(x, y)$ is an L^∞-function in y and harmonic in x, and it is $O(|x|^{1-d})$ as $|x| \to +\infty$. Therefore we readily see that $\mathcal{D}_D[\phi]$ and $\mathcal{S}_D[\phi]$ are well-defined and harmonic in $\mathbb{R}^d \setminus \partial D$. Let us list their behavior at $+\infty$.

Lemma 2.3. *The following holds:*

(i) $\mathcal{D}_D[\phi](x) = O(|x|^{1-d})$ *as* $|x| \to +\infty$.
(ii) $\mathcal{S}_D[\phi](x) = O(|x|^{2-d})$ *as* $|x| \to +\infty$ *when* $d \geq 3$.
(iii) *If* $d = 2$, *we have*

$$\mathcal{S}_D[\phi](x) = \frac{1}{2\pi} \int_{\partial D} \phi(y) \, d\sigma(y) \log |x| + O(|x|^{-1}) \quad \text{as } |x| \to +\infty .$$

(iv) *If* $\int_{\partial D} \phi(y) \, d\sigma = 0$, *then* $\mathcal{S}_D[\phi](x) = O(|x|^{1-d})$ *as* $|x| \to +\infty$ *for* $d \geq 2$.

Proof. The first three properties are fairly obvious from the definitions. Let us show (iv). If $\int_{\partial D} \phi(y) \, d\sigma = 0$, then

$$\mathcal{S}_D[\phi](x) = \int_{\partial D} [\Gamma(x, y) - \Gamma(x, y_0)]\phi(y) d\sigma(y) ,$$

where $y_0 \in D$. Since

$$|\Gamma(x,y) - \Gamma(x,y_0)| \leq C|x|^{1-d} \quad \text{as } |x| \to +\infty \text{ and } y \in \partial D \qquad (2.7)$$

for some constant C, $\mathcal{S}_D[\phi](x) = O(|x|^{1-d})$ as $|x| \to +\infty$. $\qquad\qquad\qquad\square$

Lemma 2.2 shows that if $u \in W^{1,2}(D)$ is harmonic, then for any $x \in D$,

$$u(x) = \mathcal{D}_D[u|_{\partial D}](x) - \mathcal{S}_D\left[\frac{\partial u}{\partial \nu}\bigg|_{\partial D}\right](x) . \qquad (2.8)$$

To solve the Dirichlet and Neumann problems, where either u or $\partial u/\partial \nu$ on ∂D is prescribed, we need to understand well the subtle behaviors of the functions $\mathcal{D}_D[\phi](x \pm t\nu_x)$ and $\nabla \mathcal{S}_D[\phi](x \pm t\nu_x)$ for $x \in \partial D$ as $t \to 0^+$. A detailed discussion of the behavior near the boundary ∂D of $\mathcal{D}_D[\phi]$ and $\nabla \mathcal{S}_D[\phi]$ for a $\mathcal{C}^2$-domain D and a density $\phi \in L^2(\partial D)$ is given below. We shall follow [77].

Let $\langle \, , \, \rangle$ denote the scalar product in $\mathbb{R}^d$. For ease of notation we will sometimes use the dot for the scalar product in $\mathbb{R}^d$. Assume that D is a bounded $\mathcal{C}^2$-domain. Then we have the bound

$$\left|\frac{\langle x-y, \nu_x \rangle}{|x-y|^d}\right| \leq C\frac{1}{|x-y|^{d-2}} \quad \text{for } x,y \in \partial D, x \neq y , \qquad (2.9)$$

which shows that there exists a positive constant C depending only on D such that

$$\int_{\partial D}\left(\frac{|\langle x-y, \nu_x \rangle|}{|x-y|^d} + \frac{|\langle x-y, \nu_y \rangle|}{|x-y|^d}\right) d\sigma(y) \leq C , \qquad (2.10)$$

and

$$\int_{|y-x|<\epsilon}\left(\frac{|\langle x-y, \nu_x \rangle|}{|x-y|^d} + \frac{|\langle x-y, \nu_y \rangle|}{|x-y|^d}\right) d\sigma(y) \leq C \int_0^{\epsilon} \frac{1}{r^{d-2}} r^{d-2}\, dr \qquad (2.11)$$
$$\leq C\epsilon ,$$

for any $x \in \partial D$, by integration in polar coordinates.

Introduce the operator $\mathcal{K}_D : L^2(\partial D) \to L^2(\partial D)$ given by

$$\mathcal{K}_D[\phi](x) = \frac{1}{\omega_d}\int_{\partial D}\frac{\langle y-x, \nu_y \rangle}{|x-y|^d}\phi(y)\, d\sigma(y) . \qquad (2.12)$$

The estimate (2.10) proves that this operator is bounded. In fact, for $\phi, \psi \in L^2(\partial D)$, we estimate

$$\left| \int_{\partial D} \int_{\partial D} \frac{\langle y - x, \nu_y \rangle}{|x - y|^d} \phi(y) \, \psi(x) \, d\sigma(y) \, d\sigma(x) \right| \tag{2.13}$$

via the inequality $2ab \leq a^2 + b^2$. Then, by (2.10), (2.13) is dominated by

$$C \left(||\phi||^2_{L^2(\partial D)} + ||\psi||^2_{L^2(\partial D)} \right) .$$

Replacing ϕ, ψ, by $t\phi, (1/t)\psi$, we see that (2.13) is bounded by

$$C \left(t^2 ||\phi||^2_{L^2(\partial D)} + \frac{1}{t^2} ||\psi||^2_{L^2(\partial D)} \right) ;$$

minimizing over $t \in]0, +\infty[$, via elementary calculus, we see that (2.13) is dominated by $C||\phi||_{L^2(\partial D)} ||\psi||_{L^2(\partial D)}$, proving that $\mathcal{K}_D$ is a bounded operator on $L^2(\partial D)$.

On the other hand, it is easily checked that the operator defined by

$$\mathcal{K}_D^*[\phi](x) = \frac{1}{\omega_d} \int_{\partial D} \frac{\langle x - y, \nu_x \rangle}{|x - y|^d} \phi(y) \, d\sigma(y) , \tag{2.14}$$

is the L^2-adjoint of $\mathcal{K}_D$. We refer to $\mathcal{K}_D^*$ as the Neumann-Poincaré operator.

It is now important to ask about the compactness of these operators; see Sect. 1.6. Indeed, to apply the Fredholm theory for solving the Dirichlet and Neumann problems for the Laplace equation, we will need the following lemma.[1]

Lemma 2.4. *If D is a bounded $\mathcal{C}^2$-domain, then the operators $\mathcal{K}_D$ and $\mathcal{K}_D^*$ are compact operators in $L^2(\partial D)$.*

Proof. It suffices to prove that $\mathcal{K}_D$ is compact in $L^2(\partial D)$ to assert that $\mathcal{K}_D^*$ is compact as well.

Given $\epsilon > 0$, set $\Gamma_\epsilon(x) = \Gamma(x, 0)$ if $|x| > \epsilon$, $\Gamma_\epsilon(x) = 0$ otherwise, and define

$$\mathcal{K}_D^\epsilon[\phi](x) = \int_{\partial D} \frac{\partial \Gamma_\epsilon}{\partial \nu_y}(x - y)\phi(y) \, d\sigma(y) .$$

Then

$$\int_{\partial D} \int_{\partial D} \left| \frac{\partial \Gamma_\epsilon}{\partial \nu_y}(x - y) \right|^2 d\sigma(x) \, d\sigma(y) < +\infty ,$$

hence the operator norm of $\mathcal{K}_D^\epsilon$ on $L^2(\partial D)$ satisfies

---

[1]Note that $\mathcal{C}^{1,\alpha}$-regularity, for some $0 < \alpha < 1$, is enough to ensure Lemma 2.4 to hold.

$$\|\mathcal{K}_D^\epsilon\| \leq \left\| \frac{\partial \Gamma_\epsilon}{\partial \nu} \right\|_{L^2(\partial D \times \partial D)} .$$

Let $\{\phi_p\}_{p=1}^{+\infty}$ be an orthonormal basis for $L^2(\partial D)$. It is an easy consequence of Fubini's theorem that if $\psi_{pq}(x,y) = \phi_p(x)\phi_q(y)$, then $\{\psi_{pq}\}_{p,q=1}^{+\infty}$ is an orthonormal basis for $L^2(\partial D \times \partial D)$. Hence we can write

$$\frac{\partial \Gamma_\epsilon}{\partial \nu_y}(x-y) = \sum_{p,q=1}^{+\infty} \langle \frac{\partial \Gamma_\epsilon}{\partial \nu}, \psi_{pq} \rangle \psi_{pq}(x,y) .$$

Here $\langle, \rangle$ denotes the L^2-product. For $N \in \mathbb{N}, N \geq 2$, let

$$\mathcal{K}_D^{\epsilon,N}[\phi](x) = \sum_{p+q \leq N} \int_{\partial D} \langle \frac{\partial \Gamma_\epsilon}{\partial \nu}, \psi_{pq} \rangle \psi_{pq}(x,y)\phi(y)\, d\sigma(y) .$$

It is clear that the range of $\mathcal{K}_D^{\epsilon,N}$ lies in the span of $\phi_1, \ldots, \phi_N$, so $\mathcal{K}_D^{\epsilon,N}$ is of finite rank. Moreover

$$\left\| \mathcal{K}_D^\epsilon - \mathcal{K}_D^{\epsilon,N} \right\| \leq \left\| \frac{\partial \Gamma_\epsilon}{\partial \nu} - \sum_{p+q \leq N} \langle \frac{\partial \Gamma_\epsilon}{\partial \nu}, \psi_{pq} \rangle \psi_{pq} \right\|_{L^2(\partial D \times \partial D)} \longrightarrow 0 \quad \text{as } N \to +\infty ,$$

and then $\mathcal{K}_D^\epsilon$ is compact. On the other hand,

$$\mathcal{K}_D[\phi](x) = \frac{1}{\omega_d} \int_{|y-x|>\epsilon} \frac{\langle y-x, \nu_y \rangle}{|x-y|^d} \phi(y)\, d\sigma(y)$$
$$+ \frac{1}{\omega_d} \int_{|y-x|<\epsilon} \frac{\langle y-x, \nu_y \rangle}{|x-y|^d} \phi(y)\, d\sigma(y) ,$$
$$= \mathcal{K}_D^\epsilon[\phi](x) + \frac{1}{\omega_d} \int_{|y-x|<\epsilon} \frac{\langle y-x, \nu_y \rangle}{|x-y|^d} \phi(y)\, d\sigma(y) ,$$

and then, by the estimate (2.11) the operator norm of $\mathcal{K}_D - \mathcal{K}_D^\epsilon$ tends to zero as $\epsilon \to 0$, so $\mathcal{K}_D$ is compact. $\square$

In the special case of the unit sphere, we may simplify the expressions defining the operators $\mathcal{K}_D$ and $\mathcal{K}_D^*$.

Lemma 2.5. *(i) Suppose that D is a two dimensional disk with radius r_0. Then,*

$$\frac{\langle x-y, \nu_x \rangle}{|x-y|^2} = \frac{1}{2r_0} \quad \forall\, x,y \in \partial D, x \neq y ,$$

and therefore, for any $\phi \in L^2(\partial D)$,

$$K_D^*[\phi](x) = K_D[\phi](x) = \frac{1}{4\pi r_0} \int_{\partial D} \phi(y) \, d\sigma(y) \,, \qquad (2.15)$$

for all $x \in \partial D$.

(ii) For $d \geq 3$, if D is a sphere with radius r_0, then, we have

$$\frac{\langle x - y, \nu_x \rangle}{|x - y|^d} = \frac{1}{2r_0} \frac{1}{|x - y|^{d-2}} \quad \forall \, x, y \in \partial D, x \neq y \,,$$

and for any $\phi \in L^2(\partial D)$ and $x \in \partial D$,

$$K_D^*[\phi](x) = K_D[\phi](x) = \frac{(2 - d)}{2r_0} S_D[\phi](x) \,. \qquad (2.16)$$

We also remark that if the disk D of radius r_0 is centered at the origin, then one can easily see that for each integer n

$$S_D[e^{in\theta}](x) = \begin{cases} -\dfrac{r_0}{2|n|} \left(\dfrac{r}{r_0}\right)^{|n|} e^{in\theta} & \text{if } |x| = r < r_0 \,, \\[3mm] -\dfrac{r_0}{2|n|} \left(\dfrac{r_0}{r}\right)^{|n|} e^{in\theta} & \text{if } |x| = r > r_0 \,, \end{cases} \qquad (2.17)$$

and hence

$$\frac{\partial}{\partial r} S_D[e^{in\theta}](x) = \begin{cases} -\dfrac{1}{2} \left(\dfrac{r}{r_0}\right)^{|n|-1} e^{in\theta} & \text{if } |x| = r < r_0 \,, \\[3mm] \dfrac{1}{2} \left(\dfrac{r_0}{r}\right)^{|n|+1} e^{in\theta} & \text{if } |x| = r > r_0 \,. \end{cases} \qquad (2.18)$$

It follows from (2.15) that

$$K_D^*[e^{in\theta}] = 0 \quad \forall n \neq 0 \,. \qquad (2.19)$$

We also get

$$D_D[e^{in\theta}](x) = \begin{cases} \dfrac{1}{2} \left(\dfrac{r}{r_0}\right)^{|n|} e^{in\theta} & \text{if } |x| = r < r_0 \,, \\[3mm] -\dfrac{1}{2} \left(\dfrac{r_0}{r}\right)^{|n|} e^{in\theta} & \text{if } |x| = r > r_0 \,. \end{cases}$$

Turning now to the behavior of the double layer potential at the boundary, we first establish that the double layer potential with constant density has a jump.

Lemma 2.6. *If D is a bounded C^2-domain, then $\mathcal{D}_D[1](x) = 0$ for $x \in \mathbb{R}^d \setminus \overline{D}$, $\mathcal{D}_D[1](x) = 1$ for $x \in D$, and $\mathcal{K}_D[1](x) = 1/2$ for $x \in \partial D$.*

Proof. The first result follows immediately from Green's formula, since $\Gamma(x, y)$ is in $C^\infty(\overline{D})$ and harmonic in D as a function of y when $x \in \mathbb{R}^d \setminus \overline{D}$. As for the second result, given $x \in D$, let $\epsilon > 0$ be small enough so that $\overline{B}_\epsilon \subset D$, where B_ϵ is the ball of center x and radius ϵ. We can apply Green's formula to $\Gamma(x, y)$ on the domain $D \setminus \overline{B}_\epsilon$ to obtain

$$0 = \mathcal{D}_D[1](x) - \frac{\epsilon^{1-d}}{\omega_d} \int_{\partial B_\epsilon} d\sigma(y)$$
$$= \mathcal{D}_D[1](x) - 1 \,.$$

Now we prove the third equation. Given $x \in \partial D$, again let B_ϵ be the ball of center x and radius ϵ. Set $\partial D_\epsilon = \partial D \setminus (\partial D \cap B_\epsilon)$, $\partial B'_\epsilon = \partial B_\epsilon \cap D$, and $\partial B''_\epsilon = \{y \in \partial B_\epsilon : \nu_x \cdot y < 0\}$. (Thus $\partial B''_\epsilon$ is the hemisphere of ∂B_ϵ lying on the same side of the tangent plane to ∂D at x.) A further application of Green's formula shows that

$$0 = \frac{1}{\omega_d} \int_{\partial D_\epsilon} \frac{\langle y - x, \nu_y \rangle}{|x - y|^d} \, d\sigma(y) + \int_{\partial B'_\epsilon} \frac{\partial \Gamma}{\partial \nu_y}(x, y) \, d\sigma(y) \,.$$

Thus

$$\frac{1}{\omega_d} \int_{\partial D_\epsilon} \frac{\langle y - x, \nu_y \rangle}{|x - y|^d} \, d\sigma(y) = - \int_{\partial B'_\epsilon} \frac{\partial \Gamma}{\partial \nu_y}(x, y) \, d\sigma(y) = \frac{\epsilon^{1-d}}{\omega_d} \int_{\partial B'_\epsilon} d\sigma(y) \,.$$

On the one hand, we now have

$$\int_{\partial D} \frac{\langle y - x, \nu_y \rangle}{|x - y|^d} \, d\sigma(y) = \lim_{\epsilon \to 0} \int_{\partial D_\epsilon} \frac{\langle y - x, \nu_y \rangle}{|x - y|^d} \, d\sigma(y) \,.$$

Since ∂D is C^2, the distance between the tangent plane to ∂D at x and the points on ∂D at a distance ϵ from x is $O(\epsilon^2)$, so

$$\int_{\partial B'_\epsilon} d\sigma(y) = \int_{\partial B''_\epsilon} d\sigma(y) + O(\epsilon^2) \cdot O(\epsilon^{d-1}) = \frac{\omega_d \epsilon^{d-1}}{2} + O(\epsilon^{d+1}) \,,$$

and the desired result follows. $\square$

Lemma 2.6 can be extended to general densities $\phi \in L^2(\partial D)$. For convenience we introduce the following notation. For a function u defined on $\mathbb{R}^d \setminus \partial D$, we denote

$$u|_{\pm}(x) := \lim_{t \to 0^+} u(x \pm t\nu_x), \quad x \in \partial D \,,$$

and

$$\frac{\partial u}{\partial \nu_x}\bigg|_\pm (x) := \lim_{t \to 0^+} \langle \nabla u(x \pm t\nu_x), \nu_x \rangle \,, \qquad x \in \partial D \,,$$

if the limits exist. Here ν_x is the outward unit normal to ∂D at x.

We relate in the next lemma the traces $\mathcal{D}_D|_\pm$ of the double-layer potential to the operator $\mathcal{K}_D$ defined by (2.12).

Lemma 2.7. *If D is a bounded $\mathcal{C}^2$-domain, then for $\phi \in L^2(\partial D)$*

$$(\mathcal{D}_D[\phi])\big|_\pm (x) = \left(\mp\frac{1}{2}I + \mathcal{K}_D\right)[\phi](x) \quad a.e. \ x \in \partial D \,. \tag{2.20}$$

Proof. First we consider a density $f \in \mathcal{C}^0(\partial D)$. If $x \in \partial D$ and $t < 0$ is sufficiently small, then $x + t\nu_x \in D$, so by Lemma 2.6,

$$\mathcal{D}_D[f](x + t\nu_x) = f(x) + \int_{\partial D} \frac{\partial \Gamma}{\partial \nu_y}(x + t\nu_x, y)(f(y) - f(x)) \, d\sigma(y) \,. \tag{2.21}$$

To prove that the second integral is continuous as $t \to 0^-$, given $\epsilon > 0$ let $\delta > 0$ be such that $|f(y) - f(x)| < \epsilon$ whenever $|y - x| < \delta$. Then, we have

$$\int_{\partial D} \frac{\partial \Gamma}{\partial \nu_y}(x + t\nu_x, y)(f(y) - f(x)) \, d\sigma(y) - \int_{\partial D} \frac{\partial \Gamma}{\partial \nu_y}(x, y)(f(y) - f(x)) \, d\sigma(y)$$

$$= \int_{\partial D \cap B_\delta} \frac{\partial \Gamma}{\partial \nu_y}(x + t\nu_x, y)(f(y) - f(x)) \, d\sigma(y)$$

$$- \int_{\partial D \cap B_\delta} \frac{\partial \Gamma}{\partial \nu_y}(x, y)(f(y) - f(x)) \, d\sigma(y)$$

$$+ \int_{\partial D \setminus B_\delta} \left(\frac{\partial \Gamma}{\partial \nu_y}(x + t\nu_x, y) - \frac{\partial \Gamma}{\partial \nu_y}(x, y) \right) (f(y) - f(x)) \, d\sigma(y)$$

$$= I_1 + I_2 + I_3 \,.$$

Here B_δ is the ball of center x and radius δ. It easily follows from (2.10) that $|I_2| \le C\epsilon$. Since

$$\left| \frac{\partial \Gamma}{\partial \nu_y}(x + t\nu_x, y) - \frac{\partial \Gamma}{\partial \nu_y}(x, y) \right| \le C \frac{|t|}{|x - y|^d} \quad \forall \, y \in \partial D \,,$$

we get $|I_3| \le CM|t|$, where M is the maximum of f on ∂D. To estimate I_1, we assume that $x = 0$ and near the origin, D is given by $y = (y', y_d)$ with $y_d > \varphi(y')$, where φ is a $\mathcal{C}^2$-function such that $\varphi(0) = 0$ and $\nabla \varphi(0) = 0$. With the local coordinates, we can show that

$$\left| \frac{\partial \Gamma}{\partial \nu_y}(x + t\nu_x, y) \right| \le C \frac{|\varphi(y')| + |t|}{(|y'|^2 + |t|^2)^{d/2}} \,,$$

and hence $|I_1| \leq C\epsilon$. A combination of the above estimates yields

$$\limsup_{t \to 0^-} \left| \int_{\partial D} \frac{\partial \Gamma}{\partial \nu_y}(x + t\nu_x, y)(f(y) - f(x)) \, d\sigma(y) \right.$$
$$\left. - \int_{\partial D} \frac{\partial \Gamma}{\partial \nu_y}(x, y)(f(y) - f(x) \, d\sigma(y) \right| \leq C\epsilon \ .$$

Since ϵ is arbitrary, we obtain that

$$(\mathcal{D}_D[f])\big|_-(x) = f(x) + \int_{\partial D} \frac{\partial \Gamma}{\partial \nu_y}(x, y)(f(y) - f(x)) \, d\sigma(y)$$
$$= \left(\frac{1}{2}I + \mathcal{K}_D \right)[f](x) \quad \text{for } x \in \partial D \ .$$

If $t > 0$, the argument is the same except that

$$\int_{\partial D} \frac{\partial \Gamma}{\partial \nu_y}(x + t\nu_x, y) \, d\sigma(y) = 0 \ ,$$

and hence we write

$$\mathcal{D}_D[f](x + t\nu_x) = \int_{\partial D} \frac{\partial \Gamma}{\partial \nu_y}(x + t\nu_x, y)(f(y) - f(x)) \, d\sigma(y), \quad x \in \partial D \ ,$$

instead of (2.21). We leave the rest of the proof to the reader.

Next, consider $\phi \in L^2(\partial D)$. We first note that by (2.10), $\lim_{t \to 0^+} \mathcal{D}_D[\phi](x \pm t\nu_x)$ exists and

$$\left\| \limsup_{t \to 0^+} \mathcal{D}_D[\phi](x \pm t\nu_x) \right\|_{L^2(\partial D)} \leq C\|\phi\|_{L^2(\partial D)} \ ,$$

for some positive constant C independent of ϕ.

To handle the general case, let ϵ be given and choose a function $f \in C^0(\partial D)$ satisfying $\|\phi - f\|_{L^2(\partial D)} < \epsilon$. Then

$$\left| \mathcal{D}_D[\phi](x \pm t\nu_x) - \left(\mp \frac{1}{2}I + \mathcal{K}_D \right)[\phi](x) \right|$$
$$\leq \left| \mathcal{D}_D[f](x \pm t\nu_x) - \left(\mp \frac{1}{2}I + \mathcal{K}_D \right)[f](x) \right| + \left| \mathcal{D}_D[\phi - f](x \pm t\nu_x) \right|$$
$$+ \left| \left(\mp \frac{1}{2}I + \mathcal{K}_D \right)[\phi - f](x) \right| \ .$$

For $\lambda > 0$, let

$$A_\lambda = \left\{ x \in \partial D : \limsup_{t \to 0^+} \left| \mathcal{D}_D[\phi](x \pm t\nu_x) - (\mp \frac{1}{2}I + \mathcal{K}_D)[\phi](x) \right| > \lambda \right\} \ .$$

For a set E let $|E|$ denote its Lebesgue measure. Then

$$
\begin{aligned}
|A_\lambda| &\le \left|\left\{|\mathcal{D}_D[\phi - f]| > \frac{\lambda}{3}\right\}\right| + \left|\left\{|\phi - f| > \frac{2\lambda}{3}\right\}\right| + \left|\left\{|\mathcal{K}_D[\phi - f]| > \frac{\lambda}{3}\right\}\right| \\
&\le (\frac{3}{\lambda})^2 \left(\|\phi - f\|_{L^2(\partial D)}^2 + \frac{1}{4}\|\phi - f\|_{L^2(\partial D)}^2 + \|\mathcal{K}_D[\phi - f]\|_{L^2(\partial D)}^2\right) \\
&\le C(\frac{3}{\lambda})^2 \epsilon^2 .
\end{aligned}
$$

Here we have used the L^2-boundedness of $\mathcal{K}_D$ which is an obvious consequence of Lemma 2.4. Since ϵ is arbitrary, $|A_\lambda| = 0$ for all $\lambda > 0$. This implies that

$$
\lim_{t \to 0+} \mathcal{D}_D[\phi](x \pm t\nu_x) = (\mp\frac{1}{2}I + \mathcal{K}_D)[\phi](x) \quad \text{a.e. } x \in \partial D ,
$$

and completes the proof. □

In a similar way, we can describe the behavior of the gradient of the single layer potential at the boundary. The following lemma reveals the connection between the traces $\partial \mathcal{S}_D/\partial \nu|_\pm$ and the operator $\mathcal{K}_D^*$ defined by (2.14).

Lemma 2.8. *If D is a bounded C^2-domain, then for $\phi \in L^2(\partial D)$*

$$
\frac{\partial}{\partial T}\mathcal{S}_D[\phi]\Big|_+ (x) = \frac{\partial}{\partial T}\mathcal{S}_D[\phi]\Big|_- (x) \quad \text{a.e. } x \in \partial D , \tag{2.22}
$$

where $\partial/\partial T$ is the tangential derivative and

$$
\frac{\partial}{\partial \nu}\mathcal{S}_D[\phi]\Big|_\pm (x) = \left(\pm\frac{1}{2}I + \mathcal{K}_D^*\right)[\phi](x) \quad \text{a.e. } x \in \partial D . \tag{2.23}
$$

It is worth emphasizing that the signs in (2.20) and (2.23) are opposite.
We now consider the integral equations

$$
\left(\frac{1}{2}I + \mathcal{K}_D\right)[\phi] = f \quad \text{and} \quad \left(\frac{1}{2}I - \mathcal{K}_D^*\right)[\psi] = g , \tag{2.24}
$$

for $f, g \in L^2(\partial D)$.

By the trace formulas (2.23) and (2.20) for the single- and double-layer potentials, it is easily seen that if ϕ and ψ are solutions to these equations then $\mathcal{D}_D[\phi]$ solves the Dirichlet problem with Dirichlet data f:

$$
\begin{cases}
\Delta U = 0 & \text{in } D , \\
U = f & \text{on } \partial D ,
\end{cases}
$$

and $-\mathcal{S}_D[\psi]$ solves the Neumann problem with Neumann data g:

$$
\begin{cases}
\Delta V = 0 & \text{in } D, \\
\dfrac{\partial V}{\partial \nu} = g & \text{on } \partial D,
\end{cases}
$$

if g and ψ satisfy $\int_{\partial D} g\, d\sigma = \int_{\partial D} \psi\, d\sigma = 0$.

In view of Lemma 2.4, we can apply the Fredholm theory to study the solvability of the two integral equations in (2.24).

2.1.3 Invertibility of $\lambda I - \mathcal{K}_D^*$

Let D be a bounded domain, and let

$$
L_0^2(\partial D) := \left\{ \phi \in L^2(\partial D) : \int_{\partial D} \phi\, d\sigma = 0 \right\}.
$$

Let $\lambda \neq 0$ be a real number. Of particular interest for solving the transmission problem for the Laplacian would be the invertibility of the operator $\lambda I - \mathcal{K}_D^*$ on $L^2(\partial D)$ or $L_0^2(\partial D)$ for $|\lambda| \geq 1/2$. The case $|\lambda| = 1/2$ corresponds to the integral equations in (2.24).

To further motivate this subsection, suppose that D has conductivity $0 < k \neq 1 < +\infty$. Consider the transmission problem

$$
\begin{cases}
\nabla \cdot \left(1 + (k-1)\chi(D) \right) \nabla u = 0 & \text{in } \mathbb{R}^d, \\
u(x) - H(x) \to 0 & \text{as } |x| \to +\infty,
\end{cases}
$$

where H is a harmonic function in $\mathbb{R}^d$. It can be shown that this problem can be reduced to solving the integral equation

$$
(\lambda I - \mathcal{K}_D^*)[\phi] = \frac{\partial H}{\partial \nu} \quad \text{on } \partial D,
$$

where $\lambda = (k+1)/(2(k-1))$.

First, it was proved by Kellogg in [114] that the eigenvalues of $\mathcal{K}_D^*$ on $L^2(\partial D)$ lie in $]-1/2, 1/2]$. The following injectivity result holds.

Lemma 2.9. *Let λ be a real number and let D be a bounded C^2-domain. The operator $\lambda I - \mathcal{K}_D^*$ is one to one on $L_0^2(\partial D)$ if $|\lambda| \geq 1/2$, and for $\lambda \in]-\infty, -1/2] \cup]1/2, +\infty[$, $\lambda I - \mathcal{K}_D^*$ is one to one on $L^2(\partial D)$.*

Proof. The argument is by contradiction. Let $\lambda \in]-\infty, -1/2] \cup]1/2, +\infty[$, and assume that $\phi \in L^2(\partial D)$ satisfies $(\lambda I - \mathcal{K}_D^*)[\phi] = 0$ and ϕ is not identically zero. Since $\mathcal{K}_D[1] = 1/2$ by Green's formula, we have

$$0 = \int_{\partial D} (\lambda I - \mathcal{K}_D^*)[\phi] \, d\sigma = \int_{\partial D} \phi (\lambda - \mathcal{K}_D[1]) \, d\sigma$$

and thus $\int_{\partial D} \phi \, d\sigma = 0$. Hence $\mathcal{S}_D[\phi](x) = O(|x|^{1-d})$ and $\nabla \mathcal{S}_D[\phi](x) = O(|x|^{-d})$ at infinity for $d \geq 2$. Since ϕ is not identically zero, both of the following numbers cannot be zero:

$$A = \int_D |\nabla \mathcal{S}_D[\phi]|^2 \, dx \text{ and } B = \int_{\mathbb{R}^d \setminus \overline{D}} |\nabla \mathcal{S}_D[\phi]|^2 \, dx \ .$$

In fact, if both of them are zero, then $\mathcal{S}_D[\phi] = $ constant in D and in $\mathbb{R}^d \setminus \overline{D}$. Hence $\phi = 0$ by

$$\left. \frac{\partial}{\partial \nu} \mathcal{S}_D[\phi] \right|_+ - \left. \frac{\partial}{\partial \nu} \mathcal{S}_D[\phi] \right|_- = \phi \quad \text{on } \partial D \ ,$$

which is a contradiction.

On the other hand, using the divergence theorem and (2.23), we have

$$A = \int_{\partial D} (-\frac{1}{2}I + \mathcal{K}_D^*)[\phi] \, \mathcal{S}_D[\phi] \, d\sigma \text{ and } B = - \int_{\partial D} (\frac{1}{2}I + \mathcal{K}_D^*)[\phi] \, \mathcal{S}_D[\phi] \, d\sigma \ .$$

Since $(\lambda I - \mathcal{K}_D^*)[\phi] = 0$, it follows that

$$\lambda = \frac{1}{2} \frac{B - A}{B + A} \ .$$

Thus, $|\lambda| < 1/2$, which is a contradiction and so, for $\lambda \in]-\infty, -\frac{1}{2}] \cup]\frac{1}{2}, +\infty[$, $\lambda I - \mathcal{K}_D^*$ is one to one on $L^2(\partial D)$.

If $\lambda = 1/2$, then $A = 0$ and hence $\mathcal{S}_D[\phi] = $ constant in D. Thus $\mathcal{S}_D[\phi]$ is harmonic in $\mathbb{R}^d \setminus \partial D$, behaves like $O(|x|^{1-d})$ as $|x| \to +\infty$ (since $\phi \in L_0^2(\partial D)$), and is constant on ∂D. By (2.23), we have $\mathcal{K}_D^*[\phi] = (1/2) \phi$, and hence

$$B = - \int_{\partial D} \phi \, \mathcal{S}_D[\phi] \, d\sigma = C \int_{\partial D} \phi \, d\sigma = 0 \ ,$$

which forces us to conclude that $\phi = 0$. This proves that $(1/2) I - \mathcal{K}_D^*$ is one to one on $L_0^2(\partial D)$. $\qquad \square$

We now turn to the surjectivity of the operator $\lambda I - \mathcal{K}_D^*$ on $L^2(\partial D)$ or $L_0^2(\partial D)$. Since D is a bounded $\mathcal{C}^2$-domain, as shown in Lemma 2.4, the operators $\mathcal{K}_D$ and $\mathcal{K}_D^*$ are compact operators in $L^2(\partial D)$. Therefore, the surjectivity of $\lambda I - \mathcal{K}_D^*$ holds, by applying the Fredholm alternative.

2.1.4 Symmetrization of $\mathcal{K}_D^*$

Lemma 2.9 shows that the spectrum of $\mathcal{K}_D^*$ lies in the interval $]-1/2, 1/2]$. In this subsection we symmetrize the non-self-adjoint operator $\mathcal{K}_D^*$. We first look into the kernel of $\mathcal{S}_D$. This will be needed for proving Lemma 2.10.

If $d \geq 3$, then it is known that $\mathcal{S}_D : L^2(\partial D) \to W_1^2(\partial D)$ has a bounded inverse [31, Theorem 2.26].

Suppose now that $d = 2$. If $\phi_0 \in \mathrm{Ker}(\mathcal{S}_D)$, then the function u defined by

$$u(x) := \mathcal{S}_D[\phi_0](x), \quad x \in \mathbb{R}^2$$

satisfies $u = 0$ on ∂D. Therefore, $u(x) = 0$ for all $x \in D$. It then follows from (2.23) that

$$\mathcal{K}_D^*[\phi_0] = \frac{1}{2}\phi_0 \quad \text{on } \partial D . \tag{2.25}$$

If $\phi_0 \in L_0^2(\partial D)$, then $u(x) \to 0$ as $|x| \to \infty$, and hence $u(x) = 0$ for $x \in \mathbb{R}^2 \backslash D$ as well. Thus $\phi_0 = 0$. The eigenfunctions of (2.25) make a one dimensional subspace of $L^2(\partial D)$, which means that $\mathrm{Ker}(\mathcal{S}_D)$ is of at most one dimension.

Let $(\phi_e, a) \in L^2(\partial D) \times \mathbb{R}$ denote the solution of the system

$$\begin{cases} \mathcal{S}_D[\phi_e] + a = 0 , \\ \int_{\partial D} \phi_e = 1 , \end{cases} \tag{2.26}$$

then it can be shown that $\mathcal{S}_D : L^2(\partial D) \to W_1^2(\partial D)$ has a bounded inverse if and only if $a \neq 0$; see again [31, Theorem 2.26].

The following Lemma holds.

Lemma 2.10. *Let $d \geq 2$. The operator $\mathcal{S}_D$ is self-adjoint and $-\mathcal{S}_D \geq 0$ on $L^2(\partial D)$.*

Proof. Let $d \geq 3$. Let $\phi \in L^2(\partial D)$ and define

$$u(x) = \mathcal{S}_D[\phi](x), \quad x \in \mathbb{R}^d . \tag{2.27}$$

Then we have

$$\int_D |\nabla u|^2 \, dx = \int_{\partial D} u\left(-\frac{1}{2}\phi + \mathcal{K}_D^*[\phi]\right) d\sigma ,$$

and by Lemma 2.3 (ii)

$$\int_{\mathbb{R}^d \backslash \overline{D}} |\nabla u|^2 \, dx = -\int_{\partial D} u\left(\frac{1}{2}\phi + \mathcal{K}_D^*[\phi]\right) d\sigma .$$

Summing up the above two identities we find

$$\int_{\mathbb{R}^d} |\nabla u|^2 \, dx = -\int_{\partial D} u\phi d\sigma$$
$$= \langle \phi, -\mathcal{S}_D[\phi] \rangle_{L^2(\partial D)} \,.$$

Thus $-\mathcal{S}_D \geq 0$ and $\mathcal{S}_D$ is self-adjoint. In the two-dimensional case, we use instead of ϕ, $\phi - (\int_{\partial D} \phi)\phi_e$, where ϕ_e is defined (2.26). Since $\phi - (\int_{\partial D} \phi)\phi_e$ has mean zero it follows by using Lemma 2.3 (iv) that

$$\langle \phi - (\int_{\partial D} \phi)\phi_e, -\mathcal{S}_D[\phi - (\int_{\partial D} \phi)\phi_e] \rangle_{L^2(\partial D)} \geq 0 \,,$$

and thus, $\langle \phi, -\mathcal{S}_D[\phi] \rangle_{L^2(\partial D)} \geq 0$. This completes the proof. $\qquad\square$

By Lemma 2.10, there exists a unique square root of $-\mathcal{S}_D$ which we denote by $\sqrt{-\mathcal{S}_D}$; furthermore, $\sqrt{-\mathcal{S}_D}$ is self-adjoint and $\sqrt{-\mathcal{S}_D} \geq 0$.

The following lemma can be proved by Green's formulas.

Lemma 2.11. *We have:*

(i) If u is a solution of $\Delta u = 0$ in D, then

$$\mathcal{S}_D\left[\frac{\partial u}{\partial \nu}\Big|_-\right](x) = \mathcal{D}_D\left[u\big|_-\right](x), \quad x \in \mathbb{R}^d \setminus \overline{D} \,. \qquad (2.28)$$

(ii) If u is a solution of

$$\begin{cases} \Delta u = 0 & \text{in } \mathbb{R}^d \setminus \overline{D} \,, \\ u(x) \to 0, & |x| \to \infty \,, \end{cases} \qquad (2.29)$$

then

$$\mathcal{S}_D\left[\frac{\partial u}{\partial \nu}\Big|_+\right](x) = \mathcal{D}_D\left[u\big|_+\right](x), \quad x \in D \,.$$

The following result is well-known. It shows that $\mathcal{K}_D\mathcal{S}_D$ is self-adjoint on $L^2(\partial D)$.

Lemma 2.12. *The following Calderón's identity (also known as Plemelj's symmetrization principle) holds:*

$$\mathcal{S}_D\mathcal{K}_D^* = \mathcal{K}_D\mathcal{S}_D \quad \text{on } L^2(\partial D) \,. \qquad (2.30)$$

Proof. Let $\phi \in L^2(\partial D)$. If we put $u = \mathcal{S}_D[\phi]$ in (2.28), we have

$$-\frac{1}{2}\mathcal{S}_D[\phi](x) + \mathcal{S}_D\mathcal{K}_D^*[\phi](x) = \mathcal{D}_D\mathcal{S}_D[\phi](x), \quad x \in \mathbb{R}^d \setminus \overline{D} \,.$$

By taking the limit as $x \to \partial D$ from outside D, we obtain (2.30) from Lemma 2.11 by using the jump relation of the double layer potential.   □

We now recall the following result of symmetrization.

Lemma 2.13. *Let M be a Hilbert-Schmidt operator. If there exists a strictly positive bounded self-adjoint operator R such that $R^2 M$ is self-adjoint, then there is a bounded self-adjoint Hilbert-Schmidt operator A such that*

$$AR = RM .\tag{2.31}$$

We use this result and (2.30) to show that, in the two-dimensional case, there is a bounded self-adjoint operator A_D on $\mathrm{Range}(\mathcal{S}_D)$ such that

$$A_D\sqrt{-\mathcal{S}_D} = \sqrt{-\mathcal{S}_D}\mathcal{K}_D^* .\tag{2.32}$$

It is worth mentioning that $\mathcal{K}_D^*$ is a Hilbert-Schmidt operator.

By defining A_D to be 0 on $\mathrm{Ker}(\mathcal{S}_D)$, we extend A_D to $L^2(\partial D)$. We note that (2.32) still holds and the extended operator is self-adjoint in $L^2(\partial D)$. In fact, if $\phi \in \mathrm{Ker}(\mathcal{S}_D)$, then $\mathcal{K}_D^*[\phi] = \frac{1}{2}\phi$ because of (2.25), and hence $\sqrt{-\mathcal{S}_D}\mathcal{K}_D^*[\phi] = 0$. Moreover, if $\phi, \psi \in L^2(\partial D)$, then we can decompose them as $\phi = \phi_1 + \phi_2$ and $\psi = \psi_1 + \psi_2$ where $\phi_1, \psi_1 \in \mathrm{Range}(\mathcal{S}_D)$ and $\phi_2, \psi_2 \in \mathrm{Ker}(\mathcal{S}_D)$. Let $\phi_1 = \sqrt{-\mathcal{S}_D}\tilde{\phi}_1$ and $\psi_1 = \sqrt{-\mathcal{S}_D}\tilde{\psi}_1$. We then get

$$\langle A_D\phi, \psi \rangle = \langle A_D\phi_1, \psi \rangle = \langle A_D\sqrt{-\mathcal{S}_D}\tilde{\phi}_1, \psi \rangle = \langle \sqrt{-\mathcal{S}_D}\mathcal{K}_D^*\tilde{\phi}_1, \psi \rangle$$

$$= \langle \sqrt{-\mathcal{S}_D}\mathcal{K}_D^*\tilde{\phi}_1, \psi_1 \rangle = \langle A_D\phi_1, \psi_1 \rangle = \langle \phi_1, A_D\psi_1 \rangle = \langle \phi, A_D\psi \rangle ,$$

and hence A_D is self-adjoint on $L^2(\partial D)$.

We obtain the following theorem.

Theorem 2.14. *Let $d = 2$. There exists a bounded self-adjoint Hilbert-Schmidt operator A_D such that*

$$A_D\sqrt{-\mathcal{S}_D} = \sqrt{-\mathcal{S}_D}\mathcal{K}_D^* .\tag{2.33}$$

Theorem 2.14 holds true in three dimensions as well even though $\mathcal{K}_D^*$ is not any more a Hilbert-Schmidt operator.

2.1.5 Neumann Function

Let Ω be a smooth bounded domain in $\mathbb{R}^d, d \geq 2$. Let $N(x, z)$ be the Neumann function for $-\Delta$ in Ω corresponding to a Dirac mass at z. That is, N is the solution to

$$\begin{cases} -\Delta_x N(x, z) = \delta_z & \text{in } \Omega , \\ \left. \dfrac{\partial N}{\partial \nu_x} \right|_{\partial\Omega} = -\dfrac{1}{|\partial\Omega|} , \displaystyle\int_{\partial\Omega} N(x, z) \, d\sigma(x) = 0 & \text{for } z \in \Omega . \end{cases} \tag{2.34}$$

Note that the Neumann function $N(x, z)$ is defined as a function of $x \in \overline{\Omega}$ for each fixed $z \in \Omega$.

The operator defined by $N(x, z)$ is the solution operator for the Neumann problem

$$\begin{cases} \Delta U = 0 & \text{in } \Omega , \\ \left. \dfrac{\partial U}{\partial \nu} \right|_{\partial\Omega} = g . \end{cases} \tag{2.35}$$

Namely, the function U defined by

$$U(x) := \int_{\partial\Omega} N(x, z) g(z) \, d\sigma(z) \tag{2.36}$$

is the solution to (2.35) satisfying $\int_{\partial\Omega} U \, d\sigma = 0$.

Now we discuss some properties of N as a function of x and z.

Lemma 2.15 (Neumann Function). *The Neumann function N is symmetric in its arguments, that is, $N(x, z) = N(z, x)$ for $x \neq z \in \Omega$. Furthermore, it has the form*

$$N(x, z) = \begin{cases} -\dfrac{1}{2\pi} \log |x - z| + R_2(x, z) & \text{if } d = 2 , \\ \dfrac{1}{(d-2)\omega_d} \dfrac{1}{|x - z|^{d-2}} + R_d(x, z) & \text{if } d \geq 3 , \end{cases} \tag{2.37}$$

where $R_d(\cdot, z)$ belongs to $W^{\frac{3}{2},2}(\Omega)$ for any $z \in \Omega, d \geq 2$ and solves

$$\begin{cases} \Delta_x R_d(x, z) = 0 & \text{in } \Omega , \\ \left. \dfrac{\partial R_d}{\partial \nu_x} \right|_{\partial\Omega} = -\dfrac{1}{|\partial\Omega|} + \dfrac{1}{\omega_d} \dfrac{\langle x - z, \nu_x \rangle}{|x - z|^d} & \text{for } x \in \partial\Omega . \end{cases}$$

Proof. Pick $z_1, z_2 \in \Omega$ with $z_1 \neq z_2$. Let $B_r(z_p) = \{|x - z_p| < r\}$, $p = 1, 2$. Choose $r > 0$ so small that $B_r(z_1) \cap B_r(z_2) = \emptyset$. Set $N_1(x) = N(x, z_1)$ and $N_2(x) = N(x, z_2)$. We apply Green's formula in $\Omega' = \Omega \setminus B_r(z_1) \cup B_r(z_2)$ to get

$$\int_{\Omega'} \left(N_1 \Delta N_2 - N_2 \Delta N_1 \right) dx = \int_{\partial\Omega} \left(N_1 \frac{\partial N_2}{\partial \nu} - N_2 \frac{\partial N_1}{\partial \nu} \right) d\sigma$$
$$- \int_{\partial B_r(z_1)} \left(N_1 \frac{\partial N_2}{\partial \nu} - N_2 \frac{\partial N_1}{\partial \nu} \right) d\sigma - \int_{\partial B_r(z_2)} \left(N_1 \frac{\partial N_2}{\partial \nu} - N_2 \frac{\partial N_1}{\partial \nu} \right) d\sigma ,$$

where all the derivatives are with respect to the x variable with z_1 and z_2 fixed. Since N_p, $p = 1, 2$, is harmonic for $x \neq z_p$, $\partial N_1/\partial \nu = \partial N_2/\partial \nu = -1/|\partial \Omega|$, and $\int_{\partial \Omega}(N_1 - N_2)\, d\sigma = 0$, we have

$$\int_{\partial B_r(z_1)} \left(N_1 \frac{\partial N_2}{\partial \nu} - N_2 \frac{\partial N_1}{\partial \nu} \right) d\sigma + \int_{\partial B_r(z_2)} \left(N_1 \frac{\partial N_2}{\partial \nu} - N_2 \frac{\partial N_1}{\partial \nu} \right) d\sigma = 0\,.$$

$$(2.38)$$

Thanks to (2.37) which will be proved shortly, the left-hand side of (2.38) has the same limit as $r \to 0$ as the left-hand side of the following identity:

$$\int_{\partial B_r(z_1)} \left(\Gamma \frac{\partial N_2}{\partial \nu} - N_2 \frac{\partial \Gamma}{\partial \nu} \right) d\sigma + \int_{\partial B_r(z_2)} \left(N_1 \frac{\partial \Gamma}{\partial \nu} - \Gamma \frac{\partial N_1}{\partial \nu} \right) d\sigma = 0\,.$$

Since

$$\int_{\partial B_r(z_1)} \Gamma \frac{\partial N_2}{\partial \nu}\, d\sigma \to 0\,, \quad \int_{\partial B_r(z_2)} \Gamma \frac{\partial N_1}{\partial \nu}\, d\sigma \to 0 \quad \text{as } r \to 0\,,$$

and

$$\int_{\partial B_r(z_1)} N_2 \frac{\partial \Gamma}{\partial \nu}\, d\sigma \to N_2(z_1)\,, \quad \int_{\partial B_r(z_2)} N_1 \frac{\partial \Gamma}{\partial \nu}\, d\sigma \to N_1(z_2) \quad \text{as } r \to 0\,,$$

we obtain $N_2(z_1) - N_1(z_2) = 0$, or equivalently $N(z_2, z_1) = N(z_1, z_2)$ for any $z_1 \neq z_2 \in \Omega$.

Now let $R_d, d \geq 2$, be defined by (2.37). Since $R_d(\cdot, z)$ is harmonic in Ω and $\partial R_d(\cdot, z)/\partial \nu \in L^2(\partial \Omega)$, it follows from the standard elliptic regularity theory that $R_d(\cdot, z) \in W^{\frac{3}{2}, 2}(\Omega)$ for any $z \in \Omega$. This completes the proof. $\square$

Note that, because of (2.37), the formula

$$U(x) \approx -\mathcal{S}_\Omega[g](x) \quad \text{in } \Omega$$

is obtained as a first approximation of the solution to the Neumann problem (2.35).

For D, a subset of Ω, let

$$N_D[f](x) := \int_{\partial D} N(x, y) f(y)\, d\sigma(y), \quad x \in \Omega\,.$$

The following lemma relates the fundamental solution Γ to the Neumann function N. It will be used in Chap. 8.

Lemma 2.16. For $z \in \Omega$ and $x \in \partial \Omega$, let $\Gamma_z(x) := \Gamma(x, z)$ and $N_z(x) := N(x, z)$. Then

$$\left(-\frac{1}{2}I + \mathcal{K}_\Omega\right)[N_z](x) = \Gamma_z(x) \quad \textit{modulo constants,} \quad x \in \partial\Omega\,, \qquad (2.39)$$

or, to be more precise, for any simply connected smooth domain D compactly contained in Ω and for any $g \in L_0^2(\partial D)$, we have for any $x \in \partial\Omega$

$$\int_{\partial D}\left(-\frac{1}{2}I + \mathcal{K}_\Omega\right)[N_z](x)g(z)\,d\sigma(z) = \int_{\partial D}\Gamma_z(x)g(z)\,d\sigma(z)\,, \qquad (2.40)$$

or equivalently,

$$\left(-\frac{1}{2}I + \mathcal{K}_\Omega\right)\left[(N_D[g])\big|_{\partial\Omega}\right](x) = \mathcal{S}_D[g]\big|_{\partial\Omega}(x)\,. \qquad (2.41)$$

Proof. Let $f \in L_0^2(\partial\Omega)$ and define

$$u(z) := \int_{\partial\Omega}\left(-\frac{1}{2}I + \mathcal{K}_\Omega\right)[N_z](x)f(x)\,d\sigma(x), \quad z \in \Omega\,.$$

Then

$$u(z) = \int_{\partial\Omega}N(x,z)\left(-\frac{1}{2}I + \mathcal{K}_\Omega^*\right)[f](x)\,d\sigma(x)\,.$$

Therefore, $\Delta u = 0$ in Ω and

$$\frac{\partial u}{\partial\nu}\bigg|_{\partial\Omega} = (-\frac{1}{2}I + \mathcal{K}_\Omega^*)[f]\,.$$

Hence by the uniqueness modulo constants of a solution to the Neumann problem we have

$$u(z) - \mathcal{S}_\Omega[f](z) = \text{constant}, \quad z \in \Omega\,.$$

Thus if $g \in L_0^2(\partial D)$, we obtain

$$\int_{\partial\Omega}\int_{\partial D}\left(-\frac{1}{2}I + \mathcal{K}_\Omega\right)[N_z](x)g(z)f(x)\,d\sigma(z)\,d\sigma(x)$$
$$= \int_{\partial\Omega}\int_{\partial D}\Gamma_z(x)g(z)f(x)\,d\sigma(z)\,d\sigma(x)\,.$$

Since f is arbitrary, we have (2.39) or, equivalently, (2.40). This completes the proof.   □

The following simple observation is useful.

Lemma 2.17. *Let $f \in L^2(\partial\Omega)$ satisfy $(\frac{1}{2}I - \mathcal{K}_\Omega)[f] = 0$. Then f is constant.*

Proof. Let $f \in L^2(\partial\Omega)$ be such that $((1/2)I - \mathcal{K}_\Omega)[f] = 0$. Then for any $g \in L^2(\partial\Omega)$

$$\int_{\partial\Omega} (\frac{1}{2}I - \mathcal{K}_\Omega)[f](x)g(x)\,d\sigma(x) = 0 ,$$

or equivalently,

$$\int_{\partial\Omega} f(x)(\frac{1}{2}I - \mathcal{K}_\Omega^*)[g](x)\,d\sigma(x) = 0 .$$

But $\text{Range}((1/2)I - \mathcal{K}_\Omega^*) = L_0^2(\partial\Omega)$ and so, f is constant. $\square$

We mention that the Neumann function for the ball $B_R(0)$ is given, for any $x, z \in B_R(0)$, by

$$N(x, z) = \frac{1}{4\pi|x - z|} + \frac{1}{4\pi|\frac{R}{|x|}x - \frac{|x|}{R}z|}$$

$$+ \frac{1}{4\pi R} \log \frac{2}{1 - \frac{x\cdot z}{R^2} + \frac{1}{R}|\frac{|x|}{R}z - \frac{R}{|x|}x|} - \frac{1}{2\pi R} \quad \text{for } d = 3 , \quad (2.42)$$

and by

$$N(x, z) = -\frac{1}{2\pi}\left(\log|x-z| + \log\left|\frac{R}{|x|}x - \frac{|x|}{R}z\right| \right) + \frac{\log R}{\pi} \quad \text{for } d = 2 . \quad (2.43)$$

2.1.6   Transmission Problem

Let Ω be a bounded domain in $\mathbb{R}^d$ with a connected smooth boundary and conductivity equal to 1. Consider a bounded domain $D \subset\subset \Omega$ with a connected smooth boundary and conductivity $0 < k \neq 1 < +\infty$.

Let $g \in L_0^2(\partial\Omega)$, and let u and U be respectively the solutions of the Neumann problems

$$\begin{cases} \nabla \cdot \left(1 + (k - 1)\chi(D)\right)\nabla u = 0 & \text{in } \Omega , \\ \left.\dfrac{\partial u}{\partial \nu}\right|_{\partial\Omega} = g , \\ \displaystyle\int_{\partial\Omega} u(x)\,d\sigma(x) = 0 , \end{cases} \qquad (2.44)$$

and

$$\begin{cases} \Delta U = 0 \quad \text{in } \Omega \,, \\ \left. \dfrac{\partial U}{\partial \nu} \right|_{\partial \Omega} = g \,, \\ \displaystyle\int_{\partial \Omega} U(x)\, d\sigma(x) = 0 \,, \end{cases} \tag{2.45}$$

where $\chi(D)$ is the characteristic function of D. Clearly, the Lax-Milgram lemma shows that, given $g \in L_0^2(\partial \Omega)$, there exist unique u and U in $W^{1,2}(\Omega)$ which solve (2.44) and (2.45).

At this point we have all the necessary ingredients to state a decomposition formula of the steady-state voltage potential u into a harmonic part and a refraction part. This decomposition formula is unique and inherits geometric properties of the inclusion D. We refer to [29, 111] for its proof.

Theorem 2.18 (Decomposition Formula). *Suppose that D is a domain compactly contained in Ω with a connected smooth boundary and conductivity $0 < k \neq 1 < +\infty$. Then the solution u of the Neumann problem (2.44) has the representation*

$$u(x) = H(x) + \mathcal{S}_D[\phi](x), \quad x \in \Omega \,, \tag{2.46}$$

where the harmonic function H is given by

$$H(x) = -\mathcal{S}_\Omega[g](x) + \mathcal{D}_\Omega[f](x), \quad x \in \Omega \,, \quad f := u|_{\partial \Omega} \in W^2_{\frac{1}{2}}(\partial \Omega) \,, \tag{2.47}$$

and $\phi \in L_0^2(\partial D)$ satisfies the integral equation

$$\left(\frac{k+1}{2(k-1)} I - \mathcal{K}_D^* \right)[\phi] = \left. \frac{\partial H}{\partial \nu} \right|_{\partial D} \quad \text{on } \partial D \,. \tag{2.48}$$

The decomposition (2.46) into a harmonic part and a refraction part is unique. Moreover, $\forall \, n \in \mathbb{N}$, there exists a constant $C_n = C(n, \Omega, \mathrm{dist}(D, \partial \Omega))$ independent of D and the conductivity k such that

$$\|H\|_{\mathcal{C}^n(\overline{D})} \leq C_n \|g\|_{L^2(\partial \Omega)} \,. \tag{2.49}$$

Furthermore, the following holds

$$H(x) + \mathcal{S}_D[\phi](x) = 0, \quad \forall \, x \in \mathbb{R}^d \setminus \overline{\Omega} \,. \tag{2.50}$$

Another useful expression of the harmonic part H of u is given in the following lemma.

Lemma 2.19. *We have*

$$
H(x) = \begin{cases} u(x) - (k-1) \displaystyle\int_D \nabla_y \Gamma(x,y) \cdot \nabla u(y)\, dy, & x \in \Omega\,, \\[4mm] -(k-1) \displaystyle\int_D \nabla_y \Gamma(x,y) \cdot \nabla u(y)\, dy, & x \in \mathbb{R}^d \setminus \overline{\Omega}\,. \end{cases} \tag{2.51}
$$

Proof. We claim that

$$
\phi = (k-1)\frac{\partial u}{\partial \nu}\bigg|_{-}. \tag{2.52}
$$

In fact, it follows from the jump formula (2.23) and (2.46) and (2.48) that

$$
\frac{\partial u}{\partial \nu}\bigg|_{-} = \frac{\partial H}{\partial \nu} + \frac{\partial}{\partial \nu}\mathcal{S}_D\phi\bigg|_{-} = \frac{\partial H}{\partial \nu} + (-\frac{1}{2}I + \mathcal{K}_D^*)\phi = \frac{1}{k-1}\phi \quad \text{on } \partial D\,.
$$

Then (2.51) follows from (2.50) and (2.52) by Green's formula.   □

Let $g \in L_0^2(\partial \Omega)$ and

$$
U(y) := \int_{\partial \Omega} N(x,y)g(x)\, d\sigma(x)\,.
$$

Then U is the solution to the Neumann problem (2.45) and the following representation holds.

Theorem 2.20. *The solution u of (2.44) can be represented as*

$$
u(x) = U(x) - N_D[\phi](x), \quad x \in \partial \Omega\,, \tag{2.53}
$$

where ϕ is defined in (2.48).

Proof. By substituting (2.46) into (2.47), we obtain

$$
H(x) = -\mathcal{S}_\Omega[g](x) + \mathcal{D}_\Omega\Big[H|_{\partial\Omega} + (\mathcal{S}_D[\phi])|_{\partial\Omega}\Big](x), \quad x \in \Omega\,.
$$

It then follows from (2.20) that

$$
\left(\frac{1}{2}I - \mathcal{K}_\Omega\right)[H|_{\partial\Omega}] = -(\mathcal{S}_\Omega[g])|_{\partial\Omega} + \left(\frac{1}{2}I + \mathcal{K}_\Omega\right)[(\mathcal{S}_D[\phi])|_{\partial\Omega}] \quad \text{on } \partial\Omega\,. \tag{2.54}
$$

Since $U = -\mathcal{S}_\Omega[g] + \mathcal{D}_\Omega[U|_{\partial\Omega}]$ in Ω by Green's formula, we have

$$\left(\frac{1}{2}I - \mathcal{K}_\Omega\right)[U|_{\partial\Omega}] = -(\mathcal{S}_\Omega[g])|_{\partial\Omega} . \tag{2.55}$$

Since $\phi \in L_0^2(\partial D)$, it follows from (2.39) that

$$-\left(\frac{1}{2}I - \mathcal{K}_\Omega\right)[(N_D[\phi])|_{\partial\Omega}] = (\mathcal{S}_D[\phi])|_{\partial\Omega} . \tag{2.56}$$

Then, from (2.54), (2.55), and (2.56), we have

$$\left(\frac{1}{2}I - \mathcal{K}_\Omega\right)\left[H|_{\partial\Omega} - U|_{\partial\Omega} + \left(\frac{1}{2}I + \mathcal{K}_\Omega\right)[(N_D[\phi])|_{\partial\Omega}]\right] = 0 .$$

Therefore, we have from Lemma 2.17

$$H|_{\partial\Omega} - U|_{\partial\Omega} + \left(\frac{1}{2}I + \mathcal{K}_\Omega\right)[(N_D[\phi])|_{\partial\Omega}] = C \text{ (constant).} \tag{2.57}$$

Note from (2.41) that

$$(\frac{1}{2}I + \mathcal{K}_\Omega)[(N_D[\phi])|_{\partial\Omega}] = (N_D[\phi])|_{\partial\Omega} + (\mathcal{S}_D[\phi])|_{\partial\Omega} .$$

Thus we get from (2.46) and (2.57) that

$$u|_{\partial\Omega} = U|_{\partial\Omega} - (N_D[\phi])|_{\partial\Omega} + C . \tag{2.58}$$

Since all the functions entering in (2.58) belong to $L_0^2(\partial\Omega)$, we conclude that $C = 0$, and the theorem is proved. $\qquad\square$

2.2 Helmholtz Equation

Consider the scalar wave equation $\partial_t^2 U - \Delta U = 0$. We obtain a time-harmonic solution $U(x,t) = \Re e[e^{-ikt}u(x)]$ if the space-dependent part u satisfies the Helmholtz equation, $\Delta u + k^2 u = 0$.

Mathematical models for acoustical and microwave soundings of biological media involve the Helmholtz equation.

This section begins by discussing the well-known Sommerfeld radiation condition, and by deriving a fundamental solution. We then introduce the single- and double-layer potentials, and state Rellich's lemma. Then, we establish two decomposition formulas for the solution to the transmission problem. We also discuss the reciprocity property and derive the Helmholtz-Kirchhoff identity for fundamental solutions of the Helmholtz equations.

A short introduction to geometric optics is given. The implementation of a cluttered medium is described. Finally, some useful results on the wave equation are provided.

2.2.1 Fundamental Solution

A fundamental solution $\Gamma_k(x)$ to the Helmholtz operator $\Delta + k^2$ in $\mathbb{R}^d$ is a solution (in the sense of distributions) of

$$(\Delta + k^2)\Gamma_k = \delta_0 , \tag{2.59}$$

where δ_0 is the Dirac mass at 0. Solutions are not unique, since we can add to a solution any plane wave (of the form $e^{ik\theta \cdot x}, \theta \in \mathbb{R}^d : |\theta| = 1$) or any combination of such plane waves. So, we need to specify the behavior of the solutions at infinity. It is natural to look for radial solutions of the form $\Gamma_k(x) = w_k(r)$ that is subject to the extra Sommerfeld radiation condition or outgoing wave condition

$$\left| \frac{dw_k}{dr} - ikw_k \right| \leq Cr^{-(d+1)/2} \quad \text{at infinity.} \tag{2.60}$$

If $d = 3$, (2.59) becomes

$$\frac{1}{r^2}\frac{d}{dr}r^2\frac{dw_k}{dr} + k^2 w_k = 0, \quad r > 0 ,$$

whose solution is

$$w_k(r) = c_1 \frac{e^{ikr}}{r} + c_2 \frac{e^{-ikr}}{r} .$$

It is easy to check that the Sommerfeld radiation condition (2.60) leads to $c_2 = 0$ and then (2.59) leads to $c_1 = -1/(4\pi)$.

If $d = 2$, (2.59) becomes

$$\frac{1}{r}\frac{d}{dr}r\frac{dw_k}{dr} + k^2 w_k = 0, \quad r > 0 .$$

This is a Bessel equation whose solutions are not elementary functions. From Sect. 1.3, we know that the Hankel functions of the first and second kinds of order 0, $H_0^{(1)}(kr)$ and $H_0^{(2)}(kr)$, form a basis for the solution space. At infinity $(r \to +\infty)$, only $H_0^{(1)}(kr)$ satisfies the outgoing radiation condition (2.60). At the origin $(r \to 0)$, $H_0^{(1)}(kr)$ behaves like $(2i/\pi)\log(r)$. The following lemma holds.

Lemma 2.21 (Fundamental Solution). *The outgoing fundamental solution* $\Gamma_k(x)$ *to the operator* $\Delta + k^2$ *is given by*

$$\Gamma_k(x) = \begin{cases} -\dfrac{i}{4}H_0^{(1)}(k|x|) \,, & d = 2 \,, \\[2mm] -\dfrac{e^{ik|x|}}{4\pi|x|} \,, & d = 3 \,, \end{cases}$$

for $x \neq 0$, *where* $H_0^{(1)}$ *is the Hankel function of the first kind of order* 0.

The following Graf's addition formula for $d = 2$ will be useful [158].

Lemma 2.22. *For* $|x| > |y|$, *we have*

$$H_0^{(1)}(k|x - y|) = \sum_{n \in \mathbb{Z}} H_n^{(1)}(k|x|)e^{in\theta_x} J_n(k|y|)e^{-\theta_y} \,, \qquad (2.61)$$

where $x = (|x|, \theta_x)$ *and* $y = (|y|, \theta_y)$ *in polar coordinates. Here* $H_n^{(1)}$ *is the Hankel function of the first kind of order* n *and* J_n *is the Bessel function of order* n; *see (1.11) and (1.25).*

In three dimensions, the following addition formula holds for $|x| > |y|$:

$$\frac{e^{ik|x-y|}}{4\pi|x - y|} = ik \sum_{l=0}^{+\infty} \sum_{m=-l}^{l} h_l^{(1)}(k|x|)j_l(k|y|)Y_{lm}(\theta_x, \phi_x)Y_{lm}(\theta_y, \phi_y) \,, \qquad (2.62)$$

where $x = (|x|, \theta_x, \phi_x)$, $y = (|y|, \theta_y, \phi_y)$ in the spherical coordinates and Y_{lm} is the spherical harmonic function. Here, j_l and $h_l^{(1)}$ are defined by (1.29) and (1.31). Formulas (2.61) and (2.62) are particularly useful since they will allow us to introduce in Chap. 5 the notion of scattering coefficients for the solutions to the Helmholtz equation.

Another useful decomposition of Γ_k is into plane waves. The following decomposition, known as the Weyl representation of cylindrical and spherical waves holds:

$$\Gamma_k(x) = -ic_d \int_{\mathbb{R}^{d-1}} \frac{1}{\beta(\alpha)} e^{i(\beta(\alpha)|x_d| + \alpha \cdot \tilde{x})} \, d\alpha \,, \qquad (2.63)$$

where $x = (\tilde{x}, x_d)$, $\tilde{x} = (x_1, \ldots, x_{d-1})$,

$$\beta(\alpha) = \begin{cases} \sqrt{k^2 - |\alpha|^2}, & |\alpha| < k \,, \\[2mm] i\sqrt{|\alpha|^2 - k^2}, & |\alpha| \geq k \,, \end{cases}$$

and

$$c_2 = \frac{1}{4\pi}, \quad c_3 = \frac{1}{8\pi^2} .$$

As will be shown in Chap. 13, (2.63) plays a key role in diffraction tomography. From now on, we denote by $\Gamma_k(x, y) := \Gamma_k(x - y)$ for $x \neq y$.

Particular solutions to the Helmholtz equation in $\mathbb{R}^d, d = 2, 3$, are plane waves given by $e^{ik\theta \cdot x}$ where θ is a unit real vector, and cylindrical and spherical waves defined by $\Gamma_k(x, y)$ with y being the source point for respectively $d = 2$ and 3. These particular solutions will be very useful in the subsequent chapters.

2.2.2   Layer Potentials

For a bounded smooth domain D in $\mathbb{R}^d$ and $k > 0$ let $\mathcal{S}_D^k$ and $\mathcal{D}_D^k$ be the single- and double-layer potentials defined by Γ_k, that is,

$$\mathcal{S}_D^k[\phi](x) = \int_{\partial D} \Gamma_k(x, y)\phi(y) \, d\sigma(y) , \quad x \in \mathbb{R}^d ,$$

$$\mathcal{D}_D^k[\phi](x) = \int_{\partial D} \frac{\partial \Gamma_k(x, y)}{\partial \nu_y} \phi(y) \, d\sigma(y) , \quad x \in \mathbb{R}^d \setminus \partial D ,$$

for $\phi \in L^2(\partial D)$. Because $\Gamma_k - \Gamma$, where Γ is defined by (2.1), is a smooth function, we can easily prove from (2.23) and (2.20) that

$$\frac{\partial(\mathcal{S}_D^k[\phi])}{\partial \nu}\bigg|_{\pm} (x) = \left(\pm \frac{1}{2}I + (\mathcal{K}_D^k)^* \right)[\phi](x) \quad \text{a.e. } x \in \partial D , \qquad (2.64)$$

$$(\mathcal{D}_D^k[\phi])\bigg|_{\pm} (x) = \left(\mp \frac{1}{2}I + \mathcal{K}_D^k \right)[\phi](x) \quad \text{a.e. } x \in \partial D , \qquad (2.65)$$

for $\phi \in L^2(\partial D)$, where $\mathcal{K}_D^k$ is the operator defined by

$$\mathcal{K}_D^k[\phi](x) = \int_{\partial D} \frac{\partial \Gamma_k(x, y)}{\partial \nu_y} \phi(y) d\sigma(y) , \qquad (2.66)$$

and $(\mathcal{K}_D^k)^*$ is given by

$$(\mathcal{K}_D^k)^*[\phi](x) = \int_{\partial D} \frac{\partial \Gamma_k(x, y)}{\partial \nu_x} \phi(y) d\sigma(y) . \qquad (2.67)$$

Moreover, the integral operators $\mathcal{K}_D^k$ and $(\mathcal{K}_D^k)^*$ are compact on $L^2(\partial D)$.

We will need the following important result from the theory of the Helmholtz equation. It will help us to prove uniqueness of the solution to exterior Helmholtz problems. For its proof we refer to [68, Lemma 2.11] or [129, Lemma 9.8].

Lemma 2.23 (Rellich's Lemma). *Let $R_0 > 0$ and $B_R(0) = \{|x| < R\}$. Let u satisfy the Helmholtz equation $\Delta u + k^2 u = 0$ for $|x| > R_0$. Assume, furthermore, that*

$$\lim_{R \to +\infty} \int_{\partial B_R(0)} |u(x)|^2 \, d\sigma(x) = 0 \, .$$

Then, $u \equiv 0$ for $|x| > R_0$.

Note that the assertion of this lemma does not hold if k is imaginary or $k = 0$.

Now we can state the following uniqueness result for the Helmholtz equation.

Lemma 2.24. *Suppose $d = 2$ or 3. Let D be a bounded C^2-domain in $\mathbb{R}^d$. Let $u \in W^{1,2}_{\mathrm{loc}}(\mathbb{R}^d \setminus \overline{D})$ satisfy*

$$\begin{cases} \Delta u + k^2 u = 0 & in \ \mathbb{R}^d \setminus \overline{D} \, , \\ \left| \dfrac{\partial u}{\partial r} - iku \right| = O\left(r^{-(d+1)/2} \right) & as \ r = |x| \to +\infty \quad uniformly \ in \ \dfrac{x}{|x|} \, , \\ \Im m \displaystyle\int_{\partial D} \overline{u} \dfrac{\partial u}{\partial \nu} \, d\sigma = 0 \, . \end{cases}$$

Then, $u \equiv 0$ in $\mathbb{R}^d \setminus \overline{D}$.

Proof. Let $B_R(0) = \{|x| < R\}$. For R large enough, $D \subset B_R(0)$. Notice first that by multiplying $\Delta u + k^2 u = 0$ by $\overline{u}$ and integrating by parts over $B_R(0) \setminus \overline{D}$ we arrive at

$$\Im m \int_{\partial B_R(0)} \overline{u} \frac{\partial u}{\partial \nu} \, d\sigma = 0 \, ,$$

since

$$\Im m \int_{\partial D} \overline{u} \frac{\partial u}{\partial \nu} \, d\sigma = 0 \, .$$

Thus we have

$$\Im m \int_{\partial B_R(0)} \overline{u} \left(\frac{\partial u}{\partial \nu} - iku \right) d\sigma = -k \int_{\partial B_R(0)} |u|^2 \, .$$

Applying the Cauchy–Schwarz inequality,

$$\left| \Im m \int_{\partial B_R(0)} \overline{u}\left(\frac{\partial u}{\partial \nu} - iku \right) d\sigma \right| \leq \left(\int_{\partial B_R(0)} |u|^2 \right)^{1/2} \left(\int_{\partial B_R(0)} \left| \frac{\partial u}{\partial \nu} - iku \right|^2 d\sigma \right)^{1/2}$$

and using the Sommerfeld radiation condition

$$\left| \frac{\partial u}{\partial r} - iku \right| = O\left(r^{-(d+1)/2} \right) \quad \text{as } r \to +\infty ,$$

we get

$$\left| \Im m \int_{\partial B_R(0)} \overline{u}\left(\frac{\partial u}{\partial \nu} - iku \right) d\sigma \right| \leq \frac{C}{R} \left(\int_{\partial B_R(0)} |u|^2 \right)^{1/2}$$

for some positive constant C independent of R. Consequently, we obtain that

$$\left(\int_{\partial B_R(0)} |u|^2 \right)^{1/2} \leq \frac{C}{R} ,$$

which indicates by Rellich's Lemma that $u \equiv 0$ in $\mathbb{R}^d \setminus \overline{B_R(0)}$. Hence, by the unique continuation property for $\Delta + k^2$, we can conclude that $u \equiv 0$ up to the boundary ∂D. This finishes the proof.   □

2.2.3   Transmission Problem

Introduce the piecewise constant functions

$$\mu(x) = \begin{cases} \mu_0 , & x \in \Omega \setminus \overline{D} , \\ \mu_\star , & x \in D , \end{cases} \tag{2.68}$$

and

$$\varepsilon(x) = \begin{cases} \varepsilon_0 , & x \in \Omega \setminus \overline{D} , \\ \varepsilon_\star , & x \in D , \end{cases} \tag{2.69}$$

where $\mu_0, \mu_\star, \varepsilon_0$, and $\varepsilon_\star$ are positive constants.

Let $f \in W_{\frac{1}{2}}^2(\partial \Omega)$, and let u and U denote the solutions to the Helmholtz equations

$$\begin{cases} \nabla \cdot (\frac{1}{\mu}\nabla u) + \omega^2 \varepsilon u = 0 \ \text{ in } \Omega \,, \\ u = f \ \text{ on } \partial\Omega \,, \end{cases} \tag{2.70}$$

and

$$\begin{cases} \Delta U + \omega^2 \varepsilon_0 \mu_0 U = 0 \ \text{ in } \Omega \,, \\ U = f \ \text{ on } \partial\Omega \,. \end{cases} \tag{2.71}$$

In electromagnetics, ε_0 and $\varepsilon_\star$ are electrical permittivities, μ_0 and $\mu_\star$ are magnetic permeabilities, and u and U are electric potentials. In acoustics, one replaces permittivity and permeability by compressibility and volume density of mass, and the scalar electric potential by the scalar acoustic pressure.

We now present two decompositions of the solution of (2.70) similar to the representation formula (2.46) for the transmission problem for the harmonic equation. To do so, we first state the following theorem which is of importance to us for establishing our decomposition formulas. We refer the reader to [29] for its proof.

Theorem 2.25. *Let* $k_\star^2 := \omega^2 \mu_\star \varepsilon_\star$. *Suppose that* $k_0^2 := \omega^2 \mu_0 \varepsilon_0$ *is not a Dirichlet eigenvalue for* $-\Delta$ *on* D. *For each* $(F, G) \in W_1^2(\partial D) \times L^2(\partial D)$, *there exists a unique solution* $(f, g) \in L^2(\partial D) \times L^2(\partial D)$ *to the system of integral equations*

$$\begin{cases} \mathcal{S}_D^{k_\star}[f] - \mathcal{S}_D^{k_0}[g] = F \\ \dfrac{1}{\mu_\star} \dfrac{\partial(\mathcal{S}_D^{k_\star}[f])}{\partial\nu}\bigg|_- - \dfrac{1}{\mu_0} \dfrac{\partial(\mathcal{S}_D^{k_0}[g])}{\partial\nu}\bigg|_+ = G \end{cases} \quad on \ \partial D \,. \tag{2.72}$$

Furthermore, there exists a constant C *independent of* F *and* G *such that*

$$\|f\|_{L^2(\partial D)} + \|g\|_{L^2(\partial D)} \le C\left(\|F\|_{W_1^2(\partial D)} + \|G\|_{L^2(\partial D)} \right) \,. \tag{2.73}$$

The following decomposition formula holds.

Theorem 2.26 (Decomposition Formula). *Suppose that* k_0^2 *is not a Dirichlet eigenvalue for* $-\Delta$ *on* D. *Let* u *be the solution of (2.70) and* $g := \frac{\partial u}{\partial\nu}|_{\partial\Omega}$. *Define*

$$H(x) := -\mathcal{S}_\Omega^{k_0}[g](x) + \mathcal{D}_\Omega^{k_0}[f](x) \,, \quad x \in \mathbb{R}^d \setminus \partial\Omega \,, \tag{2.74}$$

and let $(\phi, \psi) \in L^2(\partial D) \times L^2(\partial D)$ *be the unique solution of*

$$\begin{cases} \mathcal{S}_D^{k_\star}[\phi] - \mathcal{S}_D^{k_0}[\psi] = H \\ \dfrac{1}{\mu_\star} \dfrac{\partial(\mathcal{S}_D^{k_\star}[\phi])}{\partial\nu}\Big|_{-} - \dfrac{1}{\mu_0} \dfrac{\partial(\mathcal{S}_D^{k_0}[\psi])}{\partial\nu}\Big|_{+} = \dfrac{1}{\mu_0} \dfrac{\partial H}{\partial\nu} \qquad \text{on } \partial D\,. \end{cases} \tag{2.75}$$

Then u can be represented as

$$u(x) = \begin{cases} H(x) + \mathcal{S}_D^{k_0}[\psi](x)\,, & x \in \Omega \setminus \overline{D}\,, \\ \mathcal{S}_D^{k_\star}[\phi](x)\,, & x \in D\,. \end{cases} \tag{2.76}$$

Moreover, there exists $C > 0$ independent of H such that

$$\|\phi\|_{L^2(\partial D)} + \|\psi\|_{L^2(\partial D)} \le C\left(\|H\|_{L^2(\partial D)} + \|\nabla H\|_{L^2(\partial D)} \right). \tag{2.77}$$

Proof. Note that u defined by (2.76) satisfies the differential equations and the transmission condition on ∂D in (2.70). Thus in order to prove (2.76), it suffices to prove that $f = u|_{\partial\Omega}$ on $\partial\Omega$. Let $\partial u/\partial\nu|_{\partial\Omega} := g$ and consider the following transmission problem:

$$\begin{cases} (\Delta + k_0^2)v = 0 & \text{in } (\Omega \setminus \overline{D}) \cup (\mathbb{R}^d \setminus \overline{\Omega})\,, \\ (\Delta + k_\star^2)v = 0 & \text{in } D\,, \\ v|_{-} - v|_{+} = 0\,, & \dfrac{1}{\mu_\star} \dfrac{\partial v}{\partial\nu}\Big|_{-} - \dfrac{1}{\mu_0} \dfrac{\partial v}{\partial\nu}\Big|_{+} = 0 \quad \text{on } \partial D\,, \\ v|_{-} - v|_{+} = f\,, & \dfrac{\partial v}{\partial\nu}\Big|_{-} - \dfrac{\partial v}{\partial\nu}\Big|_{+} = g \quad \text{on } \partial\Omega\,, \\ \left| \dfrac{\partial v}{\partial r}(x) - ik_0 v(x) \right| = O(|x|^{-(d+1)/2})\,, & |x| \to \infty\,. \end{cases} \tag{2.78}$$

We claim that (2.78) has a unique solution. In fact, if $f = g = 0$, then we can show that $v = 0$ in $\mathbb{R}^d \setminus \overline{D}$. Thus

$$v = \dfrac{\partial v}{\partial\nu}\Big|_{-} = 0 \quad \text{on } \partial D\,.$$

By the unique continuation for the operator $\Delta + k_\star^2$, we have $v = 0$ in D, and hence $v \equiv 0$ in $\mathbb{R}^d$. Note that v_p, $p = 1, 2$, defined by

$$v_1(x) = \begin{cases} u(x)\,, & x \in \Omega\,, \\ 0\,, & x \in \mathbb{R}^d \setminus \overline{\Omega}\,, \end{cases} \qquad v_2(x) = \begin{cases} H(x) + \mathcal{S}_D^{k_0}[\psi](x)\,, & x \in \Omega \setminus \overline{D}\,, \\ \mathcal{S}_D^{k_\star}[\phi](x)\,, & x \in D\,, \end{cases}$$

are two solutions of (2.78), and hence $v_1 \equiv v_2$. Here we use the fact that $(\partial\mathcal{D}_\Omega^{k_0}/\partial\nu)[f]$ does not have a jump across $\partial\Omega$.

This completes the proof of solvability of (2.78). The estimate (2.77) is a consequence of solvability and the closed graph theorem [78]. □

The following proposition is also of importance to us. We refer again to [29] for a proof.

Proposition 2.27. *For each $n \in \mathbb{N}$ there exists C_n independent of D (but depending on $\mathrm{dist}(D, \partial\Omega)$) such that*

$$\|H\|_{C^n(\overline{D})} \leq C_n \|f\|_{W_{\frac{1}{2}}^2(\partial\Omega)} .$$

We now transform the decomposition formula (2.76) into the one using Green's function and the background solution U, that is, the solution of (2.71).

Let $G_{k_0}(x, y)$ be the Dirichlet Green function for $\Delta + k_0^2$ in Ω, i.e., for each $y \in \Omega$, G_{k_0} is the solution of

$$\begin{cases} (\Delta + k_0^2)G_{k_0}(x, y) = \delta_y(x) , & x \in \Omega , \\ G_{k_0}(x, y) = 0 , & x \in \partial\Omega . \end{cases} \tag{2.79}$$

Then,

$$U(x) = \int_{\partial\Omega} \frac{\partial G_{k_0}(x, y)}{\partial \nu_y} f(y) d\sigma(y) , \quad x \in \Omega .$$

We need to introduce some more notation. For a C^2-domain $D \subset\subset \Omega$ and $\phi \in L^2(\partial D)$, let

$$G_D^{k_0}[\phi](x) := \int_{\partial D} G_{k_0}(x, y)\phi(y) \, d\sigma(y) , \quad x \in \overline{\Omega} .$$

Our second decomposition formula is the following.

Theorem 2.28. *Let ψ be the function defined in (2.75). Then*

$$\frac{\partial u}{\partial \nu}(x) = \frac{\partial U}{\partial \nu}(x) + \frac{\partial(G_D^{k_0}[\psi])}{\partial \nu}(x) , \quad x \in \partial\Omega . \tag{2.80}$$

To prove Theorem 2.28 we first observe an easy identity. If $x \in \mathbb{R}^d \setminus \Omega$ and $z \in \Omega$, then

$$\int_{\partial\Omega} \Gamma_{k_0}(x, y) \frac{\partial G_{k_0}(z, y)}{\partial \nu_y}\bigg|_{\partial\Omega} d\sigma(y) = \Gamma_{k_0}(x, z) . \tag{2.81}$$

As a consequence of (2.81), we have

$$\left(\frac{1}{2}I + (\mathcal{K}_\Omega^{k_0})^*\right)\left[\frac{\partial G_{k_0}(z,\cdot)}{\partial \nu_y}\bigg|_{\partial\Omega}\right](x) = \frac{\partial \Gamma_{k_0}(x,z)}{\partial \nu_x} , \qquad (2.82)$$

for all $x \in \partial\Omega$ and $z \in \Omega$.

Our second observation is the following.

Lemma 2.29. *If k_0^2 is not a Dirichlet eigenvalue for $-\Delta$ on Ω, then $(1/2)\,I + (\mathcal{K}_\Omega^{k_0})^* : L^2(\partial\Omega) \to L^2(\partial\Omega)$ is invertible.*

Proof. Since $(\mathcal{K}_\Omega^{k_0})^* : L^2(\partial\Omega) \to L^2(\partial\Omega)$ is compact, it suffices by the Fredholm alternative to prove that $(1/2)\,I + (\mathcal{K}_\Omega^{k_0})^* : L^2(\partial\Omega) \to L^2(\partial\Omega)$ is injective.

Suppose that $\phi \in L^2(\partial\Omega)$ and $\left((1/2)\,I + (\mathcal{K}_\Omega^{k_0})^*\right)[\phi] = 0$. Define

$$u(x) := \mathcal{S}_\Omega^{k_0}[\phi](x), \quad x \in \mathbb{R}^d \setminus \overline{\Omega} .$$

Then u is a solution of $(\Delta + k_0^2)u = 0$ in $\mathbb{R}^d \setminus \overline{\Omega}$, and satisfies the Sommerfeld radiation condition

$$\left|\frac{\partial u}{\partial r} - ik_0 u\right| = O\left(r^{-(d+1)/2}\right) \quad \text{as } r \to +\infty ,$$

and the Neumann boundary condition

$$\frac{\partial u}{\partial \nu}\bigg|_{\partial\Omega} = \left(\frac{1}{2}I + (\mathcal{K}_\Omega^{k_0})^*\right)[\phi] = 0 .$$

Therefore, by Lemma 2.24, we obtain $\mathcal{S}_\Omega^{k_0}[\phi](x) = 0$, $x \in \mathbb{R}^d \setminus \overline{\Omega}$. Since k_0^2 is not a Dirichlet eigenvalue for $-\Delta$ on Ω, we can prove that $\phi \equiv 0$. This completes the proof. $\square$

With these two observations available we are now ready to prove Theorem 2.28.

Proof of Theorem 2.28. Let $g := \partial u/\partial \nu$ and $g_0 := \partial U/\partial \nu$ on $\partial\Omega$ for convenience. By the divergence theorem, we get

$$U(x) = -\mathcal{S}_\Omega^{k_0}[g_0](x) + \mathcal{D}_\Omega^{k_0}[f](x) , \quad x \in \Omega .$$

It then follows from (2.74) that

$$H(x) = -\mathcal{S}_\Omega^{k_0}[g](x) + \mathcal{S}_\Omega^{k_0}[g_0](x) + U(x) , \quad x \in \Omega .$$

Consequently, substituting (2.76) into the above equation, we see that for $x \in \Omega$

$$H(x) = -\mathcal{S}_\Omega^{k_0}\left[\frac{\partial H}{\partial \nu}\bigg|_{\partial\Omega} + \frac{\partial(\mathcal{S}_D^{k_0}[\psi])}{\partial \nu}\bigg|_{\partial\Omega}\right](x) + \mathcal{S}_\Omega^{k_0}[g_0](x) + U(x) .$$

Therefore the jump formula (2.64) yields

$$\frac{\partial H}{\partial \nu} = -\left(-\frac{1}{2}I + (\mathcal{K}_\Omega^{k_0})^*\right)\left[\frac{\partial H}{\partial \nu}\bigg|_{\partial\Omega} + \frac{\partial(\mathcal{S}_D^{k_0}[\psi])}{\partial \nu}\bigg|_{\partial\Omega}\right] \tag{2.83}$$
$$+ \left(\frac{1}{2}I + (\mathcal{K}_\Omega^{k_0})^*\right)[g_0] \quad \text{on } \partial\Omega .$$

By (2.82), we have for $x \in \partial\Omega$

$$\frac{\partial(\mathcal{S}_D^{k_0}[\psi])}{\partial \nu}(x) = \int_{\partial D} \frac{\partial\Gamma_{k_0}(x,y)}{\partial\nu_x}\psi(y)\,d\sigma(y)$$

$$= \left(\frac{1}{2}I + (\mathcal{K}_\Omega^{k_0})^*\right)\left[\frac{\partial(G_D^{k_0}[\psi])}{\partial \nu}\bigg|_{\partial\Omega}\right](x) . \tag{2.84}$$

Thus we obtain

$$\left(-\frac{1}{2}I + (\mathcal{K}_\Omega^{k_0})^*\right)\left[\frac{\partial(\mathcal{S}_D^{k_0}[\psi])}{\partial \nu}\bigg|_{\partial\Omega}\right]$$

$$= \left(\frac{1}{2}I + (\mathcal{K}_\Omega^{k_0})^*\right)\left[\left(-\frac{1}{2}I + (\mathcal{K}_\Omega^{k_0})^*\right)\left[\frac{\partial(G_D^{k_0}[\psi])}{\partial \nu}\bigg|_{\partial\Omega}\right]\right] \quad \text{on } \partial\Omega .$$

It then follows from (2.83) that

$$\left(\frac{1}{2}I + (\mathcal{K}_\Omega^{k_0})^*\right)\left[\frac{\partial H}{\partial \nu}\bigg|_{\partial\Omega} + \left(-\frac{1}{2}I + (\mathcal{K}_\Omega^{k_0})^*\right)\left(\frac{\partial(G_D^{k_0}[\psi])}{\partial \nu}\bigg|_{\partial\Omega}\right) - g_0\right] = 0$$

on $\partial\Omega$ and hence, by Lemma 2.29, we arrive at

$$\frac{\partial H}{\partial \nu}\bigg|_{\partial\Omega} + \left(-\frac{1}{2}I + (\mathcal{K}_\Omega^{k_0})^*\right)\left[\frac{\partial(G_D^{k_0}[\psi])}{\partial \nu}\bigg|_{\partial\Omega}\right] - g_0 = 0 \quad \text{on } \partial\Omega . \tag{2.85}$$

By substituting this equation into (2.76), we get

$$\frac{\partial u}{\partial \nu} = \frac{\partial U}{\partial \nu} - \left(-\frac{1}{2}I + (\mathcal{K}_\Omega^{k_0})^*\right)\left[\frac{\partial(G_D^{k_0}[\psi])}{\partial \nu}\bigg|_{\partial\Omega}\right] + \frac{\partial(\mathcal{S}_D^{k_0}[\psi])}{\partial \nu} \quad \text{on } \partial\Omega .$$

Finally, using (2.84) we conclude that (2.80) holds and the proof is then complete. □

Observe that, by (2.64), (2.85) is equivalent to

$$\frac{\partial}{\partial \nu}\left(H + S_\Omega^{k_0}\left[\frac{\partial(G_D^{k_0}[\psi])}{\partial \nu}\bigg|_{\partial\Omega}\right] - U\right)\bigg|_- = 0 \quad \text{on } \partial\Omega \ .$$

On the other hand, by (2.81),

$$S_\Omega^{k_0}\left(\frac{\partial(G_D^{k_0}[\psi])}{\partial \nu}\bigg|_{\partial\Omega}\right)(x) = S_D^{k_0}[\psi](x) \ , \quad x \in \partial\Omega \ .$$

Thus, by (2.76), we obtain

$$H(x) + S_\Omega^{k_0}\left[\frac{\partial(G_D^{k_0}[\psi])}{\partial \nu}\bigg|_{\partial\Omega}\right](x) - U(x) = 0 \ , \quad x \in \partial\Omega \ .$$

Then, by the unique continuation for $\Delta + k_0^2$, we obtain the following Lemma.

Lemma 2.30. *We have*

$$H(x) = U(x) - S_\Omega^{k_0}\left[\frac{\partial(G_D^{k_0}[\psi])}{\partial \nu}\bigg|_{\partial\Omega}\right](x) \ , \quad x \in \Omega \ . \tag{2.86}$$

2.2.4 Reciprocity

Let μ and ε be two piecewise smooth functions such that $\mu(x) = \mu_0$ and $\varepsilon(x) = \varepsilon_0$ for $|x| \geq R_0$ for some positive R_0. Introduce the fundamental solution Φ_{k_0} to be the solution to

$$(\nabla \cdot \frac{1}{\mu}\nabla + \omega^2\varepsilon)\Phi_{k_0} = \frac{1}{\mu_0}\delta_0 \ , \tag{2.87}$$

subject to the Sommerfeld radiation condition.

An important property satisfied by the fundamental solution Φ_{k_0} is the reciprocity property. The following holds.

Lemma 2.31. *We have, for* $x \neq y$,

$$\Phi_{k_0}(x, y) = \Phi_{k_0}(y, x) \ . \tag{2.88}$$

Identity (2.88) means that the wave recorded at x when there is a time-harmonic source at y is equal to the wave recorded at y when there is a time-harmonic source at x.

Proof. We consider the equations satisfied by the fundamental solution with the source at y_2 and with the source at y_1 (with $y_1 \neq y_2$):

$$(\nabla_x \cdot \frac{1}{\mu} \nabla_x + \omega^2 \varepsilon)\Phi_{k_0}(x, y_2) = \frac{1}{\mu_0}\delta_{y_2},$$

$$(\nabla_x \cdot \frac{1}{\mu} \nabla_x + \omega^2 \varepsilon)\Phi_{k_0}(x, y_1) = \frac{1}{\mu_0}\delta_{y_1}.$$

We multiply the first equation by $\Phi_{k_0}(x, y_1)$ and subtract the second equation multiplied by $\Phi_{k_0}(x, y_2)$:

$$\nabla_x \cdot \frac{\mu_0}{\mu}\Big[\Phi_{k_0}(x, y_1)\nabla_x\Phi_{k_0}(x, y_2) - \Phi_{k_0}(x, y_2)\nabla_x\Phi_{k_0}(x, y_1)\Big]$$

$$= -\Phi_{k_0}(x, y_2)\delta_{y_1} + \Phi_{k_0}(x, y_1)\delta_{y_2}$$

$$= -\Phi_{k_0}(y_1, y_2)\delta_{y_1} + \Phi_{k_0}(y_2, y_1)\delta_{y_2}.$$

We next integrate over the ball B_R of center 0 and radius R which contains both y_1 and y_2 and use the divergence theorem:

$$\int_{\partial B_R} \nu \cdot \Big[\Phi_{k_0}(x, y_1)\nabla_x\Phi_{k_0}(x, y_2) - \Phi_{k_0}(x, y_2)\nabla_x\Phi_{k_0}(x, y_1)\Big] d\sigma(x)$$

$$= -\Phi_{k_0}(y_1, y_2) + \Phi_{k_0}(y_2, y_1),$$

where ν is the unit outward normal to the ball B_R, which is $\nu = x/|x|$.

If $x \in \partial B_R$ and $R \to \infty$, then we have by the Sommerfeld radiation condition:

$$\nu \cdot \nabla_x\Phi_{k_0}(x, y) = ik_0\Phi_{k_0}(x, y) + O\Big(\frac{1}{R^{(d+1)/2}}\Big).$$

Therefore, as $R \to \infty$,

$$-\Phi_{k_0}(y_1, y_2) + \Phi_{k_0}(y_2, y_1)$$

$$= ik_0 \int_{\partial B_R} \Big[\Phi_{k_0}(x, y_1)\Phi_{k_0}(x, y_2) - \Phi_{k_0}(x, y_2)\Phi_{k_0}(x, y_1)\Big] d\sigma(x)$$

$$= 0,$$

which is the desired result. □

2.2.5 The Helmholtz-Kirchhoff Theorem

The Helmholtz-Kirchhoff theorem plays a key role in understanding the resolution limit in imaging with waves. The following holds.

Lemma 2.32. *We have*

$$\int_{|y|=R} \left(\frac{\partial \overline{\Gamma_{k_0}}}{\partial \nu}(x,y) \Gamma_{k_0}(z,y) - \overline{\Gamma_{k_0}}(x,y) \frac{\partial \Gamma_{k_0}}{\partial \nu}(z,y) \right) d\sigma(y) = 2i \Im m \Gamma_{k_0}(x,z) \,,$$

$$(2.89)$$

which yields

$$\lim_{R \to +\infty} \int_{|y|=R} \overline{\Gamma_{k_0}}(x,y) \Gamma_{k_0}(z,y) \, d\sigma(y) = -\frac{1}{k_0} \Im m \Gamma_{k_0}(x,z) \,, \qquad (2.90)$$

by using the radiation outgoing condition.

Identity (2.90) is valid even in inhomogeneous media. The following identity holds, which as we will see shows that the sharper the behavior of the imaginary part of the fundamental solution Φ_{k_0} around the source is, the higher is the resolution.

Theorem 2.33. *Let Φ_{k_0} be the fundamental solution defined in (2.87). We have*

$$\lim_{R \to +\infty} \int_{|y|=R} \overline{\Phi_{k_0}}(x,y) \Phi_{k_0}(z,y) \, d\sigma(y) = -\frac{1}{k_0} \Im m \Phi_{k_0}(x,z) \,. \qquad (2.91)$$

Proof. The proof is based essentially on the second Green's identity and the Sommerfeld radiation condition. Let us consider

$$(\nabla_y \cdot \tfrac{1}{\mu} \nabla_y + \omega^2 \varepsilon) \Phi_{k_0}(y,x_2) = \frac{1}{\mu_0} \delta_{x_2} \,,$$
$$(\nabla_y \cdot \tfrac{1}{\mu} \nabla_y + \omega^2 \varepsilon) \Phi_{k_0}(y,x_1) = \frac{1}{\mu_0} \delta_{x_1} \,.$$

We multiply the first equation by $\overline{\Phi_{k_0}}(y,x_1)$ and we subtract the second equation multiplied by $\Phi_{k_0}(y,x_2)$:

$$\nabla_y \frac{\mu_0}{\mu} \cdot \left[\overline{\Phi_{k_0}}(y,x_1) \nabla_y \Phi_{k_0}(y,x_2) - \Phi_{k_0}(y,x_2) \nabla_y \overline{\Phi_{k_0}}(y,x_1) \right]$$
$$= -\Phi_{k_0}(y,x_2) \delta_{x_1} + \overline{\Phi_{k_0}}(y,x_1) \delta_{x_2}$$
$$= -\Phi_{k_0}(x_1,x_2) \delta_{x_1} + \overline{\Phi_{k_0}}(x_1,x_2) \delta_{x_2} \,,$$

using the reciprocity property $\Phi_{k_0}(x_1,x_2) = \Phi_{k_0}(x_2,x_1)$.

We integrate over the ball B_R and we use the divergence theorem:

$$\int_{\partial B_R} \nu \cdot \left[\overline{\Phi_{k_0}}(y,x_1) \nabla_y \Phi_{k_0}(y,x_2) - \Phi_{k_0}(y,x_2) \nabla_y \overline{\Phi_{k_0}}(y,x_1) \right] d\sigma(y)$$
$$= -\Phi_{k_0}(x_1,x_2) + \overline{\Phi_{k_0}}(x_1,x_2) \,.$$

This equality can be viewed as an application of the second Green's identity. The Green's function also satisfies the Sommerfeld radiation condition

$$\lim_{|y|\to\infty} |y| \Big(\frac{y}{|y|} \cdot \nabla_y - ik_0 \Big) \Phi_{k_0}(y, x_1) = 0 \, ,$$

uniformly in all directions $y/|y|$. Using this property, we substitute $ik_0\Phi_{k_0}(y, x_2)$ for $\nu \cdot \nabla_y \Phi_{k_0}(y, x_2)$ in the surface integral over ∂B_R, and $-ik_0\overline{\Phi_{k_0}}(y, x_1)$ for $\nu \cdot \nabla_y \overline{\Phi_{k_0}}(y, x_1)$, and we obtain the desired result.   □

2.2.6   Geometric Optics

Eikonal and Transport Equations

Geometric optics is a high-frequency asymptotics for the solution of the Helmholtz equation. We look for an approximate expression as $\delta \to 0$ for $\Phi_{k_0/\delta}(x, y)$ solution of the Helmholtz equation

$$(\nabla \cdot \frac{1}{\mu}\nabla + (\frac{\omega}{\delta})^2\varepsilon)\Phi_{k_0/\delta} = \frac{1}{\mu_0}\delta_0 \, , \tag{2.92}$$

subject to the Sommerfeld radiation condition. Here δ is a small parameter. In the particular case when $\mu = \mu_0$ and $\varepsilon = \varepsilon_0$, then we have

$$\Gamma_{k_0/\delta}(x, y) = -\frac{1}{4\pi|x - y|}e^{i\frac{k_0}{\delta}|x-y|}, \quad d = 3 \, , \tag{2.93}$$

and

$$\Gamma_{k_0/\delta}(x, y) \approx \frac{\sqrt{\delta}e^{i\pi/4}}{2\sqrt{2\pi k_0}}\frac{e^{i\frac{k_0}{\delta}|x-y|}}{\sqrt{|x - y|}}, \quad d = 2 \, , \tag{2.94}$$

which exhibits a smooth amplitude term and a rapid phase. Motivated by this observation, in the general case when μ and ε smoothly varying, we look for an expansion of the form:

$$\Phi_{k_0/\delta}(x, y) = -e^{i\frac{\omega}{\delta}T(x,y)} \sum_{j=0}^{\infty} \frac{\delta^j a_j(x, y)}{k_0^j} \, .$$

If we substitute this ansatz, known as the WKB expansion, into the Helmholtz equation (2.92) for $x \neq y$ and collect the terms with the same powers in δ, then we obtain:

$$O\left(\frac{1}{\delta^2}\right): \qquad |\nabla_x \mathcal{T}|^2 - \mu(x)\varepsilon(x) = 0 \ ,$$

$$O\left(\frac{1}{\delta}\right): \qquad \frac{2}{\mu}\nabla_x \mathcal{T} \cdot \nabla_x a_0 + a_0 \nabla_x \cdot \frac{1}{\mu}\nabla_x \mathcal{T} = 0 \ .$$

The first equation is the eikonal equation for the quantity $\mathcal{T}$, that turns out to be the travel time from x to y, and the second equation is the transport equation for the amplitude a_0. These equations can be solved by the method of characteristics.

In these conditions the geometric optics approximation of the fundamental solution

$$\Phi_{k_0/\delta}(x,y) \approx -a(x,y)e^{i\frac{\omega}{\delta}\mathcal{T}(x,y)} \tag{2.95}$$

is valid when $\delta \ll 1$, where $\mathcal{T}$ is the travel time that can be defined by Fermat's principle as

$$\begin{aligned}
\mathcal{T}(x,y) = \inf\Big\{ T \ : \ &\exists (X_t)_{t\in[0,T]} \in \mathcal{C}^1 \, , \ X_0 = x \, , \ X_T = y \, , \\
&\left|\frac{dX_t}{dt}\right| = \varepsilon(X_t)\mu(X_t) \Big\} \ .
\end{aligned} \tag{2.96}$$

The curves X_t that minimize the functional in (2.96) are called rays. Here, we make a simple geometric assumption: $\mu(x)$ and $\varepsilon(x)$ are smooth and there is a unique ray between any pair of points in the region of interest.

Note that, in the homogeneous case, we have $\mathcal{T}(x,y) = \sqrt{\varepsilon_0\mu_0}|x-y|$. Hence, the ray connecting x and y is the straight line going from x to y.

WKB Expansion for Helmholtz Equation in the Presence of an Inclusion

Suppose that $d = 2$ and that μ and ε are given by (2.68) and (2.69), respectively, for some smooth convex inclusion D. For $y \in \mathbb{R}^2 \setminus \overline{D}$, write

$$\Phi_{k_0/\delta}(x,y) = \begin{cases} \Gamma_{k_0/\delta}(x,y) + u_y^{(s)}(x), & x \in \mathbb{R}^2 \setminus \overline{D} \, , \\ u_y^{(t)}(x), & x \in D \, , \end{cases}$$

where $u_y^{(s)}$ and $u_y^{(t)}$ are the scattered and transmitted waves. Using Green's formula, we get

$$\Phi_{k_0/\delta}(x,y) = \Gamma_{k_0/\delta}(x,y)$$
$$+ \int_{\partial D} \left(\frac{\partial \Gamma_{k_0/\delta}}{\partial \nu}(z,x) u_y^{(s)}(z) - \Gamma_{k_0/\delta}(z,x) \frac{\partial u_y^{(s)}}{\partial \nu}(z) \right) d\sigma(z) ,$$

where ν is the outward normal to ∂D. Using a WKB approximation, we write

$$u^{(s)}(z) \approx e^{i\pi/4} a_y^{(s)}(z) \frac{e^{i\frac{\omega}{\delta}\phi_s(z)}}{\sqrt{|z-y|}} \quad \text{and} \quad u^{(t)}(z) \approx e^{i\pi/4} a_y^{(t)}(z) \frac{e^{i\frac{\omega}{\delta}\phi_t(z)}}{\sqrt{|z-y|}} ,$$

where ϕ_t and ϕ_s satisfy the eikonal equations

$$|\nabla \phi_t|^2 = \varepsilon_* \mu_* \quad \text{in } D ,$$
$$|\nabla \phi_s|^2 = \varepsilon_0 \mu_0 \quad \text{in } \mathbb{R}^2 \setminus \overline{D} .$$

Moreover, it follows from the transmission conditions on ∂D that

$$\phi_s(z) = \phi_t(z) = \sqrt{\varepsilon_0 \mu_0} |z-y| \quad \text{on } \partial D ,$$
$$a^{(s)} + \frac{\sqrt{\delta}}{2\sqrt{2\pi k_0}} = a^{(t)} \quad \text{on } \partial D , \tag{2.97}$$
$$a^{(s)} \frac{\partial \phi_s}{\partial \nu} + \frac{\sqrt{\delta}}{2\sqrt{2\pi\omega}} \frac{(z-y)\cdot\nu(z)}{|y-z|} = \frac{\mu_0}{\mu_*} a^{(t)} \frac{\partial \phi_t}{\partial \nu} \quad \text{on } \partial D .$$

Using the eikonal equations for ϕ_s and ϕ_t and (2.97) it follows that

$$\left| \frac{\partial \phi_s}{\partial \nu} \right|^2 = \varepsilon_0 \mu_0 \left| \frac{(z-y)\cdot\nu(z)}{|y-z|} \right|^2 \tag{2.98}$$

and

$$\left| \frac{\partial \phi_t}{\partial \nu} \right|^2 = \varepsilon_* \mu_* - \varepsilon_0 \mu_0 \left| \frac{(z-y)\cdot\tau(z)}{|y-z|} \right|^2 , \tag{2.99}$$

where $\tau(z)$ is the tangential vector to ∂D at z.

In order to select the sign of $\frac{\partial \phi_s}{\partial \nu}$ (i.e., the physically correct solution), the radiation condition can be used. Following [93, 113], by using (2.97) we arrive at the high-frequency asymptotic expansions for $u_y^{(s)}$ and $\frac{\partial u_y^{(s)}}{\partial \nu}$ on ∂D as $k_0/\delta \to +\infty$

$$u_y^{(s)}(z) \approx e^{i\pi/4} a_y^{(s)}(z) \frac{e^{i\frac{k_0}{\delta}|z-y|}}{\sqrt{|z-y|}} \tag{2.100}$$

and

$$\frac{\partial u_y^{(s)}}{\partial \nu}(z) \approx -ie^{i\pi/4}\frac{k_0}{\delta}a_y^{(s)}(z)\frac{(z-y)\cdot \nu(z)}{|z-y|}\frac{e^{i\frac{k_0}{\delta}|z-y|}}{\sqrt{|z-y|}} \tag{2.101}$$

if $(z-y)\cdot \nu(z) < 0$, where $a_y^{(s)}$ is the amplitude, and

$$u_y^{(s)}(z) \approx \frac{\partial u_y^{(s)}}{\partial \nu}(z) \approx 0 \quad \text{if } (z-y)\cdot \nu(z) \geq 0 . \tag{2.102}$$

Using (2.94) together with

$$\frac{\partial \Gamma_{k_0/\delta}}{\partial \nu}(z,y) \approx i\frac{\sqrt{k_0/\delta}e^{i\pi/4}}{2\sqrt{2\pi}}\frac{(z-y)\cdot \nu(z)}{|z-y|}\frac{e^{i\frac{k_0}{\delta}|z-y|}}{\sqrt{|z-y|}} ,$$

it follows that

$$\Phi_{k_0/\delta}(x,y) - \Gamma_{k_0/\delta}(x,y) \approx i\frac{\sqrt{k_0/\delta}e^{i\pi/4}}{2\sqrt{2\pi}}\int_{\partial D_{\text{illum},y}} a_y^{(s)}(z)\left(\frac{(z-y)\cdot \nu(z)}{|z-y|}\right.$$
$$\left.+\frac{(z-x)\cdot \nu(z)}{|z-x|}\right)\frac{e^{i\frac{k_0}{\delta}(|z-y|+|x-z|)}}{\sqrt{|z-y||x-z|}}\,d\sigma(z) , \tag{2.103}$$

where $\partial D_{\text{illum},y} = \{z \in \partial D : (z-y)\cdot \nu(z) < 0\}$. Equation (2.103) shows that $\Phi_{k_0/\delta}(x,y) - \Gamma_{k_0/\delta}(x,y)$ in the high-frequency regime depends only on the boundary of the target that is illuminated by the incident wave $\Gamma_{k_0/\delta}(\cdot,y)$.

2.3   Cluttered Medium

Let Ω be a bounded smooth domain. Let $y \in \mathbb{R}^d \setminus \overline{\Omega}$. We consider the Helmholtz equation:

$$\left(\Delta_x + k_0^2(1 + \nu_{\text{noise}}(x))\right)\Phi_{k_0}(x,y) = \delta_y \quad \text{in } \mathbb{R}^d ,$$

subject to the Sommerfeld radiation condition. Here, $\nu_{\text{noise}}(x)$ is a stationary random process with Gaussian statistics, mean zero, and given covariance function. The windowing is supposed to be the multiplication by a characteristic function of a compact domain within Ω. The random process $\nu_{\text{noise}}(x)$ describes the random fluctuations of the coefficient of reflection in the medium. Since the coefficient of the equation is random, the fundamental solution $\Phi_{k_0}(x,y)$ is itself random. The relation between the statistics of the fluctuations of the coefficient of reflection and the statistics of $\Phi_{k_0}(x, y)$ is highly nontrivial and nonlinear. In particular cluttered noise, that is the difference between the random fundamental solution and the background homogeneous fundamental solution, is not an additive white noise.

In order to simulate ν_{noise} (and Φ_{k_0}), we first generate on a grid of points that covers the support of the windowing function a realization of a stationary Gaussian random process using the method described in Sect. 1.8.6. Then we apply the windowing function.

2.4 Wave Equation

We briefly consider the wave equation. Let $y \in \mathbb{R}^3$ and let

$$U_y(x, t) := -\frac{\delta_0(t - |x - y|)}{4\pi|x - y|} \quad \text{for } x \neq y . \tag{2.104}$$

The function U_y is the outgoing fundamental solution (also called retarded fundamental solution) to the wave equation in three dimensions:

$$(\partial_t^2 - \Delta)U_y(x, t) = -\delta_y(x)\delta_0(t) \quad \text{in } \mathbb{R}^3 \times \mathbb{R} . \tag{2.105}$$

Moreover, U_y satisfies the conditions: $U_y(x, t) = \partial_t U_y(x, t) = 0$ for $x \neq y$ and $t < 0$. The function U_y corresponds to a spherical wave generated at the source point y and propagating at speed 1.

In the two-dimensional case, the fundamental solution is given by

$$U_y(x, t) := -\frac{H(t - |x - y|)}{2\pi\sqrt{t^2 - |x - y|^2}} \quad \text{for } |x - y| \neq t ,$$

where H is the Heaviside step function.

It can be shown that the time-harmonic fundamental solution Γ_ω (with $\varepsilon_0 = \mu_0 = 1$) satisfies

$$\Gamma_\omega(x, y) = \sqrt{2\pi}\mathcal{F}(U_y(x, t))(x, \omega), \quad d = 2, 3 ,$$

where the Fourier transform is taken in t variable.

Bibliography and Discussion

The decomposition formula in Theorem 2.18 was proved in [108, 109, 111]. It seems to inherit geometric properties of the inclusion. Based on Theorem 2.18, Kang and Seo proved global uniqueness results for the inverse conductivity problem with one measurement when the conductivity inclusion is a disk or a ball in three-dimensional space [108, 110]. Uniqueness for conductivity inclusions with elliptic shapes with a finite number of measurements is still an open problem. Suppose that (the admittivity) $k = \sigma + i\omega\varepsilon$, where ω is the frequency. Using, in this case, multi-frequency measurements may be the way to solve the uniqueness problem. In this direction, we refer to the recent works on shape identification and classification in electrolocation using multi-frequency measurements [7, 10]. The symmetrization of the Neumann-Poincaré operator in Theorem 2.14 is due to Khavinson et al. [115, proof of Theorem 1]. The book by Nédélec [135] is an excellent reference on integral equations method. A complete treatment of the Helmholtz equation as well as the full time-harmonic Maxwell equations is provided there. For WKB expansions for the Helmholtz equation at high-frequencies, we refer the reader to the seminal work [113]. The derivation of (2.98) and (2.99) follows [93], where another way of selecting the physically correct solution is given. It is based on the limiting absorption principle. The implementation of the clutter noise is from [116]. A random travel time model can be used to model the effect of the clutter on the wave propagation.

Part II
Small Volume Expansions and Concept of Generalized Polarization Tensors

Chapter 3
Small Volume Expansions

In their most general forms imaging problems are severely ill-posed and nonlinear. These are the main obstacles to find non-iterative reconstruction algorithms. If, however, in advance we have additional structural information about the profile of the material property, then we may be able to determine specific features about the conductivity distribution with a satisfactory resolution. One such type of knowledge could be that the conducting body consists of a smooth background containing a number of unknown small inclusions with a significantly different conductivity. These inclusions might represent potential tumors or small defects.

Over the last 10 years or so, a considerable amount of interesting work has been dedicated to the imaging of such low volume fraction inclusions. The method of asymptotic expansions provides a useful framework to accurately and efficiently reconstruct the location and geometric features of the inclusions in a stable way, even for moderately noisy data.

Using the method of matched asymptotic expansions we formally derive the first-order perturbations due to the presence of the inclusions in the conductivity case. These perturbations are of dipole-type. A rigorous proof of these expansions is based on layer potential techniques. The concept of polarization tensor (PT) is the basic building block for the asymptotic expansion of the boundary perturbations. It is then important from an imaging point of view to precisely characterize the PT and derive some of its properties, such as symmetry, positivity, and optimal bounds on its elements, for developing efficient algorithms to reconstruct conductivity inclusions of small volume.

We then provide the leading-order term in this asymptotic formula of the solution to the Helmholtz equation in the presence of small electromagnetic inclusions. The leading-order term is the sum of a (polarized) magnetic dipole and an electric point source.

It is worth emphasizing that all the problems considered in this chapter are singularly perturbed problems. As it will be shown later, derivatives of

H. Ammari et al., *Mathematical and Statistical Methods for Multistatic Imaging*, 97
Lecture Notes in Mathematics 2098, DOI 10.1007/978-3-319-02585-8_3,
© Springer International Publishing Switzerland 2013

the solution to the perturbed problem are not, inside the inclusion, close to those of the background solution. Consequently, the far-field expansions are not uniform in the whole background domain. Nevertheless, inner expansions of the solution inside the inclusion are provided. An example of a regularly perturbed problem is the Born approximation. See (12.11).

The asymptotic expansions are first provided for bounded domains. We consider a small inclusion inside a bounded domain. A boundary condition (Neumann or Dirichlet) is applied and the perturbations of the (Dirichlet or Neumann) boundary data are derived. Then the asymptotic expansions are extended to dipole sources in the free space. The perturbations of the field at a receiver placed away from the inclusion are derived. Finally, an extension of the asymptotic approach to time-domain measurements is described. It will be shown that after truncating the high-frequency component of the measured wave, the perturbation due to the inclusion is a wavefront emitted by a dipolar source at the location of the inclusion. Such a formula will be useful for designing time-reversal techniques for inclusion localization.

3.1 Conductivity Problem

In this section we derive an asymptotic expansion of the voltage potentials in the presence of a diametrically small inclusion with conductivity different from the background conductivity.

Let $g \in L_0^2(\partial\Omega)$. Consider the solution u of

$$\begin{cases} \nabla \cdot \left(1 + (k-1)\chi(D)\right)\nabla u = 0 & \text{in } \Omega, \\ \dfrac{\partial u}{\partial \nu}\bigg|_{\partial\Omega} = g, \quad \displaystyle\int_{\partial\Omega} u \, d\sigma = 0. \end{cases} \qquad (3.1)$$

Let U be the background solution, that is, the solution to

$$\begin{cases} \Delta U = 0 & \text{in } \Omega, \\ \dfrac{\partial U}{\partial \nu}\bigg|_{\partial\Omega} = g, \quad \displaystyle\int_{\partial\Omega} U \, d\sigma = 0. \end{cases} \qquad (3.2)$$

The following asymptotic expansion expresses the fact that the conductivity inclusion can be modeled by a dipole.

Theorem 3.1 (Voltage Boundary Perturbations). *Suppose that $\dot{D} = \delta B + z$, and let u be the solution of (3.1), where $0 < k \neq 1 < +\infty$. Denote $\lambda := (k+1)/(2(k-1))$. The following pointwise asymptotic expansion on $\partial\Omega$ holds for $d = 2, 3$:*

$$u(x) = U(x) - \delta^d \nabla U(z) \cdot M(\lambda, B)\nabla_z N(x, z) + O(\delta^{d+1}), \qquad (3.3)$$

where the remainder $O(\delta^{d+1})$ is dominated by $C\delta^{d+1}\|g\|_{L^2(\partial\Omega)}$ for some C independent of $x \in \partial\Omega$. Here U is the background solution, $N(x, z)$ is the Neumann function, that is, the solution to (2.34), $M(\lambda, B) = (m_{ij})_{i,j=1}^d$ is the polarization tensor (PT) given by

$$m_{ij} = \int_{\partial B} (\lambda I - \mathcal{K}_B^*)^{-1}[\nu_i](\xi)\, \xi_j\, d\sigma(\xi), \qquad (3.4)$$

where $\nu = (\nu_1, \ldots, \nu_d)$ is the outward unit normal to ∂B and $\xi = (\xi_1, \ldots, \xi_d)$.

3.1.1 Formal Derivations

To reveal the nature of the perturbations in the solution u to (3.1) that are due to the presence of the inclusion D, we introduce the local variables $\xi = (y - z)/\delta$ for $y \in \Omega$, and set $\hat{u}(\xi) = u(z + \delta\xi)$. We expect that $u(y)$ will differ appreciably from $U(y)$ for y near z, but it will differ little from $U(y)$ for y far from z. Therefore, using the method of matched asymptotic expansions, we represent the field u by two different expansions, an inner expansion for y near z, and an outer expansion for y far from z. The outer expansion must be U to leading order, so we write:

$$u(y) = U(y) + \delta^{\tau_1} U_1(y) + \delta^{\tau_2} U_2(y) + \cdots, \quad \text{for } |y - z| \gg \delta,$$

where $0 < \tau_1 < \tau_2 < \ldots$ and $U_1, U_2, \ldots,$ are to be found.

We write the inner expansion as

$$\hat{u}(\xi) = u(z + \delta\xi) = \hat{u}_0(\xi) + \delta\hat{u}_1(\xi) + \delta^2\hat{u}_2(\xi) + \cdots, \quad \text{for } |\xi| = O(1),$$

where $\hat{u}_0, \hat{u}_1, \ldots$ are to be found. We suppose that the functions $\hat{u}_j, j = 0, 1, \ldots,$ are defined not just in the domain obtained by stretching Ω, but everywhere in $\mathbb{R}^d$.

Evidently, the functions $\hat{u}_i$ are not defined uniquely, and the question of how to choose them now arises. Thus, there is an arbitrariness in the choice of the coefficients of both the outer and the inner expansions. In order to determine the functions $U_i(y)$ and $\hat{u}_i(\xi)$, we have to equate the inner and the outer expansions in some overlap domain within which the stretched variable ξ is large and $y - z$ is small. In this domain the matching conditions are:

$$U(y) + \delta^{\tau_1} U_1(y) + \delta^{\tau_2} U_2(y) + \cdots \approx \hat{u}_0(\xi) + \delta\hat{u}_1(\xi) + \delta^2\hat{u}_2(\xi) + \cdots.$$

If we substitute the inner expansion into the transmission problem (3.1) and formally equate coefficients of δ^{-2} and δ^{-1}, we obtain $\hat{u}_0(\xi) = U(z)$, and

$$\hat{u}_1(\xi) = \hat{v}(\frac{x-z}{\delta}) \cdot \nabla U(z),$$

where $\hat{v}$ satisfies

$$
\begin{cases}
\Delta \hat{v} = 0 & \text{in } \mathbb{R}^d \setminus \overline{B}, \\
\Delta \hat{v} = 0 & \text{in } B, \\
\hat{v}|_- - \hat{v}|_+ = 0 & \text{on } \partial B, \\
k\dfrac{\partial \hat{v}}{\partial \nu}|_- - \dfrac{\partial \hat{v}}{\partial \nu}|_+ = 0 & \text{on } \partial B, \\
\hat{v}(\xi) - \xi \to 0 & \text{as } |\xi| \to +\infty.
\end{cases}
\tag{3.5}
$$

Therefore, we arrive at the following inner asymptotic formula:

$$u(x) \approx U(z) + \delta \hat{v}(\frac{x-z}{\delta}) \cdot \nabla U(z) \quad \text{for } x \text{ near } z. \tag{3.6}$$

Clearly, $\sup_D |\nabla u(x) - \nabla U(x)|$ does not approach zero as δ goes to zero, and therefore, the problem is singulary perturbed.

Note also that

$$\hat{v}(\xi) = \xi + \mathcal{S}_B(\lambda I - \mathcal{K}_B^*)^{-1}[\nu](\xi), \quad \xi \in \mathbb{R}^d.$$

We now derive the outer expansion. From (2.51) we have

$$u(x) = H(x) + (k-1)\int_D \nabla_y \Gamma(x,y) \cdot \nabla u(y)\, dy.$$

Since

$$H(x) = -\mathcal{S}_\Omega[g](x) + \mathcal{D}_\Omega[u|_{\partial\Omega}](x) = U(x) + \mathcal{D}_\Omega[(u-U)|_{\partial\Omega}](x), \quad x \in \Omega,$$

it follows from the jump relation (2.20) that

$$(\frac{1}{2} - \mathcal{K}_\Omega)[(u-U)|_{\partial\Omega}](x) = (k-1)\int_D \nabla_y \Gamma(x,y) \cdot \nabla u(y)\, dy, \quad x \in \partial\Omega.$$

Applying Lemma 2.16, we obtain that

$$
\begin{aligned}
(u-U)(x) &= (1-k)\int_D \nabla_y N(x,y) \cdot \nabla u(y)\, dy \\
&\approx (1-k)\nabla_y N(x,z) \cdot \int_D \nabla u(y)\, dy,
\end{aligned}
$$

for $x \in \partial\Omega$. By using the inner expansion (3.6), we arrive at the outer expansion:

$$u(x) \approx U(x) + \delta^d(1-k)\nabla_y N(x,z)\Big(\int_B \nabla\hat{v}(\xi)\,d\xi\Big) \cdot \nabla U(z), \quad x \in \partial\Omega.$$

Next, we compute

$$\begin{aligned}
\int_B \nabla\hat{v}(\xi)\,d\xi &= \int_B (I + \nabla\mathcal{S}_B(\lambda I - \mathcal{K}_B^*)^{-1}[\nu](\xi))\,d\xi \\
&= |B|I + \int_{\partial B}\Big(-\frac{1}{2}I + \mathcal{K}_B^*\Big)(\lambda I - \mathcal{K}_B^*)^{-1}[\nu](\xi)\xi^T\,d\sigma(\xi) \\
&= \frac{1}{k-1}\int_{\partial B}(\lambda I - \mathcal{K}_B^*)^{-1}[\nu](\xi)\,\xi^T\,d\sigma(\xi),
\end{aligned}$$

where $|B|$ is the volume of B. Here $(\lambda I - \mathcal{K}_B^*)^{-1}$ is applied term-wisely on ν. Thus we have

$$u(x) \approx U(x) - \delta^d\nabla_y N(x,z) \cdot M(\lambda,B)\nabla U(z), \quad x \in \partial\Omega, \qquad (3.7)$$

where $M(\lambda,B)$ is the polarization tensor associated with B and the conductivity $k = (2\lambda+1)/(2\lambda-1)$ defined by (3.4).

3.1.2 Polarization Tensor

The polarization tensor M can be explicitly computed for disks and ellipses in the plane and balls and ellipsoids in three-dimensional space.

Let $|\lambda| > 1/2$ and let $k = (2\lambda+1)/(2\lambda-1)$. If B is an ellipse whose semi-axes are on the x_1- and x_2-axes and of length a and b, respectively, then its polarization tensor M takes the form

$$M(\lambda,B) = (k-1)|B| \begin{pmatrix} \dfrac{a+b}{a+kb} & 0 \\ 0 & \dfrac{a+b}{b+ka} \end{pmatrix}. \qquad (3.8)$$

For an arbitrary ellipse whose semi-axes are not aligned with the coordinate axes, one can use the identity

$$M(\lambda,\mathcal{R}B) = \mathcal{R}M(\lambda,B)\mathcal{R}^T \quad \text{for any rotation } \mathcal{R}, \qquad (3.9)$$

to compute its polarization tensor.

In the three-dimensional case, a domain for which analogous analytical expressions for the elements of its polarization tensor M are available is the ellipsoid. If the coordinate axes are chosen to coincide with the principal axes of the ellipsoid B whose equation then becomes

$$\frac{x_1^2}{a^2} + \frac{x_2^2}{b^2} + \frac{x_3^2}{c^2} = 1, \quad 0 < c \le b \le a,$$

then M takes the form

$$M(\lambda, B) = (k-1)|B| \begin{pmatrix} \dfrac{1}{(1-A_1)+kA_1} & 0 & 0 \\ 0 & \dfrac{1}{(1-A_2)+kA_2} & 0 \\ 0 & 0 & \dfrac{1}{(1-A_3)+kA_3} \end{pmatrix},$$

$$(3.10)$$

where the constants $A_1, A_2,$ and A_3 are defined by

$$A_1 = \frac{bc}{a^2} \int_1^{+\infty} \frac{1}{t^2 \sqrt{t^2 - 1 + (\frac{b}{a})^2} \sqrt{t^2 - 1 + (\frac{c}{a})^2}} \, dt,$$

$$A_2 = \frac{bc}{a^2} \int_1^{+\infty} \frac{1}{(t^2 - 1 + (\frac{b}{a})^2)^{\frac{3}{2}} \sqrt{t^2 - 1 + (\frac{c}{a})^2}} \, dt,$$

$$A_3 = \frac{bc}{a^2} \int_1^{+\infty} \frac{1}{\sqrt{t^2 - 1 + (\frac{b}{a})^2}(t^2 - 1 + (\frac{c}{a})^2)^{\frac{3}{2}}} \, dt.$$

In the special case, $a = b = c$, the ellipsoid B becomes a sphere and $A_1 = A_2 = A_3 = 1/3$. Hence the polarization tensor associated with the sphere B is given by

$$M(\lambda, B) = \frac{3(k-1)}{2+k} |B| I_3,$$

with I_3 being the 3×3 identity matrix.

We now list important properties of the PT.

Theorem 3.2 (Properties of the Polarization Tensor). *For $|\lambda| > 1/2$, let $M(\lambda, B) = (m_{ij})_{i,j=1}^d$ be the PT associated with the bounded domain B in $\mathbb{R}^d$ and the conductivity $k = (2\lambda + 1)/(2\lambda - 1)$. Then*

(i) *M is symmetric.*
(ii) *If $k > 1$, then M is positive definite, and it is negative definite if $0 < k < 1$.*
(iii) *The following optimal bounds for the PT*

$$\begin{cases} \dfrac{1}{k-1} \operatorname{Tr}(M) \le (d-1+\dfrac{1}{k})|B|, \\ (k-1) \operatorname{Tr}(M^{-1}) \le \dfrac{d-1+k}{|B|}, \end{cases} \quad (3.11)$$

hold.

Note that by making use of bounds (3.11), an accurate size estimate of B can be immediately obtained.

In the literature on effective medium theory, the bounds (3.11) are known as the Hashin-Shtrikman bounds. In the second inequality in (3.11), the equality holds if and only if B is an ellipse or an ellipsoid [106, 107].

The concept of polarization tensors appears in deriving asymptotic expansions of electrical effective properties of composite dilute media. Polarization tensors involve microstructural information beyond that contained in the volume fractions such as material contrast, inclusion shape and orientation. See [31].

3.2 Polarization Tensor of Multiple Inclusions

Our goal in this section is to investigate properties of polarization tensors associated with multiple inclusions. These results are from [32].

Let B_p for $p = 1, \ldots, P$ be a bounded smooth domain in $\mathbb{R}^d$. Throughout this section, we assume that:

(H1)   there exist positive constants C_1 and C_2 such that

$$C_1 \leq \operatorname{diam} B_p \leq C_2, \quad \text{and} \quad C_1 \leq \operatorname{dist}(B_p, B_{p'}) \leq C_2, \quad p \neq p';$$

(H2)   the conductivity of the inclusion B_p for $p = 1, \ldots, P$ is equal to some positive constant $k_p \neq 1$.

3.2.1 Definition

To begin, we prove the following theorem.

Theorem 3.3. *Let H be a harmonic function in $\mathbb{R}^d$ for $d = 2$ or 3. Let u be the solution of the transmission problem*

$$\begin{cases} \nabla \cdot \left(1 + \sum_{p=1}^{P} (k_p - 1)\chi(B_p) \right) \nabla u = 0 \quad \text{in } \mathbb{R}^d, \\ u(x) - H(x) = O(|x|^{1-d}) \quad \text{as } |x| \to +\infty. \end{cases} \tag{3.12}$$

There are unique functions $\phi^{(l)} \in L_0^2(\partial B_l)$, $l = 1, \ldots, P$, such that

$$u(x) = H(x) + \sum_{l=1}^{P} \mathcal{S}_{B_l}[\phi^{(l)}](x), \quad x \in \mathbb{R}^d. \tag{3.13}$$

The potentials $\phi^{(l)}$, $l = 1, \ldots, P$, satisfy

$$(\lambda_l I - \mathcal{K}^*_{B_l})[\phi^{(l)}] - \sum_{p \neq l} \frac{\partial(\mathcal{S}_{B_p}[\phi^{(p)}])}{\partial \nu^{(l)}}\bigg|_{\partial B_l} = \frac{\partial H}{\partial \nu^{(l)}}\bigg|_{\partial B_l} \qquad \text{on } \partial B_l, \qquad (3.14)$$

where $\nu^{(l)}$ denotes the outward unit normal to ∂B_l and

$$\lambda_l = \frac{k_l + 1}{2(k_l - 1)}.$$

Proof. It is easy to see from (2.23) that u defined by (3.13) and (3.14) is the solution of (3.12). Thus it is enough to show that the integral equation (3.14) has a unique solution.

Let $X := L_0^2(\partial B_1) \times \cdots \times L_0^2(\partial B_P)$. We prove that the operator $T : X \to X$ defined by

$$T(\phi^{(1)}, \cdots, \phi^{(P)}) = T_0(\phi^{(1)}, \cdots, \phi^{(P)}) + T_1(\phi^{(1)}, \cdots, \phi^{(P)})$$

$$:= \left((\lambda_1 I - \mathcal{K}^*_{B_1})[\phi^{(1)}], \cdots, (\lambda_P I - \mathcal{K}^*_{B_P})[\phi^{(P)}] \right)$$

$$- \left(\sum_{p \neq 1} \frac{\partial(\mathcal{S}_{B_p}[\phi^{(p)}])}{\partial \nu^{(1)}}\bigg|_{\partial B_1}, \cdots, \sum_{p \neq P} \frac{\partial(\mathcal{S}_{B_p}[\phi^{(p)}])}{\partial \nu^{(P)}}\bigg|_{\partial B_P} \right)$$

is invertible. By Lemma 2.9, T_0 is invertible on X. On the other hand, since the domains B_p are a fixed distance apart, it is easy to see that T_1 is a compact operator on X. Thus, by the Fredholm alternative, it suffices to show that T is injective on X.

If $T(\phi^{(1)}, \cdots, \phi^{(P)}) = 0$, then $u(x) := \sum_{l=1}^P \mathcal{S}_{B_l}[\phi^{(l)}](x)$, $x \in \mathbb{R}^d$, is the solution of (3.12) with $H = 0$. By the uniqueness of the solution to (3.12), we get $u \equiv 0$. In particular, $\mathcal{S}_{B_l}[\phi^{(l)}]$ is smooth across ∂B_l, $l = 1, \ldots, P$. Therefore,

$$\phi^{(l)} = \frac{\partial(\mathcal{S}_{B_l}[\phi^{(l)}])}{\partial \nu^{(l)}}\bigg|_+ - \frac{\partial(\mathcal{S}_{B_l}[\phi^{(l)}])}{\partial \nu^{(l)}}\bigg|_- = 0.$$

This completes the proof. $\square$

With the above theorem, we can proceed to introduce the polarization tensor of multiple inclusions.

Definition 3.4. Let $\phi_i^{(l)}$, $i = 1, \ldots, d$, be the solution of

$$(\lambda_l I - \mathcal{K}^*_{B_l})[\phi_i^{(l)}] - \sum_{p \neq l} \frac{\partial(\mathcal{S}_{B_p}[\phi_i^{(p)}])}{\partial \nu^{(l)}}\bigg|_{\partial B_l} = \frac{\partial x_i}{\partial \nu^{(l)}}\bigg|_{\partial B_l} \qquad \text{on } \partial B_l. \qquad (3.15)$$

Then the polarization tensor $M = (m_{ij})_{i,j=1}^d$ associated with $\cup_{p=1}^P B_p$ is defined to be

$$m_{ij} = \sum_{l=1}^P \int_{\partial B_l} \xi_j \phi_i^{(l)}(\xi)\, d\sigma(\xi). \tag{3.16}$$

3.2.2 Representation by Equivalent Ellipses

Suppose $d = 2$, and let $M = (m_{ij})_{i,j=1}^2$ be the polarization tensor of the inclusions $\cup_{p=1}^P B_p$. We define the overall conductivity $\overline{k}$ of $B = \cup_{p=1}^P B_p$ by

$$\frac{\overline{k}-1}{\overline{k}+1}\sum_{p=1}^P |B_p| := \sum_{p=1}^P \frac{k_p-1}{k_p+1}|B_p|, \tag{3.17}$$

and its *center* $\overline{z}$ by

$$\frac{\overline{k}-1}{\overline{k}+1}\overline{z}\sum_{p=1}^P |B_p| = \sum_{p=1}^P \frac{k_p-1}{k_p+1}\int_{B_p} x\,dx. \tag{3.18}$$

Note that if k_p is the same for all p then $\overline{k} = k_p$ and $\overline{z}$ is the center of mass of B.

In this section we represent the multiple inclusions $\cup_{p=1}^P B_p$ by means of an ellipse, $\mathcal{E}$, of center $\overline{z}$ with the same polarization tensor. We call $\mathcal{E}$ the equivalent ellipse of $\cup_{p=1}^P B_p$.

At this point let us review a method to find an ellipse from a given polarization tensor. Let $\mathcal{E}'$ be an ellipse whose semi-axes are on the x_1- and x_2-axes and of length a and b, respectively. Let $\mathcal{E} = R_\theta \mathcal{E}'$ where $R_\theta = \begin{pmatrix} \cos\theta & -\sin\theta \\ \sin\theta & \cos\theta \end{pmatrix}$ and $\theta \in [0, \pi]$. Let M be the polarization tensor of $\mathcal{E}$. We want to recover a, b, and θ from M knowing the conductivity $k = \overline{k}$. Recall that the polarization tensor M' for $\mathcal{E}'$ takes the form

$$M' = (k-1)|\mathcal{E}'| \begin{pmatrix} \dfrac{a+b}{a+kb} & 0 \\ 0 & \dfrac{a+b}{b+ka} \end{pmatrix},$$

and that of $\mathcal{E}$ is given by $M = R_\theta M' R_\theta^T$. Suppose that the eigenvalues of M are κ_1 and κ_2, and corresponding eigenvectors of unit length are $(e_{11}, e_{12})^T$ and $(e_{21}, e_{22})^T$. Then it can be shown that

$$a = \sqrt{\frac{p}{\pi q}}, \quad b = \sqrt{\frac{pq}{\pi}}, \quad \theta = \arctan \frac{e_{21}}{e_{11}},$$

where

$$\frac{1}{p} = \frac{k-1}{k+1} \left(\frac{1}{\kappa_1} + \frac{1}{\kappa_2} \right) \quad \text{and} \quad q = \frac{\kappa_2 - k\kappa_1}{\kappa_1 - k\kappa_2}.$$

The above calculation extends to the three-dimensional case. Based on the analytical expression (3.10), the parameters a, b, and c of an ellipsoid B can be recovered from the eigenvalues of its polarization tensor $M(\lambda, B)$.

3.3 Helmholtz Equation

Suppose that an electromagnetic medium occupies a bounded domain Ω in $\mathbb{R}^d$, with a connected C^2-boundary $\partial\Omega$. Suppose that Ω contains a small inclusion of the form $D = \delta B + z$, where $z \in \Omega$ and B is a C^2-bounded domain in $\mathbb{R}^d$ containing the origin.

Let μ_0 and ε_0 denote the permeability and the permittivity of the background medium Ω, and assume that μ_0 and ε_0 are positive constants. Let $\mu_\star$ and $\varepsilon_\star$ denote the permeability and the permittivity of the inclusion D, which are also assumed to be positive constants. Introduce the piecewise constant magnetic permeability

$$\mu_\delta(x) = \begin{cases} \mu_0 , & x \in \Omega \setminus \overline{D}, \\ \mu_\star , & x \in D. \end{cases}$$

The piecewise constant electric permittivity, $\varepsilon_\delta(x)$, is defined analogously.

Let the electric field u denote the solution to the Helmholtz equation

$$\nabla \cdot (\frac{1}{\mu_\delta} \nabla u) + \omega^2 \varepsilon_\delta u = 0 \quad \text{in } \Omega, \tag{3.19}$$

with the boundary condition $u = f \in W_{\frac{1}{2}}^2(\partial\Omega)$, where $\omega > 0$ is a given frequency.

Problem (3.19) can be written as

$$\begin{cases} (\Delta + \omega^2 \varepsilon_0 \mu_0)u = 0 & \text{in } \Omega \setminus \overline{D}, \\ (\Delta + \omega^2 \varepsilon_\star \mu_\star)u = 0 & \text{in } D, \\ \dfrac{1}{\mu_\star} \dfrac{\partial u}{\partial \nu}\Big|_- - \dfrac{1}{\mu_0} \dfrac{\partial u}{\partial \nu}\Big|_+ = 0 & \text{on } \partial D, \\ u\big|_- - u\big|_+ = 0 & \text{on } \partial D, \\ u = f & \text{on } \partial\Omega. \end{cases}$$

Assuming that

$$\omega^2 \varepsilon_0 \mu_0 \text{ is not an eigenvalue for the operator } -\Delta \text{ in } L^2(\Omega)$$

with homogeneous Dirichlet boundary conditions,   (3.20)

we can prove existence and uniqueness of a solution to (3.19) at least for δ small enough.

With the notation of Sect. 2.2, the following asymptotic formula holds.

Theorem 3.5 (Boundary Perturbations). *Suppose that (3.20) holds. Let u be the solution of (3.19) and let the function U be the background solution as before. For any $x \in \partial\Omega$,*

$$\frac{\partial u}{\partial \nu}(x) = \frac{\partial U}{\partial \nu}(x) + \delta^d \left(\nabla U(z) \cdot M(\lambda, B) \frac{\partial \nabla_z G_{k_0}(x, z)}{\partial \nu_x} \right.$$

$$\left. + k_0^2 (\frac{\varepsilon_\star}{\varepsilon_0} - 1)|B|U(z) \frac{\partial G_{k_0}(x, z)}{\partial \nu_x} \right) + O(\delta^{d+1}), \qquad (3.21)$$

where $M(\lambda, B)$ is the polarization tensor defined in (3.4) with λ given by

$$\lambda := \frac{(\mu_0/\mu_\star) + 1}{2((\mu_0/\mu_\star) - 1)}. \qquad (3.22)$$

Here G_{k_0} is the Dirichlet Green function defined by (2.79).

3.3.1 Formal Derivations

From the Lippmann-Schwinger integral representation formula

$$u(x) = U(x) + (\frac{\mu_0}{\mu_\star} - 1) \int_D \nabla u(y) \cdot \nabla_y G_{k_0}(x, y) \, dy$$

$$+ k_0^2 (\frac{\varepsilon_\star}{\varepsilon_0} - 1) \int_D u(y) G_{k_0}(x, y) \, dy, \quad x \in \Omega,$$

it follows that for any $x \in \partial\Omega$,

$$\frac{\partial u}{\partial \nu}(x) = \frac{\partial U}{\partial \nu}(x) + (\frac{\mu_0}{\mu_\star} - 1) \int_D \nabla u(y) \cdot \frac{\partial \nabla_y G_{k_0}(x, y)}{\partial \nu_x} \, dy$$

$$+ k_0^2 (\frac{\varepsilon_\star}{\varepsilon_0} - 1) \int_D u(y) \frac{\partial G_{k_0}(x, y)}{\partial \nu_x} \, dy.$$

Using a Taylor expansion of $G_{k_0}(x, y)$ for $y \in D$, we readily see that for any $x \in \partial\Omega$,

$$
\begin{aligned}
\frac{\partial u}{\partial \nu}(x) &\approx \frac{\partial U}{\partial \nu}(x) + (\frac{\mu_0}{\mu_\star} - 1)\frac{\partial \nabla_z G_{k_0}(x, z)}{\partial \nu_x} \cdot (\int_D \nabla u(y)\, dy) \\
&+ k_0^2(\frac{\varepsilon_\star}{\varepsilon_0} - 1)\frac{\partial G_{k_0}(x, z)}{\partial \nu_x}(\int_D u(y)\, dy).
\end{aligned}
\tag{3.23}
$$

Following the same lines as in the derivation of the asymptotic expansion of the voltage potentials in Sect. 3.1, one can easily check that $u(y) \approx U(z)$, for $y \in D$, and

$$
\int_D \nabla u(y)\, dy \approx \delta^d \left(\int_B \nabla \hat{v}(\xi)\, d\xi \right) \cdot \nabla U(z),
$$

where $\hat{v}$ is defined by (3.5) with $k = \mu_0/\mu_\star$. Inserting these two approximations into (3.23) leads to (3.21).

Before concluding this section, we make a remark. Consider the Helmholtz equation with the Neumann data g in the presence of the inclusion D:

$$
\begin{cases}
\nabla \cdot \dfrac{1}{\mu_\delta}\nabla u + \omega^2 \varepsilon_\delta u = 0 & \text{in } \Omega, \\
\dfrac{\partial u}{\partial \nu} = g & \text{on } \partial\Omega.
\end{cases}
\tag{3.24}
$$

Let the background solution U satisfy

$$
\begin{cases}
\Delta U + k_0^2 U = 0 & \text{in } \Omega, \\
\dfrac{\partial U}{\partial \nu} = g & \text{on } \partial\Omega.
\end{cases}
\tag{3.25}
$$

The following asymptotic expansion of the solution of the Neumann problem holds. For any $x \in \partial\Omega$, we have

$$
u(x) = U(x) + \delta^d \bigg(\nabla U(z) M(\lambda, B) \nabla_z N_{k_0}(x, z)
$$

$$
+ k_0^2(\frac{\varepsilon_\star}{\varepsilon_0} - 1)|B|U(z)N_{k_0}(x, z) \bigg) + O(\delta^{d+1}),
\tag{3.26}
$$

where N_{k_0} is the Neumann function defined by

$$
\begin{cases}
\Delta_x N_{k_0}(x, z) + k_0{}^2 N_{k_0}(x, z) = -\delta_z & \text{in } \Omega, \\
\dfrac{\partial N_{k_0}}{\partial \nu_x}\bigg|_{\partial\Omega} = 0 & \text{for } z \in \Omega.
\end{cases}
\tag{3.27}
$$

The following useful relation between the Neumann function and the fundamental solution Γ_{k_0} holds:

$$(-\frac{1}{2}I + \mathcal{K}_\Omega^{k_0})[N_{k_0}(\cdot, z)](x) = \Gamma_{k_0}(x, z), \quad x \in \partial\Omega, \; z \in \Omega. \tag{3.28}$$

3.4 Asymptotic Formulas in the Time-Domain

Consider the initial boundary value problem for the (scalar) wave equation

$$\begin{cases} \partial_t^2 u - \nabla \cdot (1 + (k-1)\chi(D))\nabla u = 0 & \text{in} \quad \Omega_T, \\ u(x,0) = u_0(x), \quad \partial_t u(x,0) = u_1(x) & \text{for} \quad x \in \Omega, \\ \dfrac{\partial u}{\partial \nu} = g & \text{on} \quad \partial\Omega_T, \end{cases} \tag{3.29}$$

where $T < +\infty$ is a final observation time, $\Omega_T = \Omega \times]0, T[$, and $\partial\Omega_T = \partial\Omega \times]0, T[$. The initial data $u_0, u_1 \in \mathcal{C}^\infty(\overline{\Omega})$, and the Neumann boundary data $g \in \mathcal{C}^\infty(0, T; \mathcal{C}^\infty(\partial\Omega))$ are subject to compatibility conditions.

Define the background solution U to be the solution of the wave equation in the absence of any inclusions. Thus U satisfies

$$\begin{cases} \partial_t^2 U - \Delta U = 0 & \text{in} \quad \Omega_T, \\ U(x,0) = u_0(x), \quad \partial_t U(x,0) = u_1(x) & \text{for} \quad x \in \Omega, \\ \dfrac{\partial U}{\partial \nu} = g & \text{on} \quad \partial\Omega_T. \end{cases}$$

For $\rho > 0$, define the operator P_ρ on tempered distributions by

$$P_\rho[\psi](x, t) = \int_{|\omega| \le \rho} e^{-i\omega t} \hat{\psi}(x, \omega) \, d\omega, \tag{3.30}$$

where $\hat{\psi}(x, \omega)$ denotes the Fourier transform of $\psi(x, t)$ in the t-variable. Clearly, the operator P_ρ truncates the high-frequency component (larger than ρ) of ψ.

The following asymptotic expansion holds as $\delta \to 0$.

Theorem 3.6 (Perturbations of Weighted Boundary Measurements). *Let $w \in \mathcal{C}^\infty(\overline{\Omega}_T)$ satisfy $(\partial_t^2 - \Delta)w(x, t) = 0$ in Ω_T with $\partial_t w(x, T) = w(x, T) = 0$ for $x \in \Omega$. Suppose that $\rho \ll 1/\delta$. Define the weighted boundary measurements*

$$I_w[U, T] := \int_{\partial\Omega_T} P_\rho[u - U](x, t) \frac{\partial w}{\partial \nu}(x, t) \, d\sigma(x) \, dt.$$

Then, for any fixed $T > diam(\Omega)$, the following asymptotic expansion for $I_w[U, T]$ holds as $\delta \to 0$:

$$I_w[U, T] \approx \delta^d \int_0^T \nabla P_\rho[U](z, t) \cdot M(\lambda, B) \nabla w(z, t) \ dt, \qquad (3.31)$$

where $M(\lambda, B)$ is defined by (3.4).

Expansion (3.31) is a weighted expansion. Pointwise expansions similar to those in Theorem 3.1 which is for the steady-state model can also be obtained. See Sect. 3.5.3.

3.5 Asymptotic Formulas for Dipole Sources in Free Space

3.5.1 Conductivity Problem

Let $y \in \mathbb{R}^d \setminus \overline{D}$ and let $u_y(x)$ be the solution to the transmission problem

$$\begin{cases} \nabla \cdot (1 + (k-1)\chi(D))\nabla u_y(x) = \delta_y(x), & x \in \mathbb{R}^2, \\ u_y(x) - \Gamma(x, y) = O(|x|^{-1}), & |x - y| \to \infty. \end{cases} \qquad (3.32)$$

Let $U_y(x) = \Gamma(x, y)$ denote the background solution. We still assume that D is of the form $D = \delta B + z$. For $y \in \partial D$ and x away from z, we can prove similarly to (3.3) that the following expansion of $u_y - U_y$ for x away from z holds:

$$(u_y - U_y)(x) = -\delta^d \nabla_z \Gamma(x, z) \cdot M(\lambda, B) \nabla_z \Gamma(y, z) + O(\delta^{d+1}).$$

Note that, because of the symmetry of the PT, the leading-order term in the above expansion satisfies the reciprocity property, i.e., $\nabla_z \Gamma(x, z) \cdot M(\lambda, B)\nabla_z \Gamma(y, z) = \nabla_y \Gamma(y, z) \cdot M(\lambda, B)\nabla_z \Gamma(x, z)$.

3.5.2 Helmholtz Equation

Suppose that D is illuminated by a time-harmonic wave generated at the point source y with the operating frequency ω. In this case, the incident field is given by

$$U_y(x) = \Gamma_{k_0}(x, y),$$

and the field perturbed in the presence of the target (inclusion) is the solution to the following transmission problem:

$$\nabla \cdot \left(\frac{1}{\mu_0} \chi(\mathbb{R}^d \setminus \overline{D}) + \frac{1}{\mu_\star} \chi(D) \right) \nabla u_y + \omega^2 \left(\varepsilon_0 \chi(\mathbb{R}^d \setminus \overline{D}) + \varepsilon_\star \chi(D) \right) u_y = \frac{1}{\mu_0} \delta_y,$$

(3.33)

and is subject to the outgoing radiation condition, or equivalently

$$
\begin{cases}
\Delta u_y + k_0^2 u_y = \delta_y & \text{in } \mathbb{R}^d \setminus \overline{D}, \\
\Delta u_y + k_\star^2 u_y = 0 & \text{in } D, \\
u_y\big|_+ - u_y\big|_- = 0 & \text{on } \partial D, \\
\dfrac{1}{\mu_0} \dfrac{\partial u_y}{\partial \nu}\bigg|_+ - \dfrac{1}{\mu_\star} \dfrac{\partial u_y}{\partial \nu}\bigg|_- = 0 & \text{on } \partial D, \\
u_y \text{ satisfies the outgoing radiation condition.}
\end{cases}
$$

(3.34)

Here, $k_\star^2 = \omega^2 \varepsilon_\star \mu_\star$.

Let u_y be the solution to (3.34) and let U_y be the solution in the absence of the target, i.e., $U_y(x) = \Gamma_{k_0}(x - y)$.

As $\delta \to 0$, the following asymptotic expansion of the perturbation of the perturbation $u_y - U_y$ due to the presence of $D = \delta B + z$ can be proved analogously to (3.21):

$$
u_y(x) - U_y(x) = -\delta^d \bigg[k_0^2 \left(\frac{\varepsilon_\star}{\varepsilon_0} - 1 \right) |B| \Gamma_{k_0}(x, z) \Gamma_{k_0}(y, z)
$$
$$
+ \nabla_z \Gamma_{k_0}(x, z) \cdot M(\lambda, B) \nabla_z \Gamma_{k_0}(y, z) \bigg] + O(\delta^{d+1}),
$$

(3.35)

where λ is given in this case by (3.22). Note that (3.35) is a dipolar approximation. Formula (3.35) shows that, at the leading-order in terms of the characteristic size, the effect of a small electromagnetic inclusion on measurements is the sum of a polarized magnetic dipole and an electric point source. Moreover, the leading-order term satisfies the reciprocity property.

3.5.3 Wave Equation

For the sake of simplicity, we only consider the three dimensional case. We set $y \in \mathbb{R}^3$ be such that $|y - z| \gg \delta$. We use a spherical wave excitation

$$U(x,t) := U_y(x,t) := -\frac{\delta_0(t - |x - y|)}{4\pi|x - y|} \quad \text{for } x \neq y. \tag{3.36}$$

The wave U_y is the outgoing (or retarded) fundamental solution to the wave equation (2.105).

Consider now for the sake of simplicity the wave equation in the whole three-dimensional space with appropriate initial conditions:

$$\begin{cases} \partial_t^2 u - \nabla \cdot \left(1 + (k-1)\chi(D)\right)\nabla u = -\delta_y(x)\delta_0(t) & \text{in } \mathbb{R}^3 \times \mathbb{R}, \\ u(x,t) = 0, \quad \partial_t u(x,t) = 0 & \text{for } x \in \mathbb{R}^3, x \neq y, t < 0. \end{cases} \tag{3.37}$$

The following theorem holds.

Theorem 3.7 (Pointwise Perturbations). *Let u be the solution to (3.37). Set U_y to be the background solution. Suppose that $\rho \ll 1/\delta$.*

(i) Let P_ρ be given by (3.30). The following outer expansion holds

$$P_\rho[u - U_y](x,t) \approx \delta^3 \int_{\mathbb{R}} \nabla P_\rho[U_z](x, t - \tau) \cdot M(\lambda, B)\nabla P_\rho[U_y](z, \tau)\, d\tau, \tag{3.38}$$

for x away from z, where $M(\lambda, B)$ is defined by (3.4) and U_y and U_z by (3.36).

(ii) The following inner approximation holds:

$$P_\rho[u - U_y](x,t) \approx -\delta\hat{v}\left(\frac{x - z}{\delta}\right) \cdot \nabla P_\rho[U_y](x,t) \quad \text{for } x \text{ near } z, \tag{3.39}$$

where $\hat{v}$ is given by (3.5) and U_y by (3.36).

Formula (3.38) shows that the perturbation due to the inclusion is in the time-domain a wavefront emitted by a dipolar source located at the point z.

Taking the Fourier transform of (3.38) in the time variable yields the expansions given in (3.35) for the perturbations resulting from the presence of a small inclusion for solutions to the Helmholtz equation at low frequencies (at wavelengths large compared to the size of the inclusion). Based on (3.38) a time-reversal technique can be used to localize the inclusion from far-field measurements.

Bibliography and Discussion

Theorem 3.1 was proven in [27, 62, 80]. The results of Sect. 3.3 are from [30, 157]. The original Hashin-Shtrikman bounds can be found in the book by Milton [131]. The Hashin-Shtrikman bounds for the polarization tensor were proved in [61, 124]. The method of representing an inclusion by means of an equivalent ellipse with the same polarization tensor is from [57]. The initial boundary-value problem for the (time-dependent) wave equation in the presence of inclusions of small volume has been considered in [17]. In that paper, the asymptotic formulas (3.38) and (3.39) are rigorously derived. Formula (3.35) was obtained in [157] (see also [29, 39]). Approximation (3.7) is uniform with respect to the conductivity contrast k [29, 137] provided that $0 \leq k \neq 1 \leq \infty$. For negative k, the polarization tensor may blow up [95]. If B is smooth, then the Neumann-Poincaré operator $\mathcal{K}_B^*$ has a discrete spectrum and consequently, there is in general a sequence of negative values of k, known as plasmonic resonances [88], for which the polarization tensor does not exist. When B is a disk, in view of (2.15), there are only two plasmonic resonances.

Chapter 4
Generalized Polarization Tensors

The aim of this chapter is to introduce the concept of generalized polarization tensors (GPTs). The GPTs are the basic building blocks for the asymptotic expansions of the boundary voltage perturbations due to the presence of small conductivity inclusions inside a conductor. The GPTs contain important geometrical information on the inclusion.

In this chapter, we first introduce the concept of GPTs. Then we prove invariance properties of GPTs under translation, rotation, and scaling. We also show that the GPTs capture high-frequency shape oscillations as well as topology. We introduce a recursive matching algorithm to reconstruct the shape of a target given its first GPTs. To handle topology changes, we implement a level set version of our recursive matching GPTs algorithm. Moreover, we prove that high-frequency oscillations of the shape of a domain are only contained in its high-order GPTs and perform a stability and resolution analysis for the reconstruction of small shape changes from noisy GPTs. It will be shown that, a particular linear combination of the GPTs is very suitable for the resolution analysis. This motivates the introduction of the concept of complex contracted GPTs. We will show in Chap. 11 that the complex GPTs have some nice properties, such as simple transformation formulas under rigid motions (simpler than those derived in this chapter for the GPTs), simple relations with the shape symmetry, and more importantly, they have invariants.

4.1 Definition and Basic Properties of the GPTs

Throughout this chapter we assume that the domains under consideration have smooth boundaries and they are two dimensional. Let $|\lambda| > 1/2$. For a multi-index $\alpha = (\alpha_1, \alpha_2) \in \mathbb{N}^2$ where $\mathbb{N}$ is the set of all positive integers and a smooth bounded domain D in $\mathbb{R}^2$, define ϕ_α by

H. Ammari et al., *Mathematical and Statistical Methods for Multistatic Imaging*, 115
Lecture Notes in Mathematics 2098, DOI 10.1007/978-3-319-02585-8_4,
© Springer International Publishing Switzerland 2013

$$\phi_\alpha(y) := (\lambda I - \mathcal{K}_D^*)^{-1}[\nu(x) \cdot \nabla x^\alpha](y), \quad y \in \partial D . \tag{4.1}$$

Here and throughout this book, we use the conventional notation: $x^\alpha = x_1^{\alpha_1} x_2^{\alpha_2}$, $|\alpha| = \alpha_1 + \alpha_2$. The generalized polarization tensors (GPTs) $M_{\alpha\beta}$ for $\alpha, \beta \in \mathbb{N}^2$ ($|\alpha|, |\beta| \geq 1$) associated with the parameter λ and the domain D are defined by

$$M_{\alpha\beta} = M_{\alpha\beta}(\lambda, D) := \int_{\partial D} y^\beta \phi_\alpha(y) d\sigma(y) . \tag{4.2}$$

We emphasize that the PT introduced in the previous chapter is the GPT $M_{\alpha\beta}$ with $|\alpha| = |\beta| = 1$.

The GPTs are the building blocks in representing the perturbation of the electrical potential in the presence of an inclusion D of conductivity k inside a background medium of conductivity 1. The parameter λ is related to k via the formula

$$\lambda = \frac{k+1}{2(k-1)} . \tag{4.3}$$

Note that the GPTs are real-valued tensors. Key properties of positivity and symmetry of the GPTs are proved in [31, Chap. 4]. The following holds.

Theorem 4.1 (Symmetry). *Suppose that $a_\alpha, \alpha \in I$, and $b_\beta, \beta \in J$, where I and J are finite index sets, are real constants such that $\sum_\alpha a_\alpha y^\alpha$ and $\sum_\beta b_\beta y^\beta$ are harmonic polynomials. Then*

$$\sum_{\alpha,\beta} a_\alpha b_\beta M_{\alpha\beta} = \sum_{\alpha,\beta} a_\alpha b_\beta M_{\beta\alpha} . \tag{4.4}$$

Theorem 4.2 (Positivity). *Suppose that $a_\alpha, \alpha \in I$, where I is a finite index set, are constants such that $f(y) = \sum_{\alpha \in I} a_\alpha y^\alpha$ is a harmonic polynomial. Let $\phi = (\lambda I - \mathcal{K}_D^*)^{-1}[\partial f / \partial \nu]$. Then*

$$\sum_{\alpha,\beta \in I} a_\alpha a_\beta M_{\alpha\beta} = \frac{k-1}{k+1}\left[k Q_D(\mathcal{S}_D[\phi] + f) + Q_{\mathbb{R}^2 \setminus \overline{D}}(\mathcal{S}_D[\phi]) + Q_D(f) \right] , \tag{4.5}$$

with the quadratic forms $Q_D(u), Q_{\mathbb{R}^2 \setminus \overline{D}}(u)$ being defined by

$$Q_D(u) := \int_D |\nabla u|^2 \, dx, \quad Q_{\mathbb{R}^2 \setminus \overline{D}}(u) := \int_{\mathbb{R}^2 \setminus \overline{D}} |\nabla u|^2 \, dx . \tag{4.6}$$

Theorem 4.2 says that if $k > 1$, then GPT's are positive definite, and they are negative definite if $0 < k < 1$. We emphasize that what is important is not the individual terms $M_{\alpha\beta}$ but their harmonic combinations. A harmonic combination of GPTs is $\sum_{\alpha,\beta} a_\alpha b_\beta M_{\alpha\beta}$ where $\sum_\alpha a_\alpha x^\alpha$ and $\sum_\beta b_\beta x^\beta$ are (real) harmonic polynomials. We call such (a_α) and (b_β) (real) harmonic coefficients.

The following uniqueness result holds.

Theorem 4.3 (Uniqueness). *If all harmonic combinations of GPTs of two domains are the same, i.e.,*

$$\sum_{\alpha,\beta} a_\alpha b_\beta M_{\alpha\beta}(\lambda_1, D_1) = \sum_{\alpha,\beta} a_\alpha b_\beta M_{\alpha\beta}(\lambda_2, D_2)$$

for all pairs (a_α), (b_β) of harmonic coefficients, then $D_1 = D_2$ and $\lambda_1 = \lambda_2$.

Theorem 4.3 says that the full knowledge of (harmonic combinations of) GPTs determines the domain D and λ. It is known that the first-order GPT, $M_{\alpha\beta}$ for $|\alpha| + |\beta| = 2$, yields the equivalent ellipse as we have seen in the previous chapter. However, it is not known analytically what kind of information on D and λ the higher-order GPTs carry. It is the purpose of this chapter to show that the GPTs contain both high-frequency and topological information on the inclusion (or cluster of inclusions).

We first recall the following result from [31, Theorem 4.13] which says that the GPTs can be estimated from above and below in terms of the harmonic moments.

Proposition 4.4. *Let $f(y) = \sum_{\alpha \in I} a_\alpha y^\alpha$ be a harmonic polynomial. Then*

$$\frac{2}{2\lambda+1} \int_D |\nabla f|^2 \leq \sum_{\alpha,\beta \in I} a_\alpha a_\beta M_{\alpha\beta}(\lambda, D) \leq \frac{2}{2\lambda-1} \int_D |\nabla f|^2 . \qquad (4.7)$$

We also recall the following monotonicity of $\sum_{\alpha,\beta} a_\alpha a_\beta M_{\alpha\beta}(\lambda, D)$ with respect to the domain [33].

Proposition 4.5. *Let $D \subsetneq D'$. Then, for all (nonzero) harmonic coefficients $(a_\alpha)_{|\alpha| \geq 1}$,*

$$\sum_{\alpha,\beta} a_\alpha a_\beta M_{\alpha\beta}(\lambda, D) < \sum_{\alpha,\beta} a_\alpha a_\beta M_{\alpha\beta}(\lambda, D') \qquad if \ \lambda > \frac{1}{2} ,$$

and

$$\sum_{\alpha,\beta} a_\alpha a_\beta M_{\alpha\beta}(\lambda, D) > \sum_{\alpha,\beta} a_\alpha a_\beta M_{\alpha\beta}(\lambda, D') \qquad if \ \lambda < -\frac{1}{2} .$$

4.2   Translation, Rotation, and Scaling Properties of the GPTs

In this section we show other properties of the GPTs which are particularly useful for shape description. Let N be a positive integer. We prove that the set of $(M_{\alpha\beta}(\lambda, D))$ for $|\alpha| + |\beta| \leq N$ is invariant under translation and rotation of D. We also provide a scaling formula for the GPTs.

4.2.1   Translation

For $T = (T_1, T_2)$, define $D_T := \{y + T : y \in D\}$ and $\partial D_T = (\partial D)_T$, and let $y^T = y + T$. For $\phi \in L^2(\partial D)$, define $\phi^T \in L^2(\partial D_T)$ as

$$\phi^T(y^T) := \phi(y), \quad \text{where } y \in \partial D.$$

Note that, for ϕ defined on ∂D, we have

$$\mathcal{K}^*_{D_T}[\phi^T](x^T) = \frac{1}{2\pi} \int_{\partial D_T} \frac{\langle x^T - \tilde{y}, \nu(x^T) \rangle}{|x^T - \tilde{y}|^2} \phi^T(\tilde{y}) \, d\sigma(\tilde{y})$$

$$= \frac{1}{2\pi} \int_{\partial D} \frac{\langle x^T - y^T, \nu(x^T) \rangle}{|x^T - y^T|^2} \phi^T(y^T) \, d\sigma(y)$$

$$= \mathcal{K}^*_D[\phi](x) \,.$$

For multi-index α and γ, let the coefficients $c^T_{\alpha\gamma}$ be such that

$$(x - T)^\alpha = \sum_\gamma c^T_{\alpha\gamma} x^\gamma, \quad \forall x \in \mathbb{R}^2 \,. \tag{4.8}$$

It is worth mentioning that $c^T_{\alpha\gamma} = 0$ if $|\gamma| > |\alpha|$.

Let $\phi_{D,\alpha}$ be the density function defined by (4.1) for a given domain D and multi-index α. Then we have for $x^T \in \partial D_T$

$$\left(\lambda I - \mathcal{K}^*_{D_T}\right)[\phi^T_{D,\alpha}](x^T) = \left(\lambda I - \mathcal{K}^*_D\right)[\phi_{D,\alpha}](x)$$

$$= \nu(x) \cdot \nabla x^\alpha \Big|_{\partial D}$$

$$= \sum_\gamma c^T_{\alpha\gamma} \nu(x^T) \cdot \nabla (x^T)^\gamma \,.$$

Hence,

$$\phi^T_{D,\alpha} = \sum_\gamma c^T_{\alpha\gamma}\phi_{D_T,\gamma} \quad \text{on } \partial D_T,$$

and the following proposition holds.

Proposition 4.6. *Let* $D_T = \{y + T : y \in D\}$. *Then,*

$$M_{\alpha\beta}(\lambda, D) = \sum_{\eta,\gamma} c^T_{\beta\eta}c^T_{\alpha\gamma}M_{\eta\gamma}(\lambda, D_T) , \tag{4.9}$$

where the coefficients $c^T_{\beta\eta}$ *and* $c^T_{\alpha\gamma}$ *are given by (4.8).*

Proof. We compute

$$M_{\alpha\beta}(\lambda, D) = \int_{\partial D} y^\beta \phi_\alpha(y) \, d\sigma(y)$$

$$= \int_{\partial D_T} (\tilde{y} - T)^\beta \phi^T_{D,\alpha}(\tilde{y}) \, d\sigma(\tilde{y})$$

$$= \int_{\partial D_T} \sum_\eta c^T_{\beta\eta}\tilde{y}^\eta \sum_\gamma c^T_{\alpha\gamma}\phi_{D_T,\gamma} \, d\sigma(\tilde{y}) ,$$

to obtain (4.9). □

For example, when $\alpha = (1,0)$ and $\beta = (2,0)$, we have $(x - T)^\alpha = x_1 - T_1$ and $(x - T)^\beta = (x_1 - T_1)^2 = x_1^2 - 2T_1 x_1 + T_1^2$, and readily get

$$M_{(1,0),(2,0)}(\lambda, D) = M_{(1,0),(2,0)}(\lambda, D_T) - 2T_1 M_{(1,0),(1,0)}(\lambda, D_T) .$$

4.2.2 Rotation

For $y \in \mathbb{R}^2$, let $y_\theta = \begin{pmatrix} \cos\theta & -\sin\theta \\ \sin\theta & \cos\theta \end{pmatrix} \begin{pmatrix} y_1 \\ y_2 \end{pmatrix}$, i.e., the rotation of y with angle θ with respect to the origin. Set $D_\theta = \{y_\theta : y \in D\}$ and

$$\phi^\theta(y_\theta) := \phi(y), \quad y \in \partial D .$$

Note that, for a density function ϕ defined on ∂D, we have

$$\mathcal{K}^*_{D_\theta}[\phi^\theta](x_\theta) = \frac{1}{2\pi} \int_{\partial D_\theta} \frac{\langle x_\theta - \tilde{y}, \nu(x_\theta)\rangle}{|x_\theta - \tilde{y}|^2} \phi^\theta(\tilde{y}) \, d\sigma(\tilde{y})$$

$$= \frac{1}{2\pi} \int_{\partial D} \frac{\langle x_\theta - y_\theta, \nu(x_\theta)\rangle}{|x_\theta - y_\theta|^2} \phi^\theta(y_\theta) \, d\sigma(y) \tag{4.10}$$

$$= \mathcal{K}^*_D[\phi](x) .$$

For multi-index α and γ, let the coefficients $r_{\alpha\gamma}^{\theta}$ be such that

$$(x_{-\theta})^{\alpha} = \sum_{\gamma} r_{\alpha\gamma}^{\theta} x^{\gamma}, \quad \forall\, x \in \mathbb{R}^2 . \tag{4.11}$$

It should be noted that $r_{\alpha\gamma}^{\theta} = 0$ if $|\gamma| \neq |\alpha|$. The following rotation formula for the GPTs can be proved in the same way as the translation formula (4.9).

Proposition 4.7. *Let $D_{\theta} = \{y_{\theta} : y \in D\}$. Then*

$$M_{\alpha\beta}(\lambda, D) = \sum_{\eta,\gamma} r_{\beta\eta}^{\theta} r_{\alpha\gamma}^{\theta} M_{\eta\gamma}(\lambda, D_{\theta}) , \tag{4.12}$$

where the coefficients $r_{\beta\eta}^{\theta}$ and $r_{\alpha\gamma}^{\theta}$ are given by (4.11).

4.2.3   Scaling

Similarly, define for a positive real s, $sD := \{sy : y \in D\}$ and set $\phi^s(sy) = \phi(y)$, $y \in \partial D$. Then, we have

$$\begin{aligned}
\mathcal{K}_{sD}^*[\phi^s](x^s) &= \frac{1}{2\pi} \int_{\partial sD} \frac{\langle x^s - \tilde{y}, \nu(x^s)\rangle}{|x^s - \tilde{y}|^2} \phi^s(\tilde{y})\, d\sigma(\tilde{y}) \\
&= \frac{1}{2\pi} \int_{\partial D} \frac{\langle x^s - y^s, \nu(x^s)\rangle}{|x^s - y^s|^2} \phi^s(y^s) s\, d\sigma(y) \\
&= \mathcal{K}_D^*[\phi](x) .
\end{aligned}$$

From

$$(s^{-1}x)^{\alpha} = \frac{1}{s^{|\alpha|}} x^{\alpha}, \quad \forall\, x \in \mathbb{R}^2 ,$$

the following holds.

Proposition 4.8. *Let $sD := \{sy : y \in D\}$ for a positive real number s. Then*

$$M_{\alpha\beta}(\lambda, D) = \frac{1}{s^{|\alpha|+|\beta|}} M_{\alpha\beta}(\lambda, sD) . \tag{4.13}$$

4.3   GPTs of Multiple Inclusions

Let $D = \cup_{p=1}^{P} D_p$ be a cluster of P well-separated inclusions. Here, D_p is a bounded simply connected domain with $\mathcal{C}^2$-boundary. To each D_p, we associate $|\lambda_p| > 1/2$ and set $\lambda = (\lambda_1, \ldots, \lambda_P)$.

Let $\mathbb{K}^*$ be the Neumann-Poincaré operator corresponding to the cluster D:

$$\mathbb{K}^* := \begin{bmatrix} \mathcal{K}_{D_1}^* & \frac{\partial}{\partial \nu^{(1)}} \mathcal{S}_{D_2} & \cdots & \frac{\partial}{\partial \nu^{(1)}} \mathcal{S}_{D_P} \\ \frac{\partial}{\partial \nu^{(2)}} \mathcal{S}_{D_1} & \mathcal{K}_{D_2}^* & \cdots & \frac{\partial}{\partial \nu^{(2)}} \mathcal{S}_{D_P} \\ \vdots & \vdots & \ddots & \vdots \\ \frac{\partial}{\partial \nu^{(P)}} \mathcal{S}_{D_1} & \frac{\partial}{\partial \nu^{(P)}} \mathcal{S}_{D_2} & \cdots & \mathcal{K}_{D_P}^* \end{bmatrix} . \tag{4.14}$$

Here, $\nu^{(p)}$ denotes the outward normal to ∂D_p and $\mathcal{S}_{D_p}$ and $\mathcal{K}_{D_p}^*$ are the single layer and Neumann-Poincaré operator associated with D_p. From [32] it is known that there exists a unique solution to

$$(\lambda I - \mathbb{K}^*)[\Phi_\alpha](y) = \partial h_\alpha(y), \quad y \in \partial D , \tag{4.15}$$

where

$$\Phi_\alpha := \begin{bmatrix} \psi_\alpha^{(1)} \\ \vdots \\ \phi_\alpha^{(P)} \end{bmatrix}, \quad \partial h_\alpha(y) := \begin{bmatrix} \nu^{(1)}(y) \cdot \nabla y^\alpha \\ \vdots \\ \nu^{(P)}(y) \cdot \nabla y^\alpha \end{bmatrix} ,$$

and the GPTs of the cluster are defined by

$$M_{\alpha\beta}(\lambda, D := \cup_p D_p) := \sum_{p=1}^{P} \int_{\partial D_p} y^\beta \phi_\alpha^{(p)}(y) \, d\sigma(y) .$$

It is proved in [15] that the spectrum of $\mathbb{K}^*$ lies in $] - 1/2, 1/2]$, and therefore, (4.15) is invertible. Moreover, it is easy to see that the properties proved in this chapter for the GPTs can be generalized to those associated with multiple connected inclusions. In fact, the translation, rotation, and scaling properties of the GPTs hold for multiple inclusions.

4.4   Shape Derivative of the GPTs

Let $D = \cup_{p=1}^{P} D_p$ be as before. For δ small, let D_δ be a δ-deformation of D, i.e., there are scalar functions $h_p \in \mathcal{C}^1(\partial D_p)$, $1 \leq p \leq P$, such that

$$\partial D_\delta := \cup_{p=1}^{P} \{\tilde{x} = x + \delta h_p(x)\nu_p(x) : x \in \partial D_p\} , \tag{4.16}$$

where ν_p is the outward unit normal vector on ∂D_p. Suppose that a_α and b_β are constants such that $H(x) = \sum_\alpha a_\alpha x^\alpha$ and $F(x) = \sum_\beta b_\beta x^\beta$ are harmonic polynomials. Then, according to [37], the perturbation of a harmonic sum of GPTs due to the shape deformation is given as follows:

$$\sum_{\alpha,\beta} a_\alpha b_\beta M_{\alpha\beta}(\lambda, D_\delta) - \sum_{\alpha,\beta} a_\alpha b_\beta M_{\alpha\beta}(\lambda, D)$$

$$= \sum_{p=1}^{P} \delta(k_p - 1) \int_{\partial D_p} h_p(x) \left[\frac{\partial u}{\partial \nu}\Big|_- \frac{\partial v}{\partial \nu}\Big|_- + \frac{1}{k_p} \frac{\partial u}{\partial T}\Big|_- \frac{\partial v}{\partial T}\Big|_- \right] (x)\, d\sigma(x) + O(\delta^2) ,$$

$$(4.17)$$

where

$$k_p = (2\lambda_p + 1)/(2\lambda_p - 1) \tag{4.18}$$

and u and v are respectively solutions to the problems:

$$\begin{cases} \Delta u = 0 & \text{in } D \cup (\mathbb{R}^2 \backslash \overline{D}) , \\ u|_+ - u|_- = 0 & \text{on } \partial D_p,\ 1 \leq p \leq P , \\ \dfrac{\partial u}{\partial \nu}\Big|_+ - k_p \dfrac{\partial u}{\partial \nu}\Big|_- = 0 & \text{on } \partial D_p,\ 1 \leq p \leq P , \\ (u - H)(x) = O(|x|^{-1}) & \text{as } |x| \to \infty , \end{cases} \tag{4.19}$$

and

$$\begin{cases} \Delta v = 0 & \text{in } D \cup (\mathbb{R}^2 \backslash \overline{D}) , \\ k_p v|_+ - v|_- = 0 & \text{on } \partial D_p,\ 1 \leq p \leq P , \\ \dfrac{\partial v}{\partial \nu}\Big|_+ - \dfrac{\partial v}{\partial \nu}\Big|_- = 0 & \text{on } \partial D_p,\ 1 \leq p \leq P , \\ (v - F)(x) = O(|x|^{-1}) & \text{as } |x| \to \infty . \end{cases} \tag{4.20}$$

The problems (4.19) and (4.20) are dual to each other. The shape derivative of GPTs can be easily derived using (4.17), see Sect. 4.6.

4.5 Stability and Resolution Analysis in the Linearized Case

4.5.1 Complex Contracted GPTs

Particularly interesting choices of harmonic coefficients are those of homogeneous harmonic polynomials: for a positive integer n and a multi-index α with $|\alpha| = n$, define (a_α^n) by

$$\sum_{|\alpha|=n} a_\alpha^n x^\alpha = r^n e^{in\theta} = (x_1 + ix_2)^n ,\qquad (4.21)$$

where $x = (r, \theta)$ in polar coordinates. Using these (complex) harmonic coefficients, we introduce for positive integers m and n

$$\mathrm{N}_{mn}^{(1)}(\lambda, D) = \sum_{|\alpha|=m}\sum_{|\beta|=n} a_\alpha^m a_\beta^n M_{\alpha\beta}(\lambda, D) ,\qquad (4.22)$$

and

$$\mathrm{N}_{mn}^{(2)}(\lambda, D) = \sum_{|\alpha|=m}\sum_{|\beta|=n} a_\alpha^m \overline{a_\beta^n} M_{\alpha\beta}(\lambda, D) .\qquad (4.23)$$

We call $\mathrm{N}_{mn}^{(1)}$ and $\mathrm{N}_{mn}^{(2)}$ the complex contracted GPTs. As will be seen in the next subsection, the stability and resolution analysis of complex contracted GPTs is much easier than the one of the GPTs. Moreover, it will be shown in Chap. 11 that $\mathrm{N}_{mn}^{(1)}$ and $\mathrm{N}_{mn}^{(2)}$ have rotation and translation properties simpler than those satisfied by the GPTs.

4.5.2 Resolution and Stability Analysis

Let D be the unit disk, $|\lambda| > 1/2$, and $k = (2\lambda + 1)/(2\lambda - 1)$. Let $F(x) = r^m e^{im\theta}$ and $H(x) = r^n e^{in\theta}$ for $m, n \in \mathbb{N}$. The solutions u_n and v_m of respectively (4.19) and (4.20) are given by

$$u_n(x) = \begin{cases} \dfrac{2}{1+k} r^n e^{in\theta}, & r < 1 , \\[2mm] \left(\dfrac{1-k}{1+k}\dfrac{1}{r^n} + r^n\right)e^{in\theta}, & r > 1 , \end{cases}$$

and

$$v_m(x) = \begin{cases} \dfrac{2k}{1+k} r^m e^{im\theta}, & r < 1 , \\[2mm] \left(\dfrac{1-k}{1+k}\dfrac{1}{r^m} + r^m\right)e^{im\theta}. & r > 1 . \end{cases}$$

Let D_δ be a δ-perturbation of D:

$$\partial D_\delta := \{\tilde x = x + \delta h(x)\nu(x)\ :\ x \in \partial D\} ,$$

where $h \in \mathcal{C}^1(\partial D)$ and δ is a small positive parameter. We use the Fourier convention

$$\hat{h}_p = \frac{1}{\sqrt{2\pi}} \int_0^{2\pi} h(\theta) e^{-ip\theta} \, d\theta, \qquad h(\theta) = \frac{1}{\sqrt{2\pi}} \sum_{p \in \mathbb{Z}} \hat{h}_p e^{ip\theta} .$$

Let $\mathbb{N}_{mn}^{(1)}(\lambda, D_\delta)$ and $\mathbb{N}_{mn}^{(1)}(\lambda, D)$ be the complex contracted GPTs associated with D_δ and D respectively. Since

$$\frac{\partial u_n}{\partial \nu}\bigg|_- \frac{\partial v_m}{\partial \nu}\bigg|_- + \frac{1}{k} \frac{\partial u_n}{\partial T}\bigg|_- \frac{\partial v_m}{\partial T}\bigg|_- = \frac{4(k-1)mn}{(k+1)^2} e^{i(m+n)\theta} ,$$

we obtain the following result.

Proposition 4.9. *We have*

$$\mathbb{N}_{mn}^{(1)}(\lambda, D_\delta) - \mathbb{N}_{mn}^{(1)}(\lambda, D) = \sqrt{2\pi} \delta \frac{mn}{\lambda^2} \hat{h}_{m+n} + O(\delta^2) \qquad (4.24)$$

as $\delta \to 0$.

Proposition 4.9 shows that high-frequency oscillations of the boundary deformation of a disk-shaped inclusion are only contained in its high-order contracted GPTs. Moreover, only $\hat{h}_p$ for p up to $2N$ can be reconstructed from the set of contracted GPTs $\mathbb{N}_{mn}^{(1)}$ for $m, n \leq N$. An asymptotic formula for $\mathbb{N}_{mn}^{(2)}(\lambda, D_\delta) - \mathbb{N}_{mn}^{(2)}(\lambda, D)$ as $\delta \to 0$ can be derived in the exactly same manner as Proposition 4.9.

Now, consider the domain $\delta' D_\delta$ with $\delta' > 0$ being small. An important problem is to reconstruct h from $\mathbb{N}_{mn}^{(1)}(\lambda, \delta' D_\delta)$ for $m, n \leq N$ in the presence of noise. Following [25], we perform from (4.24) a stability and resolution analysis for the reconstruction of h. For doing so, we introduce

$$a_{mn} = \frac{\lambda^2}{\sqrt{2\pi}(\delta')^{m+n}} \left(\mathbb{N}_{mn}^{(1)}(\lambda, \delta' D_\delta) - \mathbb{N}_{mn}^{(1)}(\lambda, \delta' D) \right) .$$

Assume that $\mathbb{N}_{mn}^{(1)}(\lambda, \delta' D_\delta)$ are corrupted with white noise. Thus,

$$a_{m,n}^{\text{meas}} = a_{mn} + \sigma_{\text{noise}} W_{m,n} ,$$

where $W_{m,n}$ denotes the noise terms and σ_{noise} models the noise magnitude.

As will be shown in Chap. 10, $W_{m,n}$ can be modeled as independent standard complex circularly symmetric Gaussian random variables such that

$$\mathbb{E}[|W_{m,n}|^2] = e^{2\kappa(m+n)} , \qquad (4.25)$$

with $\kappa := |\log \delta'|$ describing the exponential growth of the noise as a function of m, n.

It follows from the scaling property (4.13) and the expansion (4.24) that

$$a_{m,n}^{\text{meas}} = \delta mn\hat{h}_{m+n} + \sigma_{\text{noise}}W_{m,n} + \delta^2 V_{m,n}^\delta , \qquad (4.26)$$

where $V_{m,n}^\delta$ denotes the approximation error. Therefore, introducing the least-squares estimator (for $p \geq 2$):

$$\hat{h}_p^{\text{est}} = \frac{1}{\delta}\sum_{n=1}^{p-1}\frac{1}{(p-n)n}a_{p-n,n}^{\text{meas}}$$

yields by using (4.26)

$$\hat{h}_p^{\text{est}} = \hat{h}_p + \frac{\sigma_{\text{noise}}}{\delta}\tilde{W}_p + \delta\tilde{V}_p^\delta , \qquad (4.27)$$

with

$$\tilde{W}_p = \sum_{n=1}^{p-1}\frac{1}{(p-n)n}W_{p-n,n}, \qquad (4.28)$$

$$\tilde{V}_p^\delta = \sum_{n=1}^{p-1}\frac{1}{(p-n)n}V_{p-n,n}^\delta . \qquad (4.29)$$

Note that the independent standard complex circularly symmetric Gaussian random variables $\tilde{W}_p$ are such that

$$\mathbb{E}[|\tilde{W}_p|^2] = \Big[\sum_{n=1}^{p-1}\frac{1}{n^2(p-n)^2}\Big]e^{2\kappa p} \overset{p\gg 1}{\approx} \frac{\pi^2}{6p^2}e^{2\kappa p} . \qquad (4.30)$$

We assume that $\delta^2 \ll \sigma_{\text{noise}}$, which insures that the measurement errors in the contracted GPTs dominate the approximation error, and introduce the signal-to-noise ratio (SNR):

$$\text{SNR} = (\frac{\delta}{\sigma_{\text{noise}}})^2 .$$

Using (4.27) and (4.30) we obtain that the estimator $\hat{h}_p^{\text{est}}$ is unbiased:

$$\mathbb{E}[\hat{h}_p^{\text{est}}] = \hat{h}_p ,$$

and has the following variance:

$$\mathbb{E}[|\hat{h}_p^{\text{est}} - \hat{h}_p|^2] \approx \frac{\pi^2}{6p^2}\text{SNR}^{-1}e^{2\kappa p} .$$

Therefore, we arrive at the following result.

Proposition 4.10. *Suppose that*

$$N < \frac{1}{2\kappa} \log \text{SNR} , \tag{4.31}$$

and $\hat{h}_p$, for $p \leq N$, are of order 1. Then, the pth mode $\hat{h}_p$ of h, for $p \leq N$, can be resolved, i.e., $\mathbb{E}[|\hat{h}_p^{\text{est}} - \hat{h}_p|^2] < 1$.

Proposition 4.10 shows that a very high SNR is needed if one wishes to resolve the high-order modes of the perturbation h. Furthermore, since κ is a decaying function of δ', we infer an expected result, that is, it is more difficult to estimate the high-order modes of the perturbation h as the radius δ' of the inclusion gets smaller.

4.6 GPTs Matching Approach

4.6.1 Minimization Algorithm

Let D be an unknown domain, which could be a cluster of separated inclusions as in Sect. 4.4. We let $M_{\alpha\beta}(\lambda, D)$ denote the GPTs associated with $D = \cup_{p=1}^{P} D_p$ and $\lambda = (\lambda_1, \ldots, \lambda_P)$. Suppose that $M_{\alpha\beta}(\lambda, D)$ are known for all $|\alpha| + |\beta| \leq K$ for some number K. Suppose that λ is known, we reconstruct the location and the shape of D by minimizing the discrepancy between the given and simulated GPTs. In [37], a recursive algorithm to approximate the shape of D is proposed. The recursive optimization procedure is to minimize over B for $l = 3, \ldots, K$,

$$\mathcal{J}^{(l)}[B] := \frac{1}{2} \sum_{|\alpha|+|\beta| \leq l} \left| \sum_{a_\alpha, b_\beta} a_\alpha b_\beta \left(M_{\alpha\beta}(\lambda, B) - M_{\alpha\beta}(\lambda, D) \right) \right|^2 . \tag{4.32}$$

Here the coefficients (a_α) and (b_β) are such that

$$H(x) = \sum a_\alpha x^\alpha \quad \text{and} \quad F(x) = \sum b_\beta x^\beta \tag{4.33}$$

are harmonic polynomials. At step l one uses as an initial guess the result of step $l-1$. At the first step ($l=3$) one gets an equivalent ellipse as well as the location of the domain [37]. Note that using definition (4.22) of the contracted GPTs, one can see that minimizing $\mathcal{J}^{(l)}$ is equivalent to minimizing

$$\mathcal{J}_c^{(l)}[B] := \frac{1}{2} \sum_{n+m \le l} \left| \mathbb{N}_{mn}^{(1)}(\lambda, B) - \mathbb{N}_{mn}^{(1)}(\lambda, D) \right|^2 . \tag{4.34}$$

To minimize $\mathcal{J}^{(l)}[B]$ we need to compute the shape derivative, $d_S \mathcal{J}^{(l)}$, of $\mathcal{J}^{(l)}$, which it can be obtained easily using (4.17). Suppose that B has P' components, i.e., $B = \cup_{p=1}^{P'} B_p$, and the conductivity of B_p is k_p given by (4.18). Let $h = (h_1, \ldots, h_{P'})$ be the functions determining the deformation of ∂B_p, $p = 1, \ldots, P'$. Let

$$w_p^{HF}(x) = (k_p - 1) \left[\frac{\partial u}{\partial \nu} \Big|_- \frac{\partial v}{\partial \nu} \Big|_- + \frac{1}{k_p} \frac{\partial u}{\partial T} \Big|_- \frac{\partial v}{\partial T} \Big|_- \right](x), \quad x \in \partial B_p ,$$

where u and v satisfy (4.19) and (4.20) with D replaced by B, respectively. From (4.17) the shape derivative of $\mathcal{J}^{(l)}$ at B in the direction of h is given by

$$\langle d_S \mathcal{J}^{(l)}[B], h \rangle = \sum_{|\alpha|+|\beta| \le l} \delta_{HF} \sum_{p=1}^{P'} \langle w_p^{HF}, h_p \rangle_{L^2(\partial B_p)} , \tag{4.35}$$

where

$$\delta_{HF} = \sum_{a_\alpha, b_\beta} a_\alpha b_\beta \big(M_{\alpha\beta}(\lambda, B) - M_{\alpha\beta}(\lambda, D) \big) .$$

It is worth mentioning that the only information about h_p which is used in formula (4.35) is the projections onto the vector space spanned by w_p^{HF}.

If the target domain D is connected (and consequently all the domains B under consideration are connected), one can modify the earlier shape B to obtain B^{mod} for the next step by applying the gradient descent method:

$$\partial B^{\mathrm{mod}} = \partial B - \left(\frac{\mathcal{J}^{(l)}[B]}{\sum_{H,F} \left(\langle d_S \mathcal{J}^{(l)}[B], w^{HF} \rangle \right)^2} \sum_{H,F} \langle d_S \mathcal{J}^{(l)}[B], w^{HF} \rangle w^{HF} \right) \nu ,$$

$$\tag{4.36}$$

where ν is the outward unit normal to B. Here F and H are given by (4.33) with $|\alpha| + |\beta| \le l$. This least-squares solution was implemented in [37] and computational results there clearly show that fine details of the shape can be reconstructed provided that the domain is connected. Computational results are shown in the last chapter.

In the same paper the procedure is applied to detect the domain with multiple components. The results show that the process can create shapes approaching the target shape, but not changing topology. In order to be able to change topology and reconstruct domains with multiple components, we develop a level-set version of the matching GPTs procedure described below.

4.6.2 Level-Set Framework

The level set approach proposed by Osher and Sethian [142] has been well known for handling topology changes, such as breaking one component into two, merging two components into one and forming sharp corners. It has been successfully applied for solving various imaging problems [59, 147].

The main idea of the level set approach is to represent the shape as the zero level set of a continuous function and to derive an evolution equation for the level set function in order to solve the minimization problem. In fact, by allowing additional time-dependence of the level set function, we can compute the geometric motion of the shape in time by evolving the level set function. A geometric motion with normal velocity can be realized by solving the Hamilton-Jacobi equation. Optimization within the level set framework consists of choosing a velocity driving the evolution towards a minimum (or at least decreasing the objective functional we want to minimize).

Adopting this level-set framework, one can change the topology in the shape reconstruction from GPTs. Hence, one can reconstruct the cluster of inclusions $D = \cup_p D_p$ without knowing the number P of separated components of D in advance.

Initial Guess. Given $M_{\alpha\beta}$ for $|\alpha| = |\beta| = 1$ (called PT for polarization tensor), one can find an (equivalent) ellipse with the same PT (see Sect. 3.1.2) but not its location since the PT is invariant under translation. One can locate this ellipse provided that its GPTs with $|\alpha| + |\beta| = 3$ are known, and it provides a good initial guess. The method is explained in detail in [29] and [37].

Recursive Scheme. Within the level set framework, one represents ∂B as the zero level set of a continuous function ϕ so that $B = \{\phi < 0\}$.

As (4.36), one converts the minimization problem of (4.32) into a level set form by choosing the gradient ascent direction $V(x)$ on $x \in \partial D_p$ as

$$V(x) = \frac{\mathcal{J}^{(l)}[B]}{\sum_{H,F} \sum_{p=1}^{P'} \left(\langle d_S \mathcal{J}^{(l)}[B], w_p^{HF}[B]\chi(\partial D_p)\rangle \right)^2} \times \sum_{H,F} \langle d_S \mathcal{J}^{(l)}[B], w_p^{HF}[B]\chi(\partial D_p)\rangle w_p^{HF}[B](x) \,, \tag{4.37}$$

for each $p = 1, \ldots, P'$. We can simply set

$$V(x) = \sum_{H,F} \alpha_p^{HF}[B] w_p^{HF}[B](x) , \qquad (4.38)$$

where α_p^{HF} is defined by (4.37). Then we evolve ϕ by solving the Hamilton-Jacobi equation

$$\frac{\partial \phi}{\partial t} + V|\nabla \phi| = 0 , \qquad (4.39)$$

for one time step.

It is worth emphasizing that in (4.38), V is only defined on the boundary ∂B, even though under the level set framework it has to be defined on the whole domain. Since $\nu = \nabla \phi / |\nabla \phi|$, we can modify w_p^{HF} as

$$
\begin{aligned}
w_p^{HF}[B]|_- &= (k_p + \frac{1}{k_p} - 2) \left(\nabla v[B]|_- \cdot \frac{\nabla \phi}{|\nabla \phi|} \right) \left(\nabla u[B]|_- \cdot \frac{\nabla \phi}{|\nabla \phi|} \right) \\
&+ (1 - \frac{1}{k_p}) \nabla v[B]|_- \cdot \nabla u[B]|_- ,
\end{aligned}
\qquad (4.40)
$$

and

$$
\begin{aligned}
w_p^{HF}[B]|_+ &= -(k_p + \frac{1}{k_p} - 2) \left(\nabla v[B]|_+ \cdot \frac{\nabla \phi}{|\nabla \phi|} \right) \left(\nabla u[B]|_+ \cdot \frac{\nabla \phi}{|\nabla \phi|} \right) \\
&+ (k_p - 1) \nabla v[B]|_+ \cdot \nabla u[B]|_+ ,
\end{aligned}
\qquad (4.41)
$$

where $u[B]$ and $v[B]$ satisfy (4.19) and (4.20), respectively (with D replaced with B). Therefore, (4.39) for ϕ can be modified as follows:

$$
\begin{aligned}
\frac{\partial \phi}{\partial t} &+ \left(- \mathrm{sgn}(k_p + \frac{1}{k_p} - 2) \left(\nabla v[B] \cdot \frac{\nabla \phi}{|\nabla \phi|} \right) \left(\nabla u[B] \cdot \frac{\nabla \phi}{|\nabla \phi|} \right) \right. \\
&+ \frac{1}{2} \left. \left(\mathrm{sgn}(k_p + \frac{1}{k_p} - 2) + (k_p - \frac{1}{k_p}) \right) \nabla v[B] \cdot \nabla u[B] \right) = 0 ,
\end{aligned}
\qquad (4.42)
$$

where sgn is the sign function.

Adopting the level-set framework, one can reconstruct the cluster of inclusions $D = \cup_p D_p$ without knowing the number P of separated components of D in advance. If the conductivities of the inclusions are different, then we assume that their average value is known. Numerical experiments in [22, 37] show that it is more difficult to reconstruct high contrast inclusions from their GPTs.

4.7 Multipolar Asymptotic Expansions

We now show that the GPTs are the building blocks for multipolar expansions.

Let $D = \delta B + z$ be a conductivity inclusion of conductivity $k \neq 1$. Let $y \in \mathbb{R}^2 \setminus \overline{D}$ and let $u_y(x)$ be the solution to the transmission problem (3.32). Let $U_y(x) = \Gamma(x, y)$ denote the background solution. For $y' \in \partial D$ and x away from z, the following expansion formula holds:

$$\Gamma(x, y') = \Gamma(x - z, y' - z) = \sum_{|\alpha|=0}^{+\infty} \frac{(-1)^{|\alpha|}}{\alpha!} \partial^\alpha \Gamma(x, z)(y' - z)^\alpha . \qquad (4.43)$$

Substitution of this expansion into

$$u_y(x) - U_y(x) = \int_{\partial D} \Gamma(x, y')(\lambda I - \mathcal{K}_D^*)^{-1}\left[\frac{\partial\Gamma}{\partial\nu}\Big|_{\partial D}(\cdot, y)\right](y')d\sigma(y') , \qquad (4.44)$$

yields the following expansion of $u_y - U_y$ for x away from z:

$$(u_y - U_y)(x) = \sum_{|\alpha|,|\beta|=1}^{K} \frac{(-1)^{|\alpha|}}{\alpha!\beta!}\partial^\alpha\Gamma(x, z)Q_{\alpha\beta}(z)\partial^\beta\Gamma(z, y) + O((\frac{\delta}{|x - z|})^{K+2}) ,$$

with

$$Q_{\alpha\beta}(z) = \int_{\partial D} (y' - z)^\alpha(\lambda I - \mathcal{K}_D^*)^{-1}\left[\frac{\partial}{\partial\nu}(\cdot - z)^\beta\right](y')d\sigma(y') .$$

The zeroth order term with $\beta = 0$ vanishes because the differentiation $\partial/\partial\nu$; the zeroth order term corresponding to $\alpha = 0$ vanishes because $(\lambda I - \mathcal{K}_D^*)^{-1}$ maps a zero mean value function on ∂D to another zero mean value function. Using the change of variable $y' - z \mapsto \tilde{y}$, the integral term $Q_{\alpha\beta}(z)$ inside the expansion of $u_y - U_y$ above can be written as

$$Q_{\alpha\beta}(z) = \int_{\partial(\delta B)} \tilde{y}^\alpha(\lambda I - \mathcal{K}_{\delta B}^*)^{-1}[\frac{\partial}{\partial\nu}\tilde{y}^\beta]\,d\sigma(\tilde{y}) , \qquad (4.45)$$

which is independent of z. Moreover, by the definition (4.2) of the GPTs, this term is $M_{\beta\alpha}(\lambda, \delta B)$. As a result, we have

$$(u_y - U_y)(x) = \sum_{|\alpha|,|\beta|=1}^{K} \frac{1}{\alpha!\beta!}\partial^\alpha\Gamma(z, x)M_{\alpha\beta}(\lambda, \delta B)\partial^\beta\Gamma(z, y)$$

$$+ O((\frac{\delta}{|x - z|})^{K+2}) . \qquad (4.46)$$

Note that we have switched the indices α and β. Finally, the scaling property (4.13) yields

$$(u_y - U_y)(x) = \sum_{|\alpha|,|\beta|=1}^{K} \frac{\delta^{|\alpha|+|\beta|}}{\alpha!\beta!} \partial^\alpha \Gamma(z,x) M_{\alpha\beta}(\lambda, B) \partial^\beta \Gamma(z,y) + O((\frac{\delta}{|x-z|})^{K+2}) .$$

$$(4.47)$$

Bibliography and Discussion

The GPTs were first introduced in [27] and their basic properties were investigated in [28] (see also [31, Chap. 4]). Theorem 4.3 was first proved in [28]. The (complex) contracted GPTs were introduced in [34]. The translation, rotation, and scaling properties of the GPTs and the stability and resolution analysis are from [22]. An efficient algorithm for computing the contracted GPTs is presented in [60]. The GPTs matching approach is from [37]. For the use of level-set techniques in inverse problems, we refer to [59, 147]. The stability and resolution analysis follows the approach described in [25]. The notion of GPTs can be extended to nonhomogeneous conductivity distributions, see [16]. As in the homogeneous case, the GPTs are the basic building blocks for the far-field expansion of the voltage in the presence of inhomogeneous conductivity inclusion.

Chapter 5
Frequency Dependent Generalized Polarization Tensors

This chapter introduces the notion of higher-order frequency-dependent polarization tensors (FDPTs). Multipolar asymptotic expansions for wave scattered by a target of characteristic size smaller than the wavelength can be written in terms of high-order derivatives of the fundamental solution to the Helmholtz equation and high-order FDPTs. This key property is useful in imaging. It shows that the notion of FDPTs is the natural extension of the one of GPTs for wave propagation. We also introduce the notion of scattering coefficients which plays the same role as the notion of contracted generalized polarization tensors. Using polar coordinates (r, θ), the scattering coefficients can be viewed as the coefficients of the wave scattered by the target in the basis spanned by $\{H_n^{(1)}(k_0 r)e^{in\theta}\}_n$, while the contracted GPTs are the those of the perturbation induced by the target in the basis spanned by $\{r^{-n}e^{in\theta}\}_n$. For any $n \in \mathbb{Z}$, $H_n^{(1)}(k_0 r)e^{in\theta}$ is a solution to the unperturbed Helmholtz equation, while $r^{-n}e^{in\theta}$ is a solution to the unperturbed conductivity equation, for $r \neq 0$.

5.1   Definition

Let B be a smooth bounded domain of electromagnetic parameters $\varepsilon_\star$ and $\mu_\star$. Let ε_0 and μ_0 be the electromagnetic parameters of the background medium and set $k_0 = \omega\sqrt{\varepsilon_0\mu_0}$ and $k_\star = \omega\sqrt{\varepsilon_\star\mu_\star}$ with ω being the operating frequency. Let δ be a small parameter and assume that $k_0\delta$ is not a Dirichlet eigenvalue of $-\Delta$ on B.

Theorem 2.25 shows that there exists a unique pair $(\psi_\alpha, \varphi_\alpha)$ in $L^2(\partial B) \times L^2(\partial B)$ solution to the system of integral equations

H. Ammari et al., *Mathematical and Statistical Methods for Multistatic Imaging*,   133
Lecture Notes in Mathematics 2098, DOI 10.1007/978-3-319-02585-8_5,
© Springer International Publishing Switzerland 2013

$$
\begin{cases}
\mathcal{S}_B^{k_\star\delta}[\varphi_\alpha] - \mathcal{S}_B^{k_0\delta}[\psi_\alpha] = x^\alpha \\
\dfrac{1}{\mu_\star}\dfrac{\partial}{\partial\nu}\mathcal{S}_B^{k_\star\delta}[\varphi_\alpha]\Big|_- - \dfrac{1}{\mu_0}\dfrac{\partial}{\partial\nu}\mathcal{S}_B^{k_0\delta}[\psi_\alpha]\Big|_+ = \dfrac{1}{\mu_0}\dfrac{\partial x^\alpha}{\partial\nu}
\end{cases}
\quad \text{on } \partial B .
\qquad (5.1)
$$

We now introduce the notion of frequency dependent polarization tensors.

Definition 5.1. With the solution $(\varphi_\alpha, \psi_\alpha)$ to (5.1), we define $W_{\alpha\beta} = W_{\alpha\beta}(B, \frac{\mu_0}{\mu_\star}, k_0\delta, k_\star\delta)$ for multi-indices α and β by

$$
W_{\alpha\beta} = \int_{\partial B} \psi_\alpha(y) y^\beta \, d\sigma(y) .
\qquad (5.2)
$$

We call $W_{\alpha\beta}$ the frequency dependent polarization tensor (FDPT).

For ease of notation, we sometimes use the notation $W_{\alpha\beta}(B)$ when the focus is only on the variation of B. It is worth emphasizing that $W_{\alpha\beta}$ depends not only on B but also k_0, $k_\star$, and δ and $\mu_0/\mu_\star$.

The following proposition from [29,30] shows the limiting behavior of $W_{\alpha\beta}$ as $\delta \to 0$ and makes the connection between $W_{\alpha\beta}$ and $M_{\alpha\beta}$.

Proposition 5.2. *The FDPT $W_{\alpha\beta}$ has the following asymptotic behavior as $\delta \to 0$: If $|\alpha| \geq 1$ and $|\beta| \geq 1$, then*

$$
W_{\alpha\beta}(B, \frac{\mu_0}{\mu_\star}, k_0\delta, k_\star\delta) \to M_{\alpha\beta}(B, \frac{\mu_0}{\mu_\star}) \quad \text{as } \delta \to 0 .
\qquad (5.3)
$$

The proof of Proposition 5.2 is more involved in two dimensions than in three dimensions because of the logarithmic singularity of the Green function [30]. Note also that if $|\alpha| = 0$ or $|\beta| = 0$, then the asymptotics of $W_{\alpha\beta}$ can be found in [29,30]. For example, in three dimensions, we have

$$
W_{(0,0,0),(0,0,0)}(B, \frac{\mu_0}{\mu_\star}, k_0\delta, k_\star\delta) = -\delta^2\omega^2\varepsilon_\star\mu_0|B| + O(\delta^3) ,
\qquad (5.4)
$$

$$
W_{\alpha,(0,0,0)}(B, \frac{\mu_0}{\mu_\star}, k_0\delta, k_\star\delta) = O(\delta^2), \quad |\alpha| = 1 ,
\qquad (5.5)
$$

$$
W_{(0,0,0),\beta}(B, \frac{\mu_0}{\mu_\star}, k_0\delta, k_\star\delta) = O(\delta^2), \quad |\beta| = 1 .
\qquad (5.6)
$$

Similar approximations can be proved in the two-dimensional case. Note that in both the two- and three-dimensional cases, $W_{\alpha\beta} = O(\delta^2)$ for $|\alpha| + |\beta| \leq 1$. On the other hand, the following asymptotic holds in two and three dimensions:

$$
W_{\alpha\beta}(B, \frac{\mu_0}{\mu_\star}, k_0\delta, k_\star\delta) = -\delta^2\omega^2\varepsilon_0\mu_0|B| + O(\delta^4) \quad \text{for } |\alpha| = 2 \text{ and } \beta = 0 .
\qquad (5.7)
$$

See [30].

5.2 Multipolar Asymptotic Expansions

Let D be an electromagnetic inclusion of electric permittivity and magnetic permeability $\varepsilon_\star$ and $\mu_\star$, respectively. Suppose that D is illuminated by a time-harmonic wave generated at the point source y with the operating frequency ω. In this case, the incident field is given by

$$U_y(x) = \Gamma_{k_0}(x, y) \, ,$$

and the field perturbed in the presence of the target is the solution to the transmission problem (3.33) and is subject to the outgoing radiation condition, or equivalently is the solution to (3.34).

Let u_y be the solution to (3.34) and let U_y be the solution in the absence of the target, i.e., $U_y(x) = \Gamma_{k_0}(x, y)$. Multipolar asymptotic expansions, as $\delta \to 0$, of the perturbation $u_y - U_y$ may be described most conveniently using the notion of the FDPTs. In fact, the following multipolar expansion of the perturbation due to the presence of $D = \delta B + z$ was obtained in [29, (12.10)]:

$$(u_y - U_y)(x) = \delta^{d-2} \sum_{p=0}^{K+1} \delta^p \sum_{|\alpha|+|\beta|=p} \frac{1}{\alpha! \beta!} W_{\alpha\beta} \partial_z^\alpha \Gamma_{k_0}(z, y) \partial_z^\beta \Gamma_{k_0}(x, z) + O(\delta^{K+d}) \, ,$$

$$(5.8)$$

where $W_{\alpha\beta}$ are the FDPTs associated with B.

The asymptotic formula (5.8) is a multipolar expansion of the scattered wave in the presence of D. It holds for x away from z and a fixed frequency ω. It remains valid for ω such that $\omega\delta \ll 1$.

It is worth emphasizing again that $W_{\alpha\beta}$ depends on δ. Indeed, the leading-order term in (5.8) is order δ^d since $W_{\alpha\beta} = O(\delta^2)$ for $|\alpha| + |\beta| \leq 1$ and all the higher-order $W_{\alpha\beta}$ are bounded which follows from Proposition 5.2.

As will be shown later, keeping this dependency in $W_{\alpha\beta}$ is convenient from an imaging point of view. We also emphasize that (5.8) holds not only for $U_y(x) = \Gamma_{k_0}(x, y)$ but also for any solution U to the Helmholtz equation satisfying the radiation condition by replacing $\Gamma_{k_0}(z, y)$ on the right-hand side of the equality (5.8) with $U(z)$.

Combining (5.3)–(5.6), together with (5.7), we can recover from (5.8) the leading-order term of the scattered wave (3.35).

The asymptotic formulas (5.3)–(5.7) show that using $W_{\alpha\beta}$ for $|\alpha|, |\beta| = 0, 1$, and $W_{(2,0),(0,0)}$, one can approximately reconstruct the volume, the equivalent ellipse, and the electric permittivity of the target. In fact, (5.7) yields an approximation of $|D| = \delta^2 |B|$, and then (5.4) yields $\varepsilon_\star$. The formula (5.3) for $|\alpha| = |\beta| = 1$ yields the polarization tensor, and hence the equivalent ellipse of the target. It is timely to mention that the location of D can be reconstructed using the method developed in [26]. We may go even further to separate out the information on $\mu_0/\mu_\star$ since we have information on both $|D|$ and the polarization tensor.

5.3   Scattering Coefficients

5.3.1   Definition

We define the scattering coefficients of an inclusion and derive some important properties of them. The scattering coefficients play the same role as the contracted generalized polarization tensors. As it will be shown in Theorem 15.1 the scattering coefficients are basically the Fourier coefficients of the far-field pattern (the scattering amplitude).

Assume that k_0^2 is not a Dirichlet eigenvalue for $-\Delta$ on D. Then, from Theorem 2.25 we know that the solution to

$$\begin{cases} \nabla \cdot \dfrac{1}{\mu}\nabla u + \omega^2 \varepsilon u = 0 \quad \text{in } \mathbb{R}^2 \, , \\ (u - U) \text{ satisfies the outgoing radiation condition,} \end{cases} \tag{5.9}$$

can be represented using the single layer potentials $\mathcal{S}_D^{k_0}$ and $\mathcal{S}_D^{k_\star}$ as follows:

$$u(x) = \begin{cases} U(x) + \mathcal{S}_D^{k_0}[\psi](x), & x \in \mathbb{R}^2 \setminus \bar{D} \, , \\ \mathcal{S}_D^{k_\star}[\varphi](x), & x \in D \, , \end{cases} \tag{5.10}$$

where the pair $(\varphi, \psi) \in L^2(\partial D) \times L^2(\partial D)$ is the unique solution to

$$\begin{cases} \mathcal{S}_D^{k_\star}[\varphi] - \mathcal{S}_D^{k_0}[\psi] = U \\ \dfrac{1}{\mu_\star}\dfrac{\partial(\mathcal{S}_D^{k_\star}[\varphi])}{\partial \nu}\bigg|_{-} - \dfrac{1}{\mu_0}\dfrac{\partial(\mathcal{S}_D^{k_0}[\psi])}{\partial \nu}\bigg|_{+} = \dfrac{1}{\mu_0}\dfrac{\partial U}{\partial \nu} \quad \text{on } \partial D \, . \end{cases} \tag{5.11}$$

Moreover, there exists a constant $C = C(k, k_0, D)$ such that

$$\|\varphi\|_{L^2(\partial D)} + \|\psi\|_{L^2(\partial D)} \leq C(\|U\|_{L^2(\partial D)} + \|\nabla U\|_{L^2(\partial D)}) \, . \tag{5.12}$$

Furthermore, there are constants δ_0 and $C = C(k, k_0, D)$ independent of δ as long as $\delta \leq \delta_0$ such that

$$\|\varphi_\delta\|_{L^2(\partial D)} + \|\psi_\delta\|_{L^2(\partial D)} \leq C(\|U\|_{L^2(\partial D)} + \|\nabla U\|_{L^2(\partial D)}) \, , \tag{5.13}$$

where $(\varphi_\delta, \psi_\delta)$ is the solution of (5.11) with $k_\star$ and k_0 respectively replaced by $\delta k_\star$ and δk_0.

Note that the following asymptotic formula holds as $|x| \to \infty$, which can be seen from (5.10) and Graf's formula (2.61):

$$u(x) - U(x) = -\frac{i}{4} \sum_{n \in \mathbb{Z}} H_n^{(1)}(k_0|x|)e^{in\theta_x} \int_{\partial D} J_n(k_0|y|)e^{-in\theta_y}\psi(y)d\sigma(y) .$$

(5.14)

Let (φ_m, ψ_m) be the solution to (5.11) with $J_m(k_0|x|)e^{im\theta_x}$ in the place of $U(x)$. We define the *scattering coefficient* as follows.

Definition 5.1. The scattering coefficients W_{nm}, $m, n \in \mathbb{Z}$, associated with the permittivity and permeability distributions ε, μ and the frequency ω (or $k_\star, k_0, D$) are defined by

$$W_{nm} = W_{nm}[\varepsilon, \mu, \omega] := \int_{\partial D} J_n(k_0|y|)e^{-in\theta_y}\psi_m(y)d\sigma(y) .$$

(5.15)

We obtain the following lemma for the size of $|W_{nm}|$.

Lemma 5.2. *There is a constant C depending on $(\varepsilon, \mu, \omega)$ such that*

$$|W_{nm}[\varepsilon, \mu, \omega]| \le \frac{C^{|n|+|m|}}{|n|^{|n|}|m|^{|m|}} \quad \textit{for all } n, m \in \mathbb{Z} .$$

(5.16)

Moreover, there exists δ_0 such that, for all $\delta \le \delta_0$,

$$|W_{nm}[\varepsilon, \mu, \delta\omega]| \le \frac{C^{|n|+|m|}}{|n|^{|n|}|m|^{|m|}}\delta^{|n|+|m|} \quad \textit{for all } n, m \in \mathbb{Z} ,$$

(5.17)

where the constant C depends on $(\varepsilon, \mu, \omega)$ but is independent of δ.

Proof. Let $U(x) = J_m(k_0|x|)e^{im\theta_x}$ and (φ_m, ψ_m) be the solution to (5.11). Since

$$J_m(t) \sim \frac{1}{\sqrt{2\pi|m|}}\left(\frac{et}{2|m|}\right)^{|m|}$$

(5.18)

as $m \to \infty$ (see (1.20)), we have

$$\|U\|_{L^2(\partial D)} + \|\nabla U\|_{L^2(\partial D)} \le \frac{C^{|m|}}{|m|^{|m|}}$$

for some constant C. Thus it follows from (5.12) that

$$\|\psi_m\|_{L^2(\partial D)} \le \frac{C^{|m|}}{|m|^{|m|}}$$

(5.19)

for another constant C. So we get (5.16) from (5.15).

On the other hand, one can see from (5.13) that (5.19) still holds for some C independent of δ as long as $\delta \leq \delta_0$ for some δ_0. Note that

$$W_{nm}[\varepsilon, \mu, \delta\omega] = \int_{\partial D} J_n(\delta k_0 |y|) e^{-in\theta_y} \psi_{m,\delta}(y) d\sigma(y) , \qquad (5.20)$$

where $(\varphi_{m,\delta}, \psi_{m,\delta})$ is the solution to (5.11) with $k_\star$ and k_0 respectively replaced by $\delta k_\star$ and δk_0 and $J_m(k_0\delta|x|)e^{im\theta_x}$ in the place of $U(x)$. So one can use (5.18) to obtain (5.17). This completes the proof. $\qquad\square$

Recall from (1.34) that the family of cylindrical waves $\{J_n(k_0|y|)e^{-in\theta_y}\}_n$ is complete. If U is given as

$$U(x) = \sum_{m\in\mathbb{Z}} a_m(U) J_m(k_0|x|) e^{im\theta_x} , \qquad (5.21)$$

where $a_m(U)$ are constants, it follows from the principle of superposition that the solution (φ, ψ) to (5.11) is given by

$$\psi = \sum_{m\in\mathbb{Z}} a_m(U) \psi_m .$$

Then one can see from (5.14) that the solution u to (5.9) can be represented as

$$u(x) - U(x) = -\frac{i}{4} \sum_{n\in\mathbb{Z}} H_n^{(1)}(k_0|x|) e^{in\theta_x} \sum_{m\in\mathbb{Z}} W_{nm} a_m(U) \quad \text{as } |x| \to \infty .$$
$$(5.22)$$

In particular, if U is given by a plane wave $e^{ik_0\xi\cdot x}$ with ξ being on the unit circle, then

$$u(x) - e^{ik_0\xi\cdot x} = -\frac{i}{4} \sum_{n\in\mathbb{Z}} H_n^{(1)}(k_0|x|) e^{in\theta_x} \sum_{m\in\mathbb{Z}} W_{nm} e^{im(\frac{\pi}{2}-\theta_\xi)} \quad \text{as } |x| \to \infty ,$$
$$(5.23)$$

where $\xi = (\cos\theta_\xi, \sin\theta_\xi)$ and $x = (|x|, \theta_x)$. In fact, from the Jacobi-Anger expansion of plane waves (1.15) it follows that

$$e^{ik_0\xi\cdot x} = \sum_{m\in\mathbb{Z}} e^{im(\frac{\pi}{2}-\theta_\xi)} J_m(k_0|x|) e^{im\theta_x} , \qquad (5.24)$$

and

$$\psi = \sum_{m\in\mathbb{Z}} e^{im(\frac{\pi}{2}-\theta_\xi)} \psi_m . \qquad (5.25)$$

Thus (5.23) holds. It is worth emphasizing that the expansion formula (5.22) or (5.23) determines uniquely the scattering coefficients W_{nm}, for $n, m \in \mathbb{Z}$.

5.3.2 Translation and Rotation Properties of the Scattering Coefficients

In this subsection we use the same notation as in Sect. 4.2.

Translation

For $T = (T_1, T_2)$, define $D_T := \{y + T : y \in D\}$ and $\partial D_T = (\partial D)_T$ and let $y^T = y + T$. For $\phi \in L^2(\partial D)$, define $\phi^T \in L^2(\partial D_T)$ as

$$\phi^T(y^T) := \phi(y), \quad \text{where } y \in \partial D.$$

For a density ϕ defined on ∂D and $k = k_0$ or k_*, we have

$$\mathcal{S}^k_{D_T}[\phi^T](x^T) = -\frac{i}{4} \int_{\partial D_T} H_0^{(1)}(k|x^T - y^T|)\phi^T(y^T)d\sigma(y^T)$$

$$= -\frac{i}{4} \int_{\partial D} H_0^{(1)}(k|x - y|)\phi(y)d\sigma(y) = \mathcal{S}^k_D[\phi](x),$$

and

$$(\mathcal{K}^k_{D_T})^*[\phi^T](x^T)$$
$$= -\frac{i}{4} \int_{\partial D_T} k\left(H_0^{(1)}\right)'(k|x^T - y^T|)\frac{\langle x^T - y^T, \nu(x^T)\rangle}{|x^T - y^T|}\phi^T(y^T)\,d\sigma(y^T)$$
$$= -\frac{i}{4} \int_{\partial D} k\left(H_0^{(1)}\right)'(k|x - y|)\frac{\langle x - y, \nu(x)\rangle}{|x - y|}\phi(y)\,d\sigma(y)$$
$$= (\mathcal{K}^k_D)^*[\phi](x).$$

From the identity

$$J_m(k_0|z|)e^{im\theta_z} = \sum_{l \in \mathbb{Z}}\left[J_{l+m}(k_0|x|)e^{i(l+m)\theta_x}\right]J_l(k_0|y|)e^{-il\theta_y}, \qquad (5.26)$$

with $z = x - y = (|z|, \theta_z)$, the solution $\phi_m[D_T](y^T)$ corresponding to D_T and $U_m(x) = J_m(k_0|x|)e^{im\theta_x}$ is

$$\phi_m[D_T](y^T) = \sum_{l \in \mathbb{Z}} J_l(k_0|T|)e^{-il(\theta_T + \pi)}\phi_{l+m}[D](y).$$

From (5.26) and the definition of the scattering coefficients, the following property holds:

Lemma 5.3. *Let $D_T = \{y + T : y \in D\}$. Then,*

$$W_{nm}[D_T] = \sum_{l \in \mathbb{Z}} [J_l(k_0|T|)]^2 \, W_{(n+l)(m+l)}[D] \ .$$

Rotation

For $y \in \mathbb{R}^2$, let $y_\theta = \begin{pmatrix} \cos\phi & -\sin\phi \\ \sin\phi & \cos\phi \end{pmatrix} \begin{pmatrix} y_1 \\ y_2 \end{pmatrix}$, i.e., the rotation of y with angle ϕ w.r.t. the origin. Set $D_\theta = \{y_\theta : y \in D\}$ and

$$\phi^\theta(y_\theta) := \phi(y), \quad y \in \partial D \ .$$

For a density ϕ defined on ∂D and $k = k_0$ or $k_\star$, we have

$$\mathcal{S}_{\tilde{D}}^k[\phi^\theta](x_\theta) = -\frac{i}{4} \int_{\partial D_\theta} H_0^{(1)}(k|x_\theta - y_\theta|)\phi^\theta(y_\theta)d\sigma(y_\theta)$$

$$= -\frac{i}{4} \int_{\partial D} H_0^{(1)}(k|x - y|)\phi(y)d\sigma(y) = \mathcal{S}_D^k[\phi](x) \ ,$$

and

$$(\mathcal{K}_{\tilde{D}}^k)^*[\phi^\theta](x_\theta) = -\frac{i}{4} \int_{\partial D_\theta} k \left(H_0^{(1)} \right)' (k|x_\theta - y_\theta|)\frac{\langle x_\theta - y_\theta, \nu(x_\theta)\rangle}{|x_\theta - y_\theta|}\phi^\theta(y_\theta) \, d\sigma(y_\theta)$$

$$= -\frac{i}{4} \int_{\partial D} k \left(H_0^{(1)} \right)' (k|x - y|)\frac{\langle x - y, \nu(x)\rangle}{|x - y|}\phi(y) \, d\sigma(y)$$

$$= (\mathcal{K}_D^k)^*[\phi](x) \ .$$

For $U_m(x) = J_m(k_0|x|)e^{im\theta_x}$, we have $U_m(x_\theta) = U_m(x)e^{im\theta}$. The solution $(\phi^\theta, \psi^\theta)(y_\theta)$ corresponding to D_θ and U_m is $e^{im\theta}(\phi, \psi)(y)$. From the definition,

$$W_{nm}(k_0, k, D_\theta) = \int_{\partial D_\theta} J_n(k_0|y_\theta|)e^{-in\theta_y}\psi^\theta(y_\theta)d\sigma(y_\theta) \ .$$

Hence, the following result holds.

Lemma 5.4. *Let $D_\theta = \{y_\theta : y \in D\}$. Then*

$$W_{nm}[D_\theta] = e^{i(m-n)\theta}W_{nm}[D] \ .$$

5.3.3 Shape Derivative of Scattering Coefficients

Let D_δ be a δ-perturbation of D:

$$\partial D_\delta := \left\{ \tilde{x} = x + h(x)\nu(x) \mid x \in \partial D \right\}, \tag{5.27}$$

where ν is the outward unit normal vector on ∂D and $h \in \mathcal{C}^1(\partial D)$ is such that $\|h\|_{\mathcal{C}^1} = O(\delta)$. Let $H(y) = J_m(k_0|y|)e^{im\theta_y}$ and $F(y) = J_n(k_0|y|)e^{-in\theta_y}$. Set u and u_δ to be solutions to (5.9) corresponding to the domain D and D_δ, respectively, where U is replaced by H. Let v be the solution corresponding to D with F in place of U. Then the perturbation of the scattering coefficients due to the shape deformation is given by the following lemma.

Lemma 5.5. *We have*

$$
W_{nm}[D_\delta] - W_{nm}[D]
$$
$$
= (\frac{\mu_0}{\mu_\star} - 1) \int_{\partial D} h(x)\left(\frac{\mu_0}{\mu_\star}\frac{\partial u}{\partial \nu}\Big|_-\frac{\partial v}{\partial \nu}\Big|_- + \frac{\partial u}{\partial T}\frac{\partial v}{\partial T}\right)(x)d\sigma(x)
$$
$$
- k_0^2(\frac{\varepsilon_\star}{\varepsilon_0} - 1) \int_{\partial D} h(x)u(x)v(x)d\sigma(x) + o(\|h\|_{\mathcal{C}^1}) .
$$

Proof. Let ψ be the solution of (5.11) with U replaced by H. We have from the jump relation and Green's formula

$$
W_{nm} = \int_{\partial D} F(y)\psi(y)d\sigma(y)
$$
$$
= \int_{\partial D} F\left(\frac{\partial}{\partial \nu}\mathcal{S}_D^{k_0}[\psi]\Big|_+ - \frac{\partial}{\partial \nu}\mathcal{S}_D^{k_0}[\psi]\Big|_-\right)d\sigma(y)
$$
$$
= \int_{\partial D}\left(F\frac{\partial}{\partial \nu}\mathcal{S}_D^{k_0}[\psi]\Big|_+ - \frac{\partial F}{\partial \nu}\mathcal{S}_D^{k_0}[\psi]\right)d\sigma(y) .
$$

Since $\int_{\partial D}\left(F\frac{\partial H}{\partial \nu} - \frac{\partial F}{\partial \nu}H\right)d\sigma = 0$, which follows from $(\Delta+k_0^2)F = (\Delta+k_0^2)H = 0$, we obtain from (5.10) that

$$
W_{nm}[D] = \int_{\partial D}\left(F\frac{\partial u}{\partial \nu}\Big|_+ - \frac{\partial F}{\partial \nu}u\right)d\sigma = \int_{\partial B_R}\left(F\frac{\partial u}{\partial \nu} - \frac{\partial F}{\partial \nu}u\right)d\sigma .
$$

Similarly, we have

$$
W_{nm}[D_\delta] = \int_{\partial B_R}\left(F\frac{\partial u_\delta}{\partial \nu} - \frac{\partial F}{\partial \nu}u_\delta\right)d\sigma .
$$

We then have, from (5.10) and the decaying property of the single-layer potential, that for R large enough

$$
W_{nm}[D_\delta] - W_{nm}[D] = \int_{\partial D} \left(F \frac{\partial u}{\partial \nu} - \frac{\partial F}{\partial \nu} u \right) d\sigma
$$

$$
= \int_{\partial B_R} \left(F \frac{\partial (u_\delta - u)}{\partial \nu} - \frac{\partial F}{\partial \nu} (u_\delta - u) \right) d\sigma
$$

$$
= \int_{\partial B_R} \left(v \frac{\partial (u_\delta - u)}{\partial \nu} - \frac{\partial v}{\partial \nu} (u_\delta - u) \right) d\sigma \ .
$$

Since u and v satisfy the Helmholtz equation with the same material profile,

$$
\int_{\partial B_R} \left(v \frac{\partial u}{\partial \nu} - \frac{\partial v}{\partial \nu} u \right) d\sigma = 0,
$$

and hence, applying the divergence theorem, we obtain that

$$
W_{nm}[D_\delta] - W_{nm}[D] = \int_{\partial B_R} \left(v \frac{\partial u_\delta}{\partial \nu} - \frac{\partial v}{\partial \nu} u_\delta \right) d\sigma
$$

$$
= \mu_0 \int_{B_R} \left(\frac{1}{\mu[D_\delta]} - \frac{1}{\mu[D]} \right) \nabla u_\delta \cdot \nabla v \, dx - \mu_0 \int_{B_R} \omega^2 (\varepsilon[D_\delta] - \varepsilon[D]) u_\delta v \, dx
$$

$$
= \mu_0 \left[\int_{D_\delta \setminus D} \left(\frac{1}{\mu[D_\delta]} - \frac{1}{\mu[D]} \right) \nabla u_\delta|_- \cdot \nabla v|_+ \, dx \right.
$$

$$
\left. + \int_{D \setminus D_\delta} \left(\frac{1}{\mu[D_\delta]} - \frac{1}{\mu[D]} \right) \nabla u_\delta|_+ \cdot \nabla v|_- \, dx \right] - \mu_0 \int_{B_R} \omega^2 (\varepsilon[D_\delta] - \varepsilon[D]) u_\delta v \, dx \ ,
$$

which by using exactly the same arguments as those in [48] yields the desired result.   □

Bibliography and Discussion

The results of this chapter on the FDPTs are from [29, 30, 38]. We only considered the two-dimensional case. However, similar results can be obtained in three dimensions. The multipolar expansion is derived in [29, 30]. In [38], it has been shown that the FDPTs can be reconstructed from multistatic measurements. Furthermore, a least-squares approach to reconstruct the FDPTs from multistatic measurements, which is similar to the one designed for the conductivity case, has been proposed. The notion of scattering coefficients has been introduced in [35]. Graf's addition formula plays a key role in solving boundary integral equations [67]. The notion of scattering coefficients can be extended to three dimensions using (1.32) and (2.62).

Part III
Multistatic Configuration

Chapter 6
Multistatic Response Matrix: Statistical Structure

In multistatic wave imaging, waves are emitted by a set of sources and they are recorded by a set of sensors in order to probe an unknown medium. The responses between each pair of source and receiver are collected and assembled in the form of the multi-static response (MSR) matrix. The indices of the MSR matrix are the index of the source and the index of the receiver. When the data are corrupted by additive noise, we study the structure of the MSR matrix using random matrix theory. We start this chapter by presenting an acquisition scheme, known as Hadamard technique, for noise reduction. Hadamard technique allows us to acquire simultaneously the elements of the MSR matrix and to reduce the noise level. The feature of this technique is to divide the variance of the noise by the number of sources. Then we investigate the statistical distributions of the singular values of the MSR matrix in the presence of point reflectors. In the presence of small inclusions, we find the statistical distribution of the angles between the left and the right singular vectors of the noisy MSR matrix with respect to those of the unperturbed one. Our results in this chapter will be useful for designing detection tests, estimating the number of point reflectors or inclusions in the medium, and localizing them.

6.1 Hadamard Technique

In the standard acquisition scheme, the response matrix is measured during a sequence of N_s experiments. In the mth experiment, $m = 1, \ldots, N_s$, the mth source generates the incident field and the N_r receivers record the scattered wave which means that they measure

$$A_{nm}^{\text{meas}} = A_{nm}^0 + W_{nm}, \quad n = 1, \ldots, N_r, \quad m = 1, \ldots, N_s ,$$

H. Ammari et al., *Mathematical and Statistical Methods for Multistatic Imaging*,   145
Lecture Notes in Mathematics 2098, DOI 10.1007/978-3-319-02585-8_6,
© Springer International Publishing Switzerland 2013

which gives the matrix

$$A^{\mathrm{meas}} = A^0 + W , \qquad (6.1)$$

where A^0 is the unperturbed response matrix and W_{nm} are independent complex Gaussian random variables with mean zero and variance $\sigma^2_{\mathrm{noise}}$ (which means that the real and imaginary parts are independent real Gaussian random variables with mean zero and variance $\sigma^2_{\mathrm{noise}}/2$).

The Hadamard technique is a noise reduction technique in the presence of additive noise that uses the structure of Hadamard matrices. It allows to acquire the elements of the MSR matrix simultaneously.

Definition 6.1. A complex Hadamard matrix H of order N_s is a $N_s \times N_s$ matrix whose elements are of modulus one and such that $H^*H = N_s I$.

Complex Hadamard matrices exist for all N_s. For instance the Fourier matrix

$$H_{nm} = \exp\left[i2\pi \frac{(n-1)(m-1)}{N_s} \right], \qquad m,n = 1,\dots,N_s, \qquad (6.2)$$

is a complex Hadamard matrix. A Hadamard matrix has maximal determinant among matrices with complex entries in the closed unit disk. More exactly Hadamard [91] proved that the determinant of any complex $N_s \times N_s$ matrix H with entries in the closed unit disk satisfies $|\det H| \leq N_s^{N_s/2}$, with equality attained by a complex Hadamard matrix.

We now describe a general multi-source acquisition scheme and show the importance of Hadamard matrices to build an optimal scheme. Let H be an invertible $N_s \times N_s$ matrix with complex entries in the closed unit disk. In the multi-source acquisition scheme, the response matrix is measured during a sequence of N_s experiments. In the mth experiment, $m = 1,\dots,N_s$, all sources generate unit amplitude time harmonic signals, the m' source generating $H_{m'm}$. This means that we use all sources to their maximal emission capacity (assumed to be one) with a specific coding of their phases. The N_r receivers record the scattered wave which means that they measure

$$B^{\mathrm{meas}}_{nm} = \sum_{m'=1}^{N_s} H_{m'm} A^0_{nm'} + W_{nm} = (A^0 H)_{nm} + W_{nm}, \qquad n = 1,\dots,N_r.$$

Collecting the recorded signals of the $m = 1,\dots,N_s$ experiments gives the matrix

$$B^{\mathrm{meas}} = A^0 H + W,$$

where A^0 is the unperturbed response matrix and W_{nm} are independent complex Gaussian random variables with mean zero and variance σ^2_{noise}. The measured response matrix A^{meas} is obtained by right multiplying the matrix B^{meas} by the matrix H^{-1}:

$$A^{\text{meas}} := B^{\text{meas}} H^{-1} = A^0 + W H^{-1}, \qquad (6.3)$$

so that we get the unperturbed matrix A^0 up to a new noise

$$A^{\text{meas}} = A^0 + \tilde{W}, \qquad \tilde{W} = W H^{-1}. \qquad (6.4)$$

The choice of the matrix H should fulfill the property that the new noise matrix $\tilde{W}$ has independent complex entries with Gaussian statistics, mean zero, and minimal variance. We have

$$\mathbb{E}\left[\overline{\tilde{W}_{nm}} \tilde{W}_{n'm'}\right] = \sum_{q,q'=1}^{N_s} \overline{(H^{-1})_{qm}} (H^{-1})_{q'm'} \mathbb{E}\left[\overline{W_{nq}} W_{n'q'}\right]$$

$$= \sigma^2_{\text{noise}} ((H^{-1})^* H^{-1})_{mm'} \delta_{nn'}.$$

This shows that we look for a complex matrix H with entries in the unit disk such that $(H^{-1})^* H^{-1} = cI$ with a minimal c. This is equivalent to require that H is unitary (up to a multiplication by a constant) and that $|\det H|$ is maximal. Using Hadamard result we know that the maximal determinant is $N_s^{N_s/2}$ and that a complex Hadamard matrix attains the maximum. Therefore a matrix H that minimizes the noise variance should be a Hadamard matrix, such as, for instance, the Fourier matrix (6.2). Note that, in the case of a linear array, the use of a Fourier matrix corresponds to an illumination in the form of plane waves with regularly sampled angles.

When the multi-source acquisition scheme is used with a Hadamard technique, we have $H^{-1} = \frac{1}{N_s} H^*$ and the new noise matrix $\tilde{W}$ in (6.4) has independent complex entries with Gaussian statistics, mean zero, and variance $\sigma^2_{\text{noise}}/N_s$:

$$\mathbb{E}\left[\overline{\tilde{W}_{nm}} \tilde{W}_{n'm'}\right] = \frac{\sigma^2_{\text{noise}}}{N_s} \delta_{mm'} \delta_{nn'}. \qquad (6.5)$$

This gain of a factor N_s in the signal-to-noise ratio is called the Hadamard advantage.

6.2 SVD of Multistatic Response Matrices

Throughout this section, we only consider the two-dimensional full-view case, where the sensor arrays englobe the reflectors or the inclusions to be imaged.

6.2.1 Point Reflectors

Suppose that $\varepsilon_0 = \mu_0 = 1$. Consider the Helmholtz equation:

$$\Delta_z \Phi_\omega(z, x) + \omega^2 \left(1 + \sum_{j=1}^{r} V_j(z)\right) \Phi_\omega(z, x) = \delta_x(z) \quad \text{in } \mathbb{R}^2 \qquad (6.6)$$

for $x \in \mathbb{R}^2$, with the Sommerfeld radiation condition imposed on Φ_ω. Here r is the number of localized reflectors, x is the location of the source, and

$$V_j(z) := \eta_j \chi(\tilde{D}_j)(z - z_j) , \qquad (6.7)$$

where, for $j = 1, \ldots, r$, $\tilde{D}_j$ is a compactly supported domain with volume $|\tilde{D}_j|$, $\chi(\tilde{D}_j)$ is the characteristic function of $\tilde{D}_j$, z_j is the center of the jth inclusion, and $\eta_j := \varepsilon_j - 1$ is the dielectric contrast (also called the strength of the point reflector at z_j).

Suppose that we have a transmitter array of N_s sources located at $\{x_1, \ldots, x_{N_s}\}$ and a receiver array of N_r elements located at $\{y_1, \ldots, y_{N_r}\}$. The $N_r \times N_s$ response matrix A describes the transmit-receive process performed at these arrays. The field received by the nth receiving element y_n when the wave is emitted from x_m is $\Phi_\omega(y_n, x_m)$. If we remove the incident field then we obtain the (n, m)-th entry of the unperturbed response matrix A^0:

$$A^0_{nm} = -\Phi_\omega(y_n, x_m) + \Gamma_\omega(y_n, x_m) . \qquad (6.8)$$

The incident field is $\Gamma_\omega(y, x_m)$.

Finally, taking into account measurement noise, the measured response matrix A^{meas} is

$$A^{\text{meas}} = A^0 + \frac{1}{\sqrt{N_s}} W , \qquad (6.9)$$

where the matrix W represents the additive measurement noise, which is a random matrix with independent and identically distributed complex entries with Gaussian statistics, mean zero and variance σ_{noise}^2. This particular scaling for the noise level is the right one to get non-trivial asymptotic regimes in the limit $N_s \to \infty$. Furthermore, it is the regime that emerges from the use of the Hadamard acquisition scheme for the response matrix.

In the Born approximation, where the volume $|\tilde{D}_j|$ of $\tilde{D}_j, j = 1, \ldots, r$, goes to zero, the measured field has approximately the following form, which follows from (3.35). We include a proof for the readers' sake.

Theorem 6.1. *We have*

$$\Phi_\omega(y_n, x_m) \approx \Gamma_\omega(y_n, x_m) - \sum_{j=1}^{r} \rho_j \Gamma_\omega(y_n, z_j) \Gamma_\omega(z_j, x_m) \qquad (6.10)$$

for $n = 1, \ldots, N_r$, $m = 1, \ldots, N_s$, where ρ_j is the coefficient of reflection defined by

$$\rho_j = \omega^2 \eta_j |\tilde{D}_j| . \qquad (6.11)$$

Proof. Suppose for simplicity that the number of reflectors is 1 ($r = 1$). Let us consider the full fundamental solution $\Phi_\omega(z, x)$ and the background fundamental solution $\Gamma_\omega(z, y)$, namely,

$$\Delta_z \Phi_\omega(z, x) + \omega^2 \Phi_\omega(z, x) = -\omega^2 V(z)\Phi_\omega(z, x) + \delta_x(z)$$
$$\Delta_z \Gamma_\omega(z, y) + \omega^2 \Gamma_\omega(z, y) = \delta_y(z) ,$$

with the radiation condition. We multiply the first equation by $\Gamma_\omega(x, y)$ and subtract the second equation multiplied by $\Phi_\omega(x, z)$:

$$\nabla_z \cdot \left[\Gamma_\omega(z, y)\nabla_z \Phi_\omega(z, x) - \Phi_\omega(z, y)\nabla_z \Gamma_\omega(z, x) \right]$$
$$= -\omega^2 V(z)\Phi_\omega(z, x)\Gamma_\omega(z, y) + \Gamma_\omega(z, y)\delta_x(z) - \Phi_\omega(z, x)\delta_y(z)$$
$$= -\omega^2 V(z)\Phi_\omega(z, x)\Gamma_\omega(z, y) + \Gamma_\omega(x, y)\delta_x(z) - \Phi_\omega(y, x)\delta_y(z)$$
$$\overset{reciprocity}{=} -\omega^2 V(z)\Phi_\omega(x, z)\Gamma_\omega(z, y) + \Gamma_\omega(x, y)\delta_x(z) - \Phi_\omega(x, y)\delta_y(z) .$$

We integrate over B_R (with R large enough so that it encloses the support of V) and send R to infinity to obtain thanks to the Sommerfeld radiation condition that

$$0 = -\omega^2 \int_{\mathbb{R}^2} \Phi_\omega(x, z)V(z)\Gamma_\omega(z, y)dz + \Gamma_\omega(x, y) - \Phi_\omega(x, y) .$$

We therefore obtain the Lippmann-Schwinger equation, which is exact:

$$\Phi_\omega(x, y) = \Gamma_\omega(x, y) - \omega^2 \int_{\mathbb{R}^2} \Phi_\omega(x, z)V(z)\Gamma_\omega(z, y)dz .$$

This equation is used as a basis for expanding the fundamental solution Φ_ω when the reflectivity V is small. If Φ_ω in the right-hand side is replaced by the background fundamental solution Γ_ω, then we obtain:

$$\Phi_\omega(x, y) \approx \Gamma_\omega(x, y) - \omega^2 \int \Gamma_\omega(x, z)V(z)\Gamma_\omega(z, y)dz , \qquad (6.12)$$

which is the (first-order) Born approximation. When the volume $|\tilde{D}_1|$ is small, the integral in (6.12) can be replaced by $-\omega^2\eta_1|\tilde{D}_1|\Gamma_\omega(x,z_1)\Gamma_\omega(z_1,y)$, which gives the desired result.   □

We introduce the normalized vector of fundamental solutions from the receiver array to the point z:

$$w(z) := \frac{1}{\left(\sum_{l=1}^{N_r}|\Gamma_\omega(z,y_l)|^2\right)^{\frac{1}{2}}}\left(\Gamma_\omega(z,y_n)\right)_{n=1,\ldots,N_r}, \qquad (6.13)$$

and the normalized vector of fundamental solutions from the transmitter array to the point z, known as the illumination vector,

$$v(z) := \frac{1}{\left(\sum_{l=1}^{N_s}|\Gamma_\omega(z,x_l)|^2\right)^{\frac{1}{2}}}\overline{\left(\Gamma_\omega(z,x_m)\right)}_{m=1,\ldots,N_s}. \qquad (6.14)$$

Using (6.10) we can then write the unperturbed response matrix approximately in the form

$$A^0 = \sum_{j=1}^{r}\sigma_j w(z_j)v(z_j)^*, \qquad (6.15)$$

with

$$\sigma_j := \rho_j\left(\sum_{n=1}^{N_r}|\Gamma_\omega(z_j,y_n)|^2\right)^{\frac{1}{2}}\left(\sum_{m=1}^{N_s}|\Gamma_\omega(z_j,x_m)|^2\right)^{\frac{1}{2}}. \qquad (6.16)$$

Here * denotes the conjugate transpose.

We assume that the arrays of transmitters and receivers are equidistributed on a disk englobing the point reflectors. Moreover, the point reflectors are at a distance from the arrays of transmitter and receivers much larger than the wavelength $2\pi/\omega$. Provided that the positions z_j of the reflectors are far from one another or well-separated (i.e., farther than the wavelength $2\pi/\omega$), the vectors $w(z_j)$, $j = 1,\ldots,r$, are approximately orthogonal to one another, as well as are the vectors $v(z_j)$, $j = 1,\ldots,r$. In fact, from the Helmholtz-Kirchhoff identity (2.90), we have

$$\frac{1}{N_r}\sum_n \Gamma_\omega(z_j,y_n)\overline{\Gamma_\omega(z_i,y_n)} \approx \frac{1}{\omega}J_0(\omega|z_i - z_j|) \qquad (6.17)$$

as $N_r \to +\infty$, where J_0 is the Bessel function of the first kind and of order zero. Moreover, $J_0(\omega|z_i - z_j|) \approx 0$ when $|z_j - z_i|$ is much larger than the wavelength. The matrix A^0 then has rank r and its nonzero singular values are σ_j, $j = 1,\ldots,r$, with the associated left and right singular vectors $w(z_j)$ and $v(z_j)$.

The following result holds.

Theorem 6.2. *Let A^0 be an $N_r \times N_s$ unperturbed response matrix with rank r. Let us denote by $\sigma_1(A^0) \geq \cdots \geq \sigma_r(A^0) > 0$ its nonzero singular values. Let W be an $N_r \times N_s$ random matrix with independent and identically distributed complex entries with Gaussian statistics, mean zero, and variance σ_{noise}^2. We define A^{meas} by (6.9). When $\gamma = N_r/N_s$ is fixed and $N_s \to \infty$, for any $j = 1, \ldots, r$, we have*

$$
\sigma_j(A^{\text{meas}}) \xrightarrow{N_s \to \infty}
\begin{cases}
\sigma_{\text{noise}} \left(\dfrac{\sigma_j^2(A^0)}{\sigma_{\text{noise}}^2} + 1 + \gamma + \gamma \dfrac{\sigma_{\text{noise}}^2}{\sigma_j^2(A^0)} \right)^{\frac{1}{2}} & \text{if } \sigma_j(A^0) > \gamma^{\frac{1}{4}} \sigma_{\text{noise}} , \\[2ex]
\sigma_{\text{noise}} (1 + \gamma^{\frac{1}{2}}) & \text{if } \sigma_j(A^0) \leq \gamma^{\frac{1}{4}} \sigma_{\text{noise}}
\end{cases}
\tag{6.18}
$$

in probability.

Theorem 6.2 shows how the singular values of the perturbed response matrix A^{meas} are related to the singular values of the unperturbed response matrix A^0. We can see that there is level repulsion for the singular values $\sigma_j(A^0)$ that are larger than the threshold value $\gamma^{1/4} \sigma_{\text{noise}}$, in the sense that $\sigma_j(A^{\text{meas}}) > \sigma_j(A^0)$. We can also observe that the singular values $\sigma_j(A^0)$ that are smaller than the threshold value $\gamma^{1/4} \sigma_{\text{noise}}$ are absorbed in the deformed quarter-circle distribution of the singular values of the noise matrix $W/\sqrt{N_s}$.

Proof. If X is an $N_r \times N_s$ matrix, then we denote by $\sigma_j(X)$, $j = 1, \ldots, N_r \wedge N_s$, the singular values of X. If X is a diagonalizable $N_s \times N_s$ matrix, then we denote by $\lambda_j(X)$, $j = 1, \ldots, N_s$, the eigenvalues of X.

Step 1. We briefly summarize some known results about the spiked population model. This is a random matrix model for $N_r \times N_s$ matrices introduced in [102].

Let r be a positive integer. Let $l_1 \geq \cdots \geq l_r > 1$ be positive real numbers. We define the $N_r \times N_r$ population covariance matrix by $\Sigma = \text{diag}(l_1, \ldots, l_r, 1, \ldots, 1)$. We consider the $N_r \times N_s$ random matrix X whose N_s columns are independent realizations of complex Gaussian vectors with mean zero and covariance Σ. We introduce the sample covariance matrix

$$
S_X = \frac{1}{N_s} X X^* .
$$

The statistical behavior of the eigenvalues $\lambda_j(S_X)$, $j = 1, \ldots, N_r$, has been obtained in [42] when $\gamma = N_r/N_s$ is fixed and $N_s \to \infty$:

Lemma 6.3. *When $\gamma = N_r/N_s$ is fixed and $N_s \to \infty$, we have for $j = 1, \ldots, r$:*

$$
\lambda_j(S_X) \xrightarrow{N_s \to \infty}
\begin{cases}
l_j + \gamma \dfrac{l_j}{l_j - 1} & \text{if } l_j > 1 + \gamma^{\frac{1}{2}} , \\[2ex]
(1 + \gamma^{\frac{1}{2}})^2 & \text{if } l_j \leq 1 + \gamma^{\frac{1}{2}} ,
\end{cases}
$$

almost surely.

We write the random matrix X as

$$X = Y + Z,$$

where Y and Z are independent, the N_s columns of Y are independent realizations of complex Gaussian vectors with mean zero and covariance $\Sigma - I$ and the N_s columns of Z are independent realizations of complex Gaussian vectors with mean zero and covariance I. In other words Z has independent and identically complex entries with mean zero and variance one. Note also that the entries Y_{nm} of Y are zero if $n \geq r+1$ (almost surely), since they are realizations of Gaussian random variables with mean zero and variance zero. Therefore:

(i) the matrix Y has the form

$$Y = \begin{pmatrix} \tilde{Y} \\ 0 \end{pmatrix},$$

where $\tilde{Y}$ is an $r \times N_s$ random matrix whose N_s columns are independent realizations of complex Gaussian vectors with mean 0 and $r \times r$ covariance matrix $\tilde{\Sigma} = \mathrm{diag}(l_1 - 1, \ldots, l_r - 1)$.

(ii) The sample covariance matrix $S_Y = \frac{1}{N_s} Y Y^*$ has the form

$$S_Y = \begin{pmatrix} \tilde{S}_{\tilde{Y}} & 0 \\ 0 & 0 \end{pmatrix},$$

where $\tilde{S}_{\tilde{Y}} = \frac{1}{N_s} \tilde{Y} \tilde{Y}^*$.

The matrix $\tilde{S}_{\tilde{Y}}$ is an $r \times r$ matrix with entries $(\tilde{S}_{\tilde{Y}})_{qq'} = \frac{1}{N_s} \sum_{m=1}^{N_s} \tilde{Y}_{qm} \overline{\tilde{Y}_{q'm}}$. By the law of large numbers we have

$$\tilde{S}_{\tilde{Y}} \xrightarrow{N_s \to \infty} \tilde{\Sigma},$$

almost surely. The almost sure convergence to zero of the Frobenius norm $\|\tilde{S}_{\tilde{Y}} - \tilde{\Sigma}\|_F$ also holds. Since we have for $j = 1, \ldots, r$

$$\sigma_j \left(\frac{1}{\sqrt{N_s}} Y \right)^2 = \lambda_j \left(\frac{1}{N_s} Y Y^* \right) = \lambda_j(S_Y) = \lambda_j(\tilde{S}_{\tilde{Y}}),$$

we find that

$$\sigma_j \left(\frac{1}{\sqrt{N_s}} Y \right) \xrightarrow{N_s \to \infty} \lambda_j(\tilde{\Sigma})^{\frac{1}{2}} = \sqrt{l_j - 1}, \tag{6.19}$$

almost surely. Note also that, for $j \geq r+1$ we have $\sigma_j \left(\frac{1}{\sqrt{N_s}} Y \right) = 0$.

Step 2. Let A^0 be the $N_r \times N_s$ rank-r unperturbed matrix whose nonzero singular values are $\sigma_1(A^0) \geq \cdots \geq \sigma_r(A^0) > 0$. Let W be an $N_r \times N_s$ random matrix with independent and identically distributed complex entries with Gaussian statistics, mean zero and variance one (i.e., the real and imaginary parts of the entries are independent and obey real Gaussian distribution with mean zero and variance $1/2$). We define A by

$$A = A^0 + \frac{1}{\sqrt{N_s}} W. \tag{6.20}$$

Lemma 6.4. *When $\gamma = N_r/N_s$ is fixed and $N_s \to \infty$, for any $j = 1, \ldots, r$,*

$$\sigma_j(A^{\mathrm{meas}}) \overset{N_s \to \infty}{\longrightarrow} \begin{cases} \left(l_j + \gamma \dfrac{l_j}{l_j - 1}\right)^{\frac{1}{2}} & \text{if } l_j > 1 + \gamma^{\frac{1}{2}}, \\ 1 + \gamma^{\frac{1}{2}} & \text{if } l_j \leq 1 + \gamma^{\frac{1}{2}}, \end{cases}$$

in probability, with $l_j = 1 + \sigma_j(A^0)^2$.

Proof. We first establish a relationship between the perturbed model (6.20) and a spiked population model. The idea to use such a relationship was proposed recently by Shabalin and Nobel [149] in another context. We consider the spiked population model X with $l_j = 1 + \sigma_j^2(A^0)$ and we introduce the decomposition $X = Y + Z$ as in the previous step. We denote by $Y = U_Y D_Y V_Y^*$ the singular value decomposition of Y. We denote by $A^0 = U_{A^0} D_{A^0} V_{A^0}^*$ the singular value decomposition of A^0.

Let us define

$$\tilde{A} = D_{A^0} + \frac{1}{\sqrt{N_s}} U_Y^* Z V_Y .$$

We have

$$U_{A^0} \tilde{A} V_{A^0}^* = A^0 + \frac{1}{\sqrt{N_s}} U_{A^0} U_Y^* Z V_Y V_{A^0}^* .$$

Since $U_{A^0} U_Y^* Z V_Y V_{A^0}^*$ has the same statistical distribution as W (the distribution of W is invariant with respect to multiplication by unitary matrices) and $\tilde{A}$ and $U_{A^0} \tilde{A} V_{A^0}^*$ have the same singular values, it is sufficient to show the lemma for $\tilde{A}$ instead of A.

We have

$$\sum_{j=1}^{N_s \wedge N_r} \left| \lambda_j(S_X)^{\frac{1}{2}} - \sigma_j(\tilde{A}) \right|^2 = \sum_{j=1}^{N_s \wedge N_r} \left| \sigma_j\left(\frac{1}{\sqrt{N_s}} X\right) - \sigma_j(\tilde{A}) \right|^2$$

$$= \sum_{j=1}^{N_s \wedge N_r} \left| \sigma_j\left(\frac{1}{\sqrt{N_s}} U_Y^* X V_Y\right) - \sigma_j(\tilde{A}) \right|^2$$

$$\leq \left\| \frac{1}{\sqrt{N_s}} U_Y^* X V_Y - \tilde{A} \right\|_F^2 \quad \text{by Lemma 6.8}$$

$$= \left\| \frac{1}{\sqrt{N_s}} U_Y^* Y V_Y + \frac{1}{\sqrt{N_s}} U_Y^* Z V_Y - D_{A^0} \right.$$

$$\left. - \frac{1}{\sqrt{N_s}} U_Y^* Z V_Y \right\|_F^2$$

$$= \left\| \frac{1}{\sqrt{N_s}} D_Y - D_{A^0} \right\|_F^2 ,$$

and hence,

$$\sum_{j=1}^{N_s \wedge N_r} \left| \lambda_j (S_X)^{\frac{1}{2}} - \sigma_j(\tilde{A}) \right|^2 \leq \sum_{j=1}^{N_s \wedge N_r} \left| \sigma_j \left(\frac{1}{\sqrt{N_s}} D_Y \right) - \sigma_j(D_{A^0}) \right|^2$$

$$= \sum_{j=1}^{N_s \wedge N_r} \left| \sigma_j \left(\frac{1}{\sqrt{N_s}} Y \right) - \sigma_j(A^0) \right|^2$$

$$= \sum_{j=1}^{r} \left| \sigma_j \left(\frac{1}{\sqrt{N_s}} Y \right) - \sqrt{l_j - 1} \right|^2$$

$$\xrightarrow{N_s \to \infty} 0 \quad \text{by (6.19).}$$

Therefore, for all $j = 1, \ldots, r$, we have $\left| \lambda_j (S_X)^{\frac{1}{2}} - \sigma_j(\tilde{A}) \right|^2 \to 0$ as $N_s \to \infty$. Lemma 6.3 gives the convergence of $\lambda_j(S_X)$, which in turn ensures the convergence of $\sigma_j(\tilde{A})$, which completes the proof of Lemma 6.4. $\square$

Step 3.   Let A^0 be the $N_r \times N_s$ rank-r deterministic matrix whose nonzero singular values are $\sigma_1(A^0) \geq \cdots \geq \sigma_r(A^0) > 0$. Let W be an $N_r \times N_s$ random matrix with independent and identically distributed complex entries with Gaussian statistics, mean zero and variance σ_{noise}^2. We can now prove the statement of Theorem 6.2.

Proof. We introduce

$$\tilde{A} = \frac{1}{\sigma_{\text{noise}}} A^{\text{meas}}, \qquad \tilde{A}^0 = \frac{1}{\sigma_{\text{noise}}} A^0, \qquad \tilde{W} = \frac{1}{\sigma_{\text{noise}}} W .$$

We have

$$\tilde{A} = \tilde{A}^0 + \frac{1}{\sqrt{N_s}} \tilde{W} ,$$

where $\tilde{A}^0$ is a $N_r \times N_s$ rank-r deterministic matrix whose nonzero singular values are $\sigma_j(\tilde{A}^0) = \sigma_j(A^0)/\sigma_{\text{noise}}$ and $\tilde{W}$ is an $N_r \times N_s$ random matrix with independent and identically distributed complex entries with Gaussian statistics, mean zero and variance one. Using Lemma 6.4 gives the limits of the singular values of $\tilde{A}$, which in turn yields the desired result since $\sigma_j(A^{\mathrm{meas}}) = \sigma_{\mathrm{noise}}\sigma_j(\tilde{A})$. □

6.2.2 Inclusions

We now consider that there are R inclusions $(D_j)_{j=1,\ldots,R}$ with parameters $0 < \mu_j < +\infty$ and $0 < \varepsilon_j < +\infty$ located in a background medium with permeability and permittivity equal to 1. Each inclusion is of the form $D_j = \tilde{D}_j + z_j$. Further, we assume that the inclusions are small and far from each other or well-separated. Then, from (3.35), the response matrix can be approximately written in the form

$$A^0_{nm} \approx \sum_{j=1}^{R} \nabla \Gamma_\omega(z_j, y_n) \cdot M(\mu_j, \tilde{D}_j) \nabla \Gamma_\omega(z_j, x_m)$$

$$+ \sum_{j=1}^{R} \omega^2(\varepsilon_j - 1)|\tilde{D}_j| \Gamma_\omega(z_j, y_n) \Gamma_\omega(z_j, x_m) ,$$

where we have used the reciprocity relation $\Gamma_\omega(x, y) = \Gamma_\omega(y, x)$. Using (3.9), it follows that the polarization tensor $M(\mu_j, \tilde{D}_j)$ is diagonalizable:

$$M(\mu_j, \tilde{D}_j) = \alpha_j a(\theta_j) a(\theta_j)^T + \beta_j a(\theta_j + \pi/2) a(\theta_j + \pi/2)^T ,$$

where $a(\theta) = (\cos\theta, \sin\theta)^T$. We can then write the matrix A^0 in the form:

$$A^0 = \sum_{j=1}^{3R} \sigma_j w_j v_j^* , \tag{6.21}$$

where

$$\sigma_{3(j-1)+1} = \rho_j \left(\sum_{n=1}^{N_r} |\Gamma_\omega(z_j, y_n)|^2 \right)^{\frac{1}{2}} \left(\sum_{m=1}^{N_s} |\Gamma_\omega(z_j, x_m)|^2 \right)^{\frac{1}{2}} ,$$

$$\sigma_{3(j-1)+2} = \alpha_j \left(\sum_{n=1}^{N_r} |a(\theta_j) \cdot \nabla \Gamma_\omega(z_j, y_n)|^2 \right)^{\frac{1}{2}} \left(\sum_{m=1}^{N_s} |a(\theta_j) \cdot \nabla \Gamma_\omega(z_j, x_m)|^2 \right)^{\frac{1}{2}} ,$$

$$\sigma_{3(j-1)+3} = \beta_j \left(\sum_{n=1}^{N_r} |a(\theta_j + \pi/2) \cdot \nabla \Gamma_\omega(z_j, y_n)|^2 \right)^{\frac{1}{2}} \left(\sum_{m=1}^{N_s} |a(\theta_j + \pi/2) \cdot \nabla \Gamma_\omega(z_j, x_m)|^2 \right)^{\frac{1}{2}} ,$$

$\rho_j = \omega^2 (\varepsilon_j - 1)|\tilde{D}_j|$, and

$$w_{3(j-1)+1} = w(z_j), \qquad v_{3(j-1)+1} = v(z_j) ,$$
$$w_{3(j-1)+2} = U(z_j, \theta_j), \qquad v_{3(j-1)+2} = V(z_j, \theta_j) ,$$
$$w_{3(j-1)+3} = U(z_j, \theta_j + \pi/2), \qquad v_{3(j-1)+3} = V(z_j, \theta_j + \pi/2) ,$$

with

$$U(z, \theta) = \frac{1}{\left(\sum_{l=1}^{N_r} |a(\theta) \cdot \nabla \Gamma_\omega(z, y_l)|^2\right)^{\frac{1}{2}}} \left(a(\theta) \cdot \nabla \Gamma_\omega(z, y_n)\right)_{n=1,\dots,N_r} ,$$
$$(6.22)$$

$$V(z, \theta) = \frac{1}{\left(\sum_{l=1}^{N_s} |a(\theta) \cdot \nabla \Gamma_\omega(z, x_l)|^2\right)^{\frac{1}{2}}} \left(a(\theta) \cdot \nabla \Gamma_\omega(z, x_m)\right)_{m=1,\dots,N_s} .$$
$$(6.23)$$

Note that (6.21) is not a priori a singular value decomposition, since the vectors w_j (and v_j) may not be orthogonal. However, as in the previous section, the orthogonality condition is guaranteed provided that:

- the positions z_j of the inclusions are far from each other (i.e., much farther than the wavelength),
- the sensors cover the surface of a disk or a sphere surrounding the search region.

The second condition ensures the orthogonality of the three vectors associated to the same inclusion (using the Helmholtz-Kirchhoff identity). The first condition ensures the orthogonality of the vectors associated to different inclusions. When these two conditions are fulfilled, the vectors w_j, $j = 1, \dots, 3R$, are approximately orthogonal to each other, as well as the vectors v_j, $j = 1, \dots, 3R$. The matrix A^0 has then rank $3R$ and its nonzero singular values are σ_j, $j = 1, \dots, 3R$, with the associated left and right singular vectors w_j and v_j.

Taking into account measurement noise, the measured response matrix A^{meas} is

$$A^{\text{meas}} = A^0 + \frac{1}{\sqrt{N_s}} W , \qquad (6.24)$$

where the matrix A^0 is the unperturbed response matrix (6.21) and the matrix W represents the additive measurement noise, which is a random matrix with independent and identically distributed complex entries with Gaussian statistics, mean zero and variance σ_{noise}^2. The singular values of the perturbed matrix A^{meas} and of the unperturbed matrix A^0 are related as described in Theorem 6.2. It is of interest to describe the statistical distribution of the angles between the left and right singular vectors $w_j(A^{\text{meas}})$ and $v_j(A^{\text{meas}})$ of the noisy matrix A^{meas} with respect to the left and right singular vectors $w_j(A^0)$ and $v_j(A^0)$ of the unperturbed matrix A^0.

Theorem 6.5. *We assume the same conditions as in Theorem 6.2 and moreover that the nonzero singular values $\sigma_j^2(A^0)$ are distinct. When $\gamma = N_r/N_s$ is fixed and $N_s \to \infty$, for any $j = 1, \ldots, 3R$ such that $\sigma_j(A^0) > \gamma^{\frac{1}{4}}\sigma_{\text{noise}}$, we have*

$$\left| w_j(A^0)^* w_j(A^{\text{meas}}) \right|^2 \overset{N_s \to \infty}{\longrightarrow} \frac{1 - \gamma \frac{\sigma_{\text{noise}}^4}{\sigma_j^4(A^0)}}{1 + \gamma \frac{\sigma_{\text{noise}}^2}{\sigma_j^2(A^0)}} \tag{6.25}$$

and

$$\left| v_j(A^0)^* v_j(A^{\text{meas}}) \right|^2 \overset{N_s \to \infty}{\longrightarrow} \frac{1 - \gamma \frac{\sigma_{\text{noise}}^4}{\sigma_j^4(A^0)}}{1 + \frac{\sigma_{\text{noise}}^2}{\sigma_j^2(A^0)}} \tag{6.26}$$

in probability.

This theorem shows that the singular vectors of the perturbed matrix A^{meas} have a deterministic angle with respect to the singular vectors of the unperturbed matrix A^0 provided the corresponding perturbed singular values emerge from the deformed quarter-circle distribution (i.e., $\sigma_j(A^0) > \gamma^{\frac{1}{4}}\sigma_{\text{noise}}$). Although we will not give the proof of the following result (which we will not use in the sequel), we also have, for any $j = 1, \ldots, 3R$ such that $\sigma_j(A^0) \le \gamma^{\frac{1}{4}}\sigma_{\text{noise}}$,

$$\left| w_j(A^0)^* w_j(A^{\text{meas}}) \right|^2 \overset{N_s \to \infty}{\longrightarrow} 0 \quad \text{and} \quad \left| v_j(A^0)^* v_j(A^{\text{meas}}) \right|^2 \overset{N_s \to \infty}{\longrightarrow} 0 \tag{6.27}$$

in probability.

Proof (of Theorem 6.5). We address only the case of the left singular vectors, since the result obtained for them can be used to obtain the equivalent result for the right singular vectors after transposition of the matrices. We use the same notations as in the proof of Theorem 6.2.

Step 1.   The statistical behavior of the eigenvectors $w_j(S_X)$, $j = 1, \ldots, N_r$ has been obtained in [46] when $l_1 > \cdots > l_r > 1$, $\gamma = N_r/N_s$ is fixed and $N_s \to \infty$:

Lemma 6.6. *When $\gamma = N_r/N_s$ is fixed and $N_s \to \infty$ we have for $j = 1, \ldots, r$:*

$$\left| w_j(S_X)^* w_j(\Sigma) \right|^2 \overset{N_s \to \infty}{\longrightarrow} \begin{cases} \dfrac{1 - \dfrac{\gamma}{(l_j - 1)^2}}{1 + \dfrac{\gamma}{l_j - 1}} & \text{if } l_j > 1 + \gamma^{\frac{1}{2}}, \\ 0 & \text{if } l_j \le 1 + \gamma^{\frac{1}{2}}, \end{cases}$$

almost surely, and for $j \neq k$

$$\left| w_j(S_X)^* w_k(\Sigma) \right|^2 \xrightarrow{N_s \to \infty} 0,$$

in probability.

Here Σ is diagonal with distinct eigenvalues so that the jth singular vector $w_j(\Sigma)$ is the vector $e^{(N_r,j)}$, that is, the N_r-dimensional vector whose entries are zero but the j-th entry which is equal to one ($e_k^{(N_r,j)} = 0$ if $k \neq j$ and $e_j^{(N_r,j)} = 1$). In fact, this result is proved in [143] in the case of real-valued spiked covariance matrices and it has been recently extended to the complex case in [46].

As shown in the proof of Theorem 6.2, we have $\tilde{S}_{\tilde{Y}} \xrightarrow{N_s \to \infty} \tilde{\Sigma}$ almost surely. Since the j-th eigenvector of $\tilde{\Sigma}$ is the vector $e^{(r,j)}$, we have by Lemma 6.9

$$\left| w_j(\tilde{S}_{\tilde{Y}})^* e^{(r,j)} \right|^2 \xrightarrow{N_s \to \infty} 1 ,$$

for all $j = 1, \ldots, r$ almost surely. We have

$$w_j(Y) = w_j\left(\frac{1}{\sqrt{N_s}} Y \right) = w_j\left(\frac{1}{N_s} Y Y^* \right) = w_j(S_Y)$$

and $w_j(S_Y)^* e^{(N_r,j)} = w_j(\tilde{S}_{\tilde{Y}})^* e^{(r,j)}$ for all $j = 1, \ldots, r$. Therefore,

$$\left| w_j(Y)^* e^{(N_r,j)} \right|^2 \xrightarrow{N_s \to \infty} 1 , \tag{6.28}$$

for all $j = 1, \ldots, r$ almost surely.

Step 2. Let A^0 be the $N_r \times N_s$ rank-r unperturbed matrix whose nonzero singular values are $\sigma_1(A^0) > \cdots > \sigma_r(A^0) > 0$. Let W be an $N_r \times N_s$ random matrix with independent and identically distributed complex entries with Gaussian statistics, mean zero and variance one. We define A by (6.20).

Lemma 6.7. *When $\gamma = N_r/N_s$ is fixed and $N_s \to \infty$, for any $j = 1, \ldots, r$ such that $l_j > 1 + \gamma^{\frac{1}{2}}$ and for any $k = 1, \ldots, r$, we have*

$$\left| w_j(A)^* w_k(A^0) \right|^2 \xrightarrow{N_s \to \infty} \begin{cases} \dfrac{1 - \frac{\gamma}{(l_j - 1)^2}}{1 + \frac{\gamma}{l_j - 1}} & \text{if } k = j , \\ 0 & \text{otherwise} , \end{cases}$$

in probability, with $l_j = 1 + \sigma_j(A^0)^2$.

Proof. We use the same notation as in the proof of Lemma 6.4. We use again the relationship between randomly perturbed low-rank matrices and spiked population models [149]. Let us fix an index $j = 1, \ldots, r$ such that $l_j > 1 + \gamma^{\frac{1}{2}}$. For $k = 1, \ldots, r$, let us denote $L_k = \left(l_k + \frac{\gamma l_k}{l_k - 1}\right)^{\frac{1}{2}}$ if $l_k > 1 + \gamma^{\frac{1}{2}}$ and $L_k = 1 + \gamma^{\frac{1}{2}}$ if $l_k \leq 1 + \gamma^{\frac{1}{2}}$. We can find $\delta > 0$ such that $L_j > 2\delta$ and $\min_{k=1,\ldots,r,\,k\neq j} |L_k - L_j| > 2\delta$. We know from the proof of Theorem 6.2 that $(\sigma_k(\tilde{A}))_{k=1,\ldots,r}$ and $(\sigma_k(\frac{1}{\sqrt{N_s}}U_Y^* X V_Y))_{k=1,\ldots,r}$ both converge to $(L_k)_{k=1,\ldots,r}$ almost surely. Therefore, for N_s large enough, we have

$$\min_{k\neq j} \left|\sigma_j(\tilde{A}) - \sigma_k(\frac{1}{\sqrt{N_s}}U_Y^* X V_Y)\right| \geq \delta \quad \text{and} \quad \sigma_j(\tilde{A}) \geq \delta \,,$$

and we can apply Lemma 6.9 which gives, for N_s large enough

$$\left|w_j(\tilde{A})^* w_j(\frac{1}{\sqrt{N_s}}U_Y^* X V_Y)\right|^2 \geq 1 - \frac{2\|\frac{1}{\sqrt{N_s}}U_Y^* X V_Y - \tilde{A}\|_F^2}{\delta^2} \,.$$

Using the same arguments as in the proof of Theorem 6.2 we find that the right-hand side converges almost surely to one, and therefore,

$$\left|w_j(\tilde{A})^* w_j(\frac{1}{\sqrt{N_s}}U_Y^* X V_Y)\right|^2 \xrightarrow{N_s \to \infty} 1 \,, \tag{6.29}$$

almost surely. The left singular vectors of $\frac{1}{\sqrt{N_s}}U_Y^* X V_Y$ are related to those of X through

$$w_j(\frac{1}{\sqrt{N_s}}U_Y^* X V_Y) = U_Y^* w_j(X) \,,$$

and therefore, (6.29) implies

$$\left|w_j(\tilde{A})^* U_Y^* w_j(X)\right|^2 \xrightarrow{N_s \to \infty} 1 \,, \tag{6.30}$$

almost surely.

By Lemma 6.6 and the fact that $w_j(S_X) = w_j(\frac{1}{N_s}XX^*) = w_j(X)$ we have

$$\left|w_j(X)^* e^{(N_r,k)}\right|^2 \xrightarrow{N_s \to \infty} \xi_j \delta_{jk} \tag{6.31}$$

for all $k = 1, \ldots, r$ in probability, where $\xi_j = (1 - \frac{\gamma}{(l_j-1)^2})/(1 + \frac{\gamma}{l_j-1})$. The matrix U_Y consists of the left singular vectors of Y and we have $U_Y e^{(N_r,k)} = w_k(Y)$. By (6.28) (in Step 1) we find that

$$\left|e^{(N_r,k)\,*} U_Y e^{(N_r,k)}\right| = \left|w_k(Y)^* e^{(N_r,k)}\right| \xrightarrow{N_s \to \infty} 1 \tag{6.32}$$

for all $k = 1, \ldots, r$. Combining (6.31) and (6.32) we obtain

$$\left|[U_Y^* w_j(X)]^* e^{(N_r,k)}\right|^2 = \left|w_j(X)^* U_Y e^{(N_r,k)}\right|^2 = \left|w_j(X)^* w_k(Y)\right|^2 \stackrel{N_s \to \infty}{\longrightarrow} \delta_{jk} \xi_j$$

for all $k = 1, \ldots, r$ in probability. Using (6.30) we obtain

$$\left|w_j(\tilde{A})^* e^{(N_r,k)}\right|^2 \stackrel{N_s \to \infty}{\longrightarrow} \delta_{jk} \xi_j$$

for all $k = 1, \ldots, r$ in probability. The vector $e^{(N_r,k)}$ is the k-th left singular vector of D_{A^0}, so

$$\left|w_j(\tilde{A})^* w_k(D_{A^0})\right|^2 \stackrel{N_s \to \infty}{\longrightarrow} \delta_{jk} \xi_j \tag{6.33}$$

for all $k = 1, \ldots, r$ in probability.

Remember that $\tilde{A} = D_{A^0} + \frac{1}{\sqrt{N_s}} U_Y^* Z V_Y$, so

$$U_{A^0} \tilde{A} V_{A^0}^* = A^0 + \frac{1}{\sqrt{N_s}} U_{A^0} U_Y^* Z V_Y V_{A^0}^* \ ,$$

and $U_{A^0} U_Y^* Z V_Y V_{A^0}^*$ has the same statistical distribution as W. As a result A^{meas} and $U_{A^0} \tilde{A} V_{A^0}^*$ have the same statistical distribution. Consequently, we have

$$w_j(A) \stackrel{in\,dist.}{=} w_j(U_{A^0} \tilde{A} V_{A^0}^*) U_{A^0} w_j(\tilde{A}) \ ,$$

and

$$\left|w_j(A)^* w_k(A^0)\right|^2 \stackrel{in\,dist.}{=} \left|w_j(\tilde{A})^* U_{A^0}^* w_k(A^0)\right|^2$$
$$= \left|w_j(\tilde{A})^* w_k(U_{A^0}^* A^0 V_{A^0})\right|^2 = \left|w_j(\tilde{A})^* w_k(D_{A^0})\right|^2 \ ,$$

which converges in probability to $\delta_{jk} \xi_j$ as $N_s \to \infty$ by (6.33). $\qquad \square$

Step 3.   This step is identical to the one of Theorem 6.2.

6.3 Two Useful Lemmas

We give a classical lemma that we use in the proof of Theorem 6.2.

Lemma 6.8. *Let X and Y be two $N_r \times N_s$ matrices. Then we have*

$$\sum_{j=1}^{N_s \wedge N_r} \left|\sigma_j(X) - \sigma_j(Y)\right|^2 \le \left\|X - Y\right\|_F^2 \ ,$$

where

$$\|X\|_F^2 = \sum_{n=1}^{N_r} \sum_{m=1}^{N_s} |X_{mn}|^2 = \sum_{j=1}^{N_s \wedge N_r} \sigma_j^2(X)$$

is the Frobenius norm.

We also state a second lemma used in the proof of Theorem 6.5.

Lemma 6.9. *Let X and Y be two $N_r \times N_s$ matrices. Let $j \leq N_s \wedge N_r$. If $\delta > 0$ is such that*

$$\min_{k \neq j} |\sigma_j(Y) - \sigma_k(X)| \geq \delta \quad and \quad \sigma_j(Y) \geq \delta \,,$$

then we have

$$|w_j(Y)^* w_j(X)|^2 + |v_j(Y)^* v_j(X)|^2 \geq 2 - \frac{2\|X - Y\|_F^2}{\delta^2} \,.$$

Bibliography and Discussion

The results of this chapter are from [18]. A proof of Lemma 6.8 can be found for instance in [97, p. 448]. A proof of Lemma 6.9 can be found for instance in [152, Theorem 4] and it comes from a more general result due to Wedin [159]. The use of the Hadamard technique allows to acquire simultaneously the elements of the MSR matrix [71]. The recovery of the MSR matrix data from electronic signals is simple and fast because the solution of the set of equations requires a simple matrix inversion [151]. In [101] it was first shown that Hadamard technique enhances the signal-to-noise ratio.

A challenging problem would be to extend the results of this chapter on the structure of the MSR matrix to the limited-view case.

Chapter 7
MSR Matrices Using Multipolar Expansions

In this chapter we analyze the structure of the MSR matrices, using the multipolar expansions (4.46) and (5.8). We show the linear dependence of the multistatic data with respect to the GPTs or the FDPTs in which geometrical features of the target are encoded in a nonlinear way. As will be shown later, a least-squares approach will allow an accurate reconstruction of the GPTs or FDPTs from multistatic data. We also clarify the link between multistatic and boundary measurements in order to justify the continuum approximation, when the numbers of receivers and transmitters tend to ∞.

7.1 Conductivity Problem

For the conductivity problem, the MSR matrix is constructed as follows. Let $\{x_n\}_{n=1}^{N_r}$ and $\{x_m\}_{m=1}^{N_s}$ model a set of electric potential point detectors and electric point sources. For the sake of simplicity, we assume that the two sets of locations coincide and $N_r = N_s = N$. We also assume that $d = 2$. The MSR matrix A is an $N \times N$ matrix whose nm-element is the difference of electric potentials with and without the conductivity inclusions:

$$A_{nm} = -u_m(x_n) + \Gamma(x_n, x_m), \quad n, m = 1, \ldots, N , \quad (7.1)$$

with Γ being the fundamental solution. Here, $u_m(x)$ is the solution to the transmission problem

$$\begin{cases} \nabla \cdot (1 + (k-1)\chi(D))\nabla u_m(x) = \delta_{x_m}(x), & x \in \mathbb{R}^2, \\ u_m(x) - \Gamma(x, x_m) = O(|x|^{-1}), & |x| \to \infty . \end{cases} \quad (7.2)$$

H. Ammari et al., *Mathematical and Statistical Methods for Multistatic Imaging*, 163
Lecture Notes in Mathematics 2098, DOI 10.1007/978-3-319-02585-8_7,
© Springer International Publishing Switzerland 2013

We still assume that D is of the form $D = \delta B + z$. As modeled above, the MSR matrix characterizes the perturbed potential field $u_m(x_n) - \Gamma(x_n, x_m)$. From (4.46) it follows that

$$A_{nm} = - \sum_{|\alpha|, |\beta| = 1}^{K} \frac{1}{\alpha! \beta!} \partial^\alpha \Gamma(z, x_n) M_{\alpha\beta}(\lambda, \delta B) \partial^\beta \Gamma(z, x_m) + O((\frac{\delta}{R})^{K+2}) ,$$

(7.3)

where R is the distance from the receiver sources to the inclusion D and $\lambda = (k+1)/[2(k-1)]$.

The MSR matrix A consisting of $u_m(x_n) - \Gamma(x_n, x_m)$ depends only on the inclusion (λ, D). However, the GPTs involved in the representation (7.3) depend on the (non-unique) characterization $(z, \delta B)$ of D.

7.1.1   Expansion for MSR using Real Contracted GPTs

In this section, we further simplify the expression of MSR using the notion of real contracted GPT (CGPT). Using CGPT, we can write the MSR matrix A as a product of a CGPT matrix with coefficient matrices, which is a very convenient form for inversion.

Let $P_m(x)$ be the complex-valued polynomial

$$P_m(x) = (x_1 + ix_2)^m := \sum_{|\alpha| = m} a_\alpha^m x^\alpha + i \sum_{|\beta| = m} b_\beta^m x^\beta .$$

(7.4)

Using polar coordinates $x = re^{i\theta}$, the above coefficients a_α^m and b_β^m can also be characterized by

$$\sum_{|\alpha| = m} a_\alpha^m x^\alpha = r^m \cos m\theta, \quad \text{and} \quad \sum_{|\beta| = m} b_\beta^m x^\beta = r^m \sin m\theta .$$

(7.5)

For a generic conductivity inclusion D with contrast λ, the associated GPT $M_{\alpha\beta}(\lambda, D)$ is defined as in (4.2). We introduce the associated real CGPT to be the following combination of GPTs using the coefficients in (7.4):

$$M_{mn}^{cc} = \sum_{|\alpha| = m} \sum_{|\beta| = n} a_\alpha^m a_\beta^n M_{\alpha\beta} ,$$

(7.6)

$$M_{mn}^{cs} = \sum_{|\alpha| = m} \sum_{|\beta| = n} a_\alpha^m b_\beta^n M_{\alpha\beta} ,$$

(7.7)

$$M^{sc}_{mn} = \sum_{|\alpha|=m} \sum_{|\beta|=n} b^m_\alpha a^n_\beta M_{\alpha\beta} , \tag{7.8}$$

$$M^{ss}_{mn} = \sum_{|\alpha|=m} \sum_{|\beta|=n} b^m_\alpha b^n_\beta M_{\alpha\beta} . \tag{7.9}$$

Therefore, in terms of the complex CGPTs, $\mathbb{N}^{(1)}_{mn}(\lambda, D)$ and $\mathbb{N}^{(2)}_{mn}(\lambda, D)$, given by (4.22) and (4.23) we have

$$\begin{aligned} \mathbb{N}^{(1)}_{mn}(\lambda, D) &= (M^{cc}_{mn} - M^{ss}_{mn}) + i(M^{cs}_{mn} + M^{sc}_{mn}) , \\ \mathbb{N}^{(2)}_{mn}(\lambda, D) &= (M^{cc}_{mn} + M^{ss}_{mn}) + i(M^{cs}_{mn} - M^{sc}_{mn}) . \end{aligned} \tag{7.10}$$

Moreover, we have that

$$\frac{(-1)^{|\alpha|}}{\alpha!} \partial^\alpha \Gamma(x, 0) = \frac{-1}{2\pi|\alpha|} \left[a^{|\alpha|}_\alpha \frac{\cos|\alpha|\theta}{r^{|\alpha|}} + b^{|\alpha|}_\alpha \frac{\sin|\alpha|\theta}{r^{|\alpha|}} \right] . \tag{7.11}$$

Recall that $\{x_r\}^N_{r=1}$ and $\{x_s\}^N_{s=1}$ denote the locations of the receivers and electric sources. Define R_r and θ_r so that the complex representation of $x_r - z$ is $R_r e^{i\theta_r}$ with z being the location of the target. Similarly define R_s and θ_s. Substituting formula (7.11) into the expression (7.3) of the MSR, we get

$$\begin{aligned} A_{rs} &= - \sum_{|\alpha|=1, |\beta|=1}^K \frac{a^{|\alpha|}_\alpha \cos|\alpha|\theta_s + b^{|\alpha|}_\alpha \sin|\alpha|\theta_s}{2\pi|\alpha|R^{|\alpha|}_s} \\ &\quad \times M_{\alpha\beta}(\lambda, \delta B) \frac{a^{|\beta|}_\beta \cos|\beta|\theta_r + b^{|\beta|}_\beta \sin|\beta|\theta_r}{2\pi|\beta|R^{|\beta|}_r} + E_{rs} \\ &= - \sum_{m,n=1}^K \underbrace{\frac{1}{2\pi m R^m_s} (\cos m\theta_s, \sin m\theta_s)}_{V_{sm}} \underbrace{\begin{pmatrix} M^{cc}_{mn} & M^{cs}_{mn} \\ M^{sc}_{mn} & M^{ss}_{mn} \end{pmatrix}}_{\mathbb{M}_{mn}} \underbrace{\begin{pmatrix} \cos n\theta_r \\ \sin n\theta_r \end{pmatrix} \frac{1}{2\pi n R^n_r}}_{(V_{rn})^T} \\ &\quad + E_{rs} . \end{aligned} \tag{7.12}$$

Here, the short-hand notations $\mathbb{M}_{mn}$ and V_{sm} represent the 2×2 and 1×2 real matrices respectively, and $(V_{rn})^T$ is the transpose. As m, n run from 1 to K, which is the truncation order of CGPT, and r, s run from 1 to N, which is the number of receivers (sources), these matrices build up the $2K \times 2K$ CGPT block matrix $\mathbb{M}$ and the $N \times 2K$ coefficient matrix V as follows:

$$
\mathbb{M} = \begin{pmatrix} \mathbb{M}_{11} & \mathbb{M}_{12} & \cdots & \mathbb{M}_{1K} \\ \mathbb{M}_{21} & \mathbb{M}_{22} & \cdots & \mathbb{M}_{2K} \\ \cdots & \cdots & \ddots & \cdots \\ \mathbb{M}_{K1} & \mathbb{M}_{K2} & \cdots & \mathbb{M}_{KK} \end{pmatrix} ; V = \begin{pmatrix} V_{11} & V_{12} & \cdots & V_{1K} \\ V_{21} & V_{22} & \cdots & V_{2K} \\ \cdots & \cdots & \ddots & \cdots \\ V_{N1} & V_{N2} & \cdots & V_{NK} \end{pmatrix} . \tag{7.13}
$$

Using these notations, the following holds.

Proposition 7.1. *The MSR matrix A can be written as*

$$
A = V\mathbb{M}V^T + E , \tag{7.14}
$$

where the matrix $E = (E_{rs}) = O(\delta^{K+2})$, for fixed R, represents the truncation error.

We emphasize again that the CGPT above is for the "shifted" inclusion δB. We note also that the dimension of A depends on the number of sources/receivers but does not depend on the expansion order K in (7.3).

Due to the symmetry of harmonic combination of GPTs (4.4), the matrix $\mathbb{M}$ is symmetric. Since A is symmetric, the truncation error E is also symmetric.

A representation of the MSR data in terms of the GPTs similar to (7.14) holds in dimension three as well.

7.2 Helmholtz Equation

In this section we consider the Helmholtz equation and analyze the structure of the MSR matrix, using the multipolar expansion (5.8). We first rewrite (5.8) as follows:

$$
\begin{aligned}
(u_y - U_y)(x) = {} & \delta^{d-2} W_{(0,0),(0,0)} \Gamma_{k_0}(z,y) \Gamma_{k_0}(x,z) \\
& + \delta^{d-2}\delta \sum_{|\alpha|+|\beta|=1} W_{\alpha\beta} \partial_z^\alpha \Gamma_{k_0}(z,y) \partial_z^\beta \Gamma_{k_0}(x,z) \\
& + \delta^{d-2} \sum_{p=2}^{K+1} \delta^p \sum_{|\alpha|+|\beta|=p} \frac{1}{\alpha!\beta!} W_{\alpha\beta} \partial_z^\alpha \Gamma_{k_0}(z,y) \partial_z^\beta \Gamma_{k_0}(x,z) \\
& + O(\delta^{K+d}) .
\end{aligned} \tag{7.15}
$$

The first two terms on the right-hand side of (7.15) are the sum of point source and dipolar approximations of the target while the third term gives a multipolar approximation of the target written in terms of higher-order derivatives of the Green function.

Recall now that we have coincident transmitter and receiver arrays $\{y_1, \ldots, y_N\}$ of $N(= N_r = N_s)$ elements, used to detect the target located at z. In the presence of the target the scattered wave induced at the n-th receiver from the scattering of an incident wave generated at y_m can be approximated using the multipolar expansion (7.15). The following proposition holds.

Proposition 7.2. *We have*

$$(u_{y_m} - U_{y_m})(y_n) = \mathcal{G}(y_n, z)\mathcal{W}\mathcal{G}(y_m, z)^T + O(\delta^{K+d}), \qquad (7.16)$$

where T denotes the transpose, $\mathcal{G}(y_n, z)$ is a row vector of size $\frac{1}{2}(K+1)(K+2)$ in dimension two and $\frac{1}{6}(K+1)(K+2)(K+3)$ in the three-dimensional case, which is given by

$$\mathcal{G}(y_n, z) = \left(\frac{1}{\alpha!}\partial_z^\alpha \Gamma_{k_0}(y_n, z)\right)_{|\alpha| \leq K}, \qquad (7.17)$$

and $\mathcal{W}$ is defined by

$$\mathcal{W} = (\mathcal{W}_{\alpha\beta})_{|\alpha|,|\beta| \leq K} = \left(\delta^{d-2+|\alpha|+|\beta|} W_{\alpha\beta}\right)_{|\alpha|,|\beta| \leq K}. \qquad (7.18)$$

If δ is small, then higher-order terms can be neglected. In this case, the analysis of the MSR matrix reduces to the one which is based on a dipolar approximation. As δ is increasing, more and more multipolar terms should be included in formula (7.15) in order to well approximate the response of the target. We also emphasize that in the approximation (7.16) there are some terms which do not appear in (7.15). But these terms are all of order $O(\delta^{K+d})$, and hence do not play a role in the approximation.

In view of (7.17), the signal space of the MSR matrix becomes richer. The set of singular vectors consists of the Green function and its high-order derivatives on the array. The significant singular values of the MSR matrix are perturbed, even those associated to the dipolar approximation. The difference between those based on a point approximation and those based on a multipolar approximation (high-order approximation) are measured in terms of the difference between the polarization tensor M and the new quantities $W_{\alpha\beta}$ for $|\alpha|, |\beta| = 0, 1$. Indeed, when δ is increasing, new significant singular values can emerge. Those are related to higher-order multipolar terms. They can be expressed in terms of $W_{\alpha\beta}$ for $|\alpha|$ or $|\beta| \geq 2$. These new singular values, which are intermediate between the larger ones and zero, contain some information on the target and give better approximation of its shape and electromagnetic parameters.

7.3 Continuum Approximation

In this section we clarify the link between boundary and multistatic measurements. Consider the boundary value problem (3.1) for the conductivity equation, where g is given. Suppose that the background medium Ω contains a small inclusion D of the form $z + \delta B$. The boundary measurements are then approximated by (3.3), where U is the background solution and N is the Neumann function given by (2.34).

On the other hand, suppose that there are two arrays of receivers and transmitters equi-distributed on $\partial\Omega$. In the continuum approximation, when the numbers of receivers and transmitters tend to ∞, the multistatic measurements are approximated by

$$A_D(x, y) = -\delta^d \nabla_z \Gamma(x, z) \cdot M(\lambda, B) \nabla_z \Gamma(y, z) \,,$$

for $x, y \in \partial\Omega$. Using the relation (2.39) between the Neumann function and the fundamental solution together with the representation formula (2.36) for the background solution it follows that the leading-order term of the boundary measurements (at $x \in \partial\Omega$) is nothing else than

$$\left(-\frac{1}{2}I + \mathcal{K}_\Omega\right)^{-1}\left[\int_{\partial\Omega} A_D(\cdot, y)\left(-\frac{1}{2}I + \mathcal{K}_\Omega\right)^{-1}[g](y)\, d\sigma(y)\right](x) \,.$$

Note that A_D depends only on D and $\mathcal{K}_\Omega$ depends only on Ω.

A similar relation can be derived for the Helmholtz equation. Consider the Helmholtz equation in Ω with Neumann data g. Let the multistatic data be acquired on $\partial\Omega$. Introduce

$$A_{D,k_0}(x, y) = -\delta^d\left[\nabla_z \Gamma_{k_0}(x, z) \cdot M(\lambda, B)\nabla_z \Gamma_{k_0}(y, z) + k_0^2(\frac{\varepsilon_\star}{\varepsilon_0} - 1)|B|\Gamma_{k_0}(x, z)\Gamma_{k_0}(y, z)\right] \,,$$

for $x, y \in \partial\Omega$, where λ is defined by (3.22) and Γ_{k_0} is the fundamental solution. Using (3.26) together with the relation (3.28) between the Neumann function N_{k_0}, given by (3.27), and Γ_{k_0}, it follows that, in terms of the continuum approximation $A_{D,k_0}(x, y)$ of the multistatic measurements on $\partial\Omega$, the leading-order term in the asymptotic expansion of the boundary measurements as the size of the inclusion goes to zero is given by

$$\left(-\frac{1}{2}I + \mathcal{K}_\Omega^{k_0}\right)^{-1}\left[\int_{\partial\Omega} A_{D,k_0}(\cdot, y)\left(-\frac{1}{2}I + \mathcal{K}_\Omega^{k_0}\right)^{-1}[g](y)\, d\sigma(y)\right](x)$$

for $x \in \partial\Omega$.

Bibliography and Discussion

The expansions of the MSR matrices for, respectively, the conductivity and Helmholtz problems in terms of the GPTs and FDPTs are from [8] and [38]. The real CGPTs have been introduced in [34]. The continuum approximation may be useful for carrying out a detailed resolution and stability analysis for imaging from multistatic data.

Part IV
Localization and Detection Algorithms

Chapter 8
Direct Imaging Functionals for Inclusions in the Continuum Approximation

In this chapter we apply the accurate asymptotic formulas derived in Chap. 3 for the purpose of identifying the location and certain properties of the inclusions. Formulas (3.3) and (3.21) model perturbations to the MSR measurements due to the presence of a small inclusion in the continuum approximation where the number of array elements $N \to +\infty$. We restrict ourselves to conductivity and electromagnetic imaging and single out simple fundamental algorithms. Using (3.3) and (3.21), least-squares solutions to the imaging problems for the conductivity and the Helmholtz equations can be computed. However, the computations are done iteratively and may be difficult because of the nonlinear dependence of the data on the location, the physical parameter, the size, and the orientation of the inclusion. Moreover, there may be considerable non-uniqueness of the minimizer in the case where all parameters of the inclusions are unknown [62].

In this chapter we construct various direct (non-iterative) reconstruction algorithms that take advantage of the smallness of the inclusions. In particular, MUltiple Signal Classification algorithm (MUSIC), backpropagation, Kirchhoff migration, and topological derivative are investigated. We investigate their stability with respect to medium and measurement noises as well as their resolution. We also discuss multifrequency imaging. In the presence of (independent and identically distributed) measurement noise summing a given imaging functional over frequencies yields an improvement in the signal-to-noise ratio. However, if some correlation between frequency-dependent measurements exists, for example because of a medium noise, then summing an imaging functional over frequencies may not be appropriate. A single-frequency imaging functional at the frequency which maximizes the signal-to-noise ratio may give a better reconstruction.

H. Ammari et al., *Mathematical and Statistical Methods for Multistatic Imaging*,
Lecture Notes in Mathematics 2098, DOI 10.1007/978-3-319-02585-8_8,
© Springer International Publishing Switzerland 2013

8.1   Direct Imaging for the Conductivity Problem

In this section one applies the asymptotic formula (3.3) for the purpose of identifying the location and certain properties of the conductivity inclusions. Two simple fundamental algorithms that take advantage of the smallness of the inclusions are singled out: projection-type algorithms and MUSIC-type algorithms. These algorithms are fast, stable, and efficient.

8.1.1   Detection of a Single Inclusion: A Projection-Type Algorithm

We briefly discuss a simple algorithm for detecting a single inclusion. The projection-type location search algorithm makes use of constant current sources. Let Ω be the background medium and let U be the background solution. One wants to apply a special type of current that makes ∇U constant in the inclusion D. The injection current $g = a \cdot \nu$ for a fixed unit vector $a \in \mathbb{R}^d$ yields $\nabla U = a$ in Ω.

Let the conductivity inclusion D be of the form $z + \delta B$. Let w be a smooth harmonic function in Ω. From (3.3) it follows that the weighted boundary measurements $I_w[U]$ satisfies

$$I_w[U] := \int_{\partial \Omega} (u - U)(x) \frac{\partial w}{\partial \nu}(x)\, d\sigma(x) \approx -\delta^d \nabla U(z) \cdot M(\lambda, B) \nabla w(z) , \quad (8.1)$$

where $\lambda = (k + 1)/(2(k - 1))$, k being the conductivity of D.

Assume for the sake of simplicity that $d = 2$ and D is a disk. Set

$$w(x) = -(1/2\pi) \log |x - y| \quad \text{for } y \in \mathbb{R}^2 \setminus \overline{\Omega}, x \in \Omega .$$

Since w is harmonic in Ω, then from (3.8) and (8.1), it follows that

$$I_w[U] \approx \frac{(k - 1)|D|}{\pi(k + 1)} \frac{(y - z) \cdot a}{|y - z|^2} , \quad y \in \mathbb{R}^2 \setminus \overline{\Omega} . \quad (8.2)$$

The first step for the reconstruction procedure is to locate the inclusion. The location search algorithm is as follows. Take two observation lines Σ_1 and Σ_2 contained in $\mathbb{R}^2 \setminus \overline{\Omega}$ given by

$$\Sigma_1 := \text{ a line parallel to } a ,$$

$$\Sigma_2 := \text{ a line normal to } a .$$

Find two points $z_i^S \in \Sigma_i, i = 1, 2$, so that

$$I_w[U](z_1^S) = 0, \quad I_w[U](z_2^S) = \max_{y \in \Sigma_2} |I_w[U](y)| \ .$$

From (8.2), one can see that the intersecting point z^S of the two lines

$$\Pi_1(z_1^S) := \{y \mid a \cdot (y - z_1^S) = 0\} \ , \tag{8.3}$$

$$\Pi_2(z_2^S) := \{y \mid (y - z_2^S) \text{ is parallel to } a\} \tag{8.4}$$

is close to the center z of the inclusion D: $|z^S - z| = O(\delta^2)$.

Once one locates the inclusion, the factor $|D|(k-1)/(k+1)$ can be estimated. As it has been said before, this information is a mixture of the conductivity and the volume. A small inclusion with high conductivity and larger inclusion with lower conductivity can have the same polarization tensor.

An arbitrary shaped inclusion can be represented by means of an equivalent ellipse (ellipsoid).

8.1.2 Detection of Multiple Inclusions: A MUSIC-Type Algorithm

Consider P well-separated inclusions $D_p = \delta B_p + z_p$ (these are a fixed distance apart), with conductivities $k_p, p = 1, \ldots, P$. Suppose for the sake of simplicity that all the domains B_p are disks. Let $y_l \in \mathbb{R}^2 \setminus \Omega$ for $l = 1, \ldots, n$ denote the source points. Set

$$U_{y_l} = w_{y_l} := -(1/2\pi) \log |x - y_l| \quad \text{for } x \in \Omega, \quad l = 1, \ldots, n \ .$$

The MUSIC-type location search algorithm for detecting multiple inclusions is as follows. For $n \in \mathbb{N}$ sufficiently large, define the response matrix $A = (A_{ll'})_{l,l'=1}^n$ by

$$A_{ll'} = I_{w_{y_l}}[U_{y_{l'}}] := \int_{\partial\Omega} (u - U_{y_{l'}})(x) \frac{\partial w_{y_l}}{\partial \nu}(x) \, d\sigma(x) \ .$$

Expansion (8.1) yields

$$A_{ll'} \approx -\sum_{p=1}^P \frac{2(k_p - 1)|D_p|}{k_p + 1} \nabla U_{y_{l'}}(z_p) \cdot \nabla U_{y_l}(z_p) \ .$$

For $j = 1, 2$, introduce

$$g^{(j)}(z^S) = \left(e_j \cdot \nabla U_{y_1}(z^S), \ldots, e_j \cdot \nabla U_{y_n}(z^S) \right)^T, \quad z^S \in \Omega,$$

where $\{e_1, e_2\}$ is an orthonormal basis of $\mathbb{R}^2$.

Lemma 8.1 (MUSIC Characterization). *There exists $n_0 > dP$ such that for any $n > n_0$ the following characterization of the location of the inclusions in terms of the range of the matrix A holds:*

$$g^{(j)}(z^S) \in \text{Range}(A) \text{ for } j = 1, 2 \text{ iff } z^S \in \{z_1, \ldots, z_P\}. \tag{8.5}$$

The MUSIC-type algorithm to determine the location of the inclusions is as follows. Let $\mathbb{P}_{\text{noise}} = I - \mathbb{P}$, where $\mathbb{P}$ is the orthogonal projection onto the range of A. Given any point $z^S \in \Omega$, form the vector $g^{(j)}(z^S)$. The MUSIC characterization (8.5) says that the point z^S coincides with the location of an inclusion if and only if $\mathbb{P}_{\text{noise}}[g^{(j)}](z^S) = 0$, $j = 1, 2$. Thus one can form an image of the inclusions by plotting, at each point z^S, the cost function

$$\mathcal{I}_{\text{MU}}(z^S) = \frac{1}{\sqrt{||\mathbb{P}_{\text{noise}}[g^{(1)}](z^S)||^2 + ||\mathbb{P}_{\text{noise}}[g^{(2)}](z^S)||^2}}.$$

The resulting plot will have large peaks at the locations of the inclusions.

Once one locates the inclusions, the factors $|D_p|(k_p - 1)/(k_p + 1)$, $p = 1, \ldots, P$, can be estimated from the significant singular values of A.

8.1.3   Detection of Multiple Inclusions: A Topological Derivative Based Algorithm

With the same notation as in the previous subsections, we apply constant current sources for imaging multiple inclusions. From the boundary data we can approximately compute H_j^{meas} defined by (2.47) for $g = e_j \cdot \nu$ with $\{e_1, e_2\}$ being an orthonormal basis. Then we obtain that

$$H^{\text{meas}}(y) := H_1^{\text{meas}} e_1 + H_2^{\text{meas}} e_2 \approx \sum_{p=1}^{P} \alpha_p \frac{(y - z_p)}{|y - z_p|^2}$$

for all $y \in \mathbb{R}^2 \setminus \overline{\Omega}$, where $\alpha_p := 2(k_p - 1)|D_p|/(k_p + 1)$.

Let z^S be a search point in Ω. We set

$$H'(y) := \frac{(y - z^S)}{|y - z^S|^2}, \quad y \in \mathbb{R}^2 \setminus \overline{\Omega}.$$

If we identify $x = (x_1, x_2)$ with $x_1 + ix_2$, then

$$H^{\text{meas}}(y) \approx \sum_{p=1}^{P} \alpha_p \frac{e^{i\theta_{yz_p}}}{|y - z_p|} \quad \text{and} \quad H'(y) = \frac{e^{i\theta_{yz^S}}}{|y - z^S|} \, ,$$

where θ_{yz_p} and θ_{yz^S} are the angles between y and z_p and y and z^S, respectively. Let B_R be the disk of radius R and centered at the origin. Assume that Ω contains the origin and R is much larger than the diam(Ω) so that $|y - z_p| \approx |y - z^S| \approx R$. Then we have

$$
\begin{aligned}
\mathcal{I}_{TD}(z^S) &:= \Re \int_{\partial B_R} \overline{H'}(y) H^{\text{meas}}(y) \, d\sigma(y) \\
&\approx \sum_{p=1}^{P} \frac{\alpha_p}{R^2} \Re \int_{\partial B_R} e^{i(\theta_{yz_p} - \theta_{yz^S})} \, d\sigma(y) \\
&\neq 0 \quad \text{iff} \quad z^S \in \{z_1, \dots, z_P\} \, .
\end{aligned}
\tag{8.6}
$$

The imaging functional $\mathcal{I}_{TD}(z^S)$ has the following interpretation. The locations of the maxima of $z^S \mapsto \mathcal{I}_{TD}(z^S)$ correspond to the points at which the insertion of an inclusion D' of conductivity k' centered at one of those points maximally decreases the misfit between the harmonic parts on ∂B_R of the computed and the true harmonic parts of the solutions. Define the misfit functional

$$\mathcal{E}(z^S, \alpha') := \frac{1}{2} \int_{\partial B_R} |H^{\text{meas}} - \alpha' H'|^2(y) \, d\sigma(y) \, ,$$

where $\alpha' := 2(k' - 1)|D'|/(k' + 1)$. We have

$$\mathcal{I}_{TD}(z^S) = -\frac{\partial \mathcal{E}}{\partial \alpha'}(z^S, \alpha')|_{\alpha'=0} \, ,$$

which show that $\mathcal{I}_{TD}(z^S)$ gives, at every search point z^S, the sensitivity of the misfit functional $\mathcal{E}(z^S, \alpha')$.

8.2 Direct Imaging Algorithms for the Helmholtz Equation at a Fixed Frequency

In this section, we design direct imaging functionals for small inclusions at a fixed frequency ω. Consider the Helmholtz equation (3.24) with the Neumann data g in the presence of the inclusion D and let the background solution U be defined by (3.25).

Let w be a smooth function such that $(\Delta + k_0^2)w = 0$ in Ω. The weighted boundary measurements $I_w[U, \omega]$ defined by

$$I_w[U, \omega] := \int_{\partial\Omega} (u - U)(x)\frac{\partial w}{\partial \nu}(x)\, d\sigma(x) \tag{8.7}$$

satisfies

$$I_w[U, \omega] = -\delta^d \Big(\nabla U(z) \cdot M(\lambda, B)\nabla w(z) + k_0^2(\frac{\varepsilon_\star}{\varepsilon_0} - 1)|B|U(z)w(z)\Big)$$
$$+o(\delta^d)\,, \tag{8.8}$$

with λ given by (3.22).

We apply the asymptotic formulas (3.21) and (8.8) for the purpose of identifying the location and certain properties of the inclusions.

Consider P well-separated inclusions $D_p = z_p + \delta B_p$, $p = 1, \ldots, P$. The magnetic permeability and electric permittivity of D_p are denoted by μ_p and ε_p, respectively. Suppose that all the domains B_p are disks. In this case, we have

$$I_w[U, \omega] \approx -\sum_{p=1}^{P} |D_p|\Big(2\frac{\mu_p - \mu_0}{\mu_0 + \mu_p}\nabla U(z) \cdot \nabla w(z) + k_0^2(\frac{\varepsilon_p}{\varepsilon_0} - 1)U(z)w(z)\Big)\,.$$

8.2.1 MUSIC-Type Algorithm

Let $(\boldsymbol{\theta}_1, \ldots, \boldsymbol{\theta}_n)$ be n unit vectors in $\mathbb{R}^d$. For $\boldsymbol{\theta} \in \{\boldsymbol{\theta}_1, \ldots, \boldsymbol{\theta}_n\}$, we assume that we are in possession of the boundary data u when the domain Ω is illuminated with the plane wave $U(x) = e^{ik_0\boldsymbol{\theta}\cdot x}$. Taking the harmonic function $w(x) = e^{-ik_0\boldsymbol{\theta}'\cdot x}$ for $\boldsymbol{\theta}' \in \{\boldsymbol{\theta}_1, \ldots, \boldsymbol{\theta}_n\}$ and using (3.8) shows that the weighted boundary measurement is approximately equal to

$$I_w[U, \omega] \approx -\sum_{p=1}^{P} |D_p|k_0^2\Big(2\frac{\mu_0 - \mu_p}{\mu_0 + \mu_p}\boldsymbol{\theta} \cdot \boldsymbol{\theta}' + \frac{\varepsilon_p}{\varepsilon_0} - 1\Big)e^{ik_0(\boldsymbol{\theta}-\boldsymbol{\theta}')\cdot z_p}\,.$$

Define the response matrix $A = (A_{ll'})_{l,l'=1}^{n} \in \mathbb{C}^{n\times n}$ by

$$A_{ll'} := I_{w_{l'}}[U_l, \omega]\,, \tag{8.9}$$

where $U_l(x) = e^{ik_0\boldsymbol{\theta}_l\cdot x}$, $w_l(x) = e^{-ik_0\boldsymbol{\theta}_l\cdot x}$, $l = 1, \ldots, n$. It is approximately given by

$$A_{ll'} \approx - \sum_{p=1}^{P} |D_p| k_0^2 \Big(2\frac{\mu_0 - \mu_p}{\mu_0 + \mu_p} \boldsymbol{\theta}_l \cdot \boldsymbol{\theta}_{l'} + \frac{\varepsilon_p}{\varepsilon_0} - 1 \Big) e^{ik_0(\boldsymbol{\theta}_l - \boldsymbol{\theta}_{l'}) \cdot z_p} , \qquad (8.10)$$

for $l, l' = 1, \ldots, n$. Introduce the n-dimensional vector fields $g^{(j)}(z^S)$, for $z^S \in \Omega$ and $j = 1, \ldots, d+1$, by

$$g^{(j)}(z^S) = \frac{1}{\sqrt{n}} \big(e_j \cdot \boldsymbol{\theta}_1 e^{ik_0 \boldsymbol{\theta}_1 \cdot z^S} , \ldots, e_j \cdot \boldsymbol{\theta}_n e^{ik_0 \boldsymbol{\theta}_n \cdot z^S} \big)^T , \quad j = 1, \ldots, d , \quad (8.11)$$

and

$$g^{(d+1)}(z^S) = \frac{1}{\sqrt{n}} \big(e^{ik_0 \boldsymbol{\theta}_1 \cdot z^S} , \ldots, e^{ik_0 \boldsymbol{\theta}_n \cdot z^S} \big)^T , \qquad (8.12)$$

where $\{e_1, \ldots, e_d\}$ is an orthonormal basis of $\mathbb{R}^d$. Let $g(z^S)$ be the $n \times d$ matrix whose columns are $g^{(1)}(z^S), \ldots, g^{(d)}(z^S)$. Then (8.10) can be written as

$$A \approx -n \sum_{p=1}^{P} |D_p| k_0^2 \Big(2\frac{\mu_0 - \mu_p}{\mu_0 + \mu_p} g(z_p)\overline{g(z_p)}^T + (\frac{\varepsilon_p}{\varepsilon_0} - 1) g^{(d+1)}(z_p)\overline{g^{(d+1)}(z_p)}^T \Big) .$$

Let $\mathbb{P}_{\text{noise}} = I - \mathbb{P}$, where $\mathbb{P}$ is the orthogonal projection onto the range of A as before. The MUSIC-type imaging functional is defined by

$$\mathcal{I}_{\text{MU}}(z^S, \omega) := \Big(\sum_{j=1}^{d+1} \| \mathbb{P}_{\text{noise}}[g^{(j)}](z^S) \|^2 \Big)^{-1/2} . \qquad (8.13)$$

This functional has large peaks only at the locations of the inclusions.

8.2.2 Backpropagation-Type Algorithms

Let $(\boldsymbol{\theta}_1, \ldots, \boldsymbol{\theta}_n)$ be n unit vectors in $\mathbb{R}^d$. A backpropagation-type imaging functional at a single frequency ω is given by

$$\mathcal{I}_{\text{BP}}(z^S, \omega) := \frac{1}{n} \sum_{l=1}^{n} e^{-2ik_0 \boldsymbol{\theta}_l \cdot z^S} I_{w_l}[U_l, \omega] , \qquad (8.14)$$

where $U_l(x) = w_l(x) = e^{ik_0 \boldsymbol{\theta}_l \cdot x}$, $l = 1, \ldots, n$. Suppose that $(\boldsymbol{\theta}_1, \ldots, \boldsymbol{\theta}_n)$ are equidistant points on the unit sphere S^{d-1}. For sufficiently large n, we have

$$\frac{1}{n}\sum_{l=1}^{n} e^{ik_0\boldsymbol{\theta}_l \cdot x} \approx 4(\frac{\pi}{k_0})^{d-2}\,\Im\left\{\Gamma_{k_0}(x,0)\right\} = \begin{cases} \text{sinc}(k_0|x|) & \text{for } d = 3\,, \\ J_0(k_0|x|) & \text{for } d = 2\,, \end{cases}$$

(8.15)

where $\text{sinc}(s) = \sin(s)/s$ is the sinc function and J_0 is the Bessel function of the first kind and of order zero.

Therefore, it follows that

$$\mathcal{I}_{\text{BP}}(z^S,\omega) \approx -\sum_{p=1}^{P}|D_p|k_0^2\left(2\frac{\mu_p-\mu_0}{\mu_0+\mu_p}+(\frac{\varepsilon_p}{\varepsilon_0}-1)\right)\times \begin{cases} \text{sinc}(2k_0|z^S - z_p|) & \text{for } d = 3, \\ J_0(2k_0|z^S - z_p|) & \text{for } d = 2\,. \end{cases}$$

These formulas show that the resolution of the imaging functional is the standard diffraction limit. It is of the order of half the wavelength $\lambda = 2\pi/k_0$.

Note that $\mathcal{I}_{\text{BP}}$ uses only the diagonal terms of the response matrix A, defined by (8.9). Using the whole matrix, we arrive at the Kirchhoff migration functional:

$$\mathcal{I}_{\text{KM}}(z^S,\omega) = \sum_{j=1}^{d+1}\overline{g^{(j)}(z^S)}\cdot Ag^{(j)}(z^S)\,,$$

(8.16)

where $g^{(j)}$ are defined by (8.11) and (8.12).

Suppose for simplicity that $P = 1$ and $\mu_\star = \mu_0$. In this case the response matrix is

$$A = -n|D|k_0^2(\frac{\varepsilon_\star}{\varepsilon_0} - 1)g^{(d+1)}(z)\overline{g^{(d+1)}(z)}^T$$

and we can prove that $\mathcal{I}_{\text{MU}}$ is a nonlinear function of $\mathcal{I}_{\text{KM}}$ [23]. In fact, we have

$$\mathcal{I}_{\text{KM}}(z^S,\omega) = -n|D|k_0^2(\frac{\varepsilon_\star}{\varepsilon_0} - 1)\left(1 - \mathcal{I}_{\text{MU}}^{-2}(z^S,\omega)\right)\,.$$

It is worth pointing out that this transformation does not improve neither the stability nor the resolution.

Moreover, in the presence of additive measurement noise with variance $k_0^2\sigma_{\text{noise}}^2$, the response matrix can be written as

$$A = -n|D|k_0^2(\frac{\varepsilon_\star}{\varepsilon_0} - 1)g^{(d+1)}(z)\overline{g^{(d+1)}(z)}^T + \sigma_{\text{noise}}k_0 W\,,$$

where W is a complex symmetric Gaussian matrix with mean zero and variance 1.

According to [23], the Signal-to-Noise Ratio (SNR) of the imaging functional $\mathcal{I}_{\text{KM}}$, defined by

$$\text{SNR}(\mathcal{I}_{\text{KM}}) = \frac{\mathbb{E}[\mathcal{I}_{\text{KM}}(z,\omega)]}{\text{Var}(\mathcal{I}_{\text{KM}}(z,\omega))^{1/2}} \,,$$

is then equal to

$$\text{SNR}(\mathcal{I}_{\text{KM}}) = \frac{nk_0|D|\,|\frac{\varepsilon_*}{\varepsilon_0} - 1|}{\sigma_{\text{noise}}} \,. \tag{8.17}$$

For the MUSIC algorithm, the peak of $\mathcal{I}_{\text{MU}}$ is affected by measurement noise. We have [79]

$$\mathcal{I}_{\text{MU}}(z,\omega) = \begin{cases} \frac{n|D|k_0|\frac{\varepsilon_*}{\varepsilon_0}-1|}{\sigma_{\text{noise}}} & \text{if } n|D|k_0|\frac{\varepsilon_*}{\varepsilon_0} - 1| \gg \sigma_{\text{noise}}, \\ 1 & \text{if } n|D|k_0|\frac{\varepsilon_*}{\varepsilon_0} - 1| \ll \sigma_{\text{noise}} \,. \end{cases}$$

Suppose now that the medium is randomly heterogeneous around a constant background. Let ε_* be the electric permittivity of the inclusion D. The coefficient of reflection is of the form $1 + (\frac{\varepsilon_*}{\varepsilon_0} - 1)\chi(D)(x) + \nu_{\text{noise}}(x)$, where 1 stands for the constant background, $(\frac{\varepsilon_*}{\varepsilon_0} - 1)\chi(D)$ stands for the localized perturbation of the coefficient of reflection due to the inclusion, and $\nu_{\text{noise}}(x)$ stands for the fluctuations of the coefficient of reflection due to clutter (i.e., medium noise). We assume that ν_{noise} is a random process with Gaussian statistics and mean zero, and that it is compactly supported within Ω.

If the random process ν_{noise} has a small amplitude, then the background solution U, i.e., the field that would be observed without the inclusion, can be approximated by

$$U(x) \approx U^{(0)}(x) - k_0^2 \int_\Omega N_{k_0}^{(0)}(x,y)\nu_{\text{noise}}(y)U^{(0)}(y)\, dy \,,$$

where $U^{(0)}$ and $N_{k_0}^{(0)}$ are respectively the background solution and the Neumann function in the constant background case. On the other hand, in the weak fluctuation regime, the phase mismatch between $N_{k_0}(x,z)$, the Neumann function in the random background, and $N_{k_0}^{(0)}(x,z^S)$ when z^S is close to z comes from the random fluctuations of the travel time between x and z which is approximately equal to the integral of $\nu_{\text{noise}}/2$ along the ray from x to z:

$$N_{k_0}(x,z) \approx N_{k_0}^{(0)}(x,z)e^{ik_0 T(x)} \,,$$

with

$$T(x) \approx \frac{|x-z|}{2} \int_0^1 \nu_{\text{noise}}\left(z + (x-z)s\right) ds \,.$$

Therefore, for any smooth function w satisfying $(\Delta + k_0^2)w = 0$ in Ω, the weighted boundary measurements $I_w[U^{(0)}, w]$, defined by (8.7), is approximately given by

$$
\begin{aligned}
I_w[U^{(0)}, w] \approx & -|D|k_0^2(\frac{\varepsilon_\star}{\varepsilon_0} - 1)e^{-\frac{k_0^2 \mathrm{Var}(T)}{2}} w(z)U^{(0)}(z) \\
& -k_0^2 \int_\Omega w(y)U^{(0)}(y)\nu_{\mathrm{noise}}(y)\, dy\ ,
\end{aligned}
\tag{8.18}
$$

provided that the correlation length of the random process ν_{noise} is small [20]. Without the medium noise,

$$
I_w[U^{(0)}, w] \approx -|D|k_0^2(\frac{\varepsilon_\star}{\varepsilon_0} - 1)w(z)U^{(0)}(z)\ .
$$

So, expansion (8.18) shows that the medium noise reduces the height of the main peak of $\mathcal{I}_{\mathrm{KM}}$ by the damping factor $e^{-k_0^2 \mathrm{Var}(T)/2}$ and on the other hand it induces random fluctuations of the associated image in the form of a speckle field due to the second term on the right-hand side of (8.18).

8.2.3 Topological Derivative Based Imaging Functional

The topological derivative based imaging functional was introduced in [20].

Let $D' = z^S + \delta'B'$, $\mu' > \mu_0$, $\varepsilon' > \varepsilon_0$, B' be chosen a priori (usually a disk), and let δ' be small. If $\mu_\star < \mu_0$ and $\varepsilon_\star < \varepsilon_0$, then we choose $\mu' < \mu_0$ and $\varepsilon' < \varepsilon_0$.

Let w be the solution of the Helmholtz equation

$$
\begin{cases}
\Delta w + k_0^2 w = 0 & \text{in } \Omega\ , \\
\dfrac{\partial w}{\partial \nu} = (-\dfrac{1}{2}I + (\mathcal{K}_\Omega^{-k_0})^*)\overline{(-\dfrac{1}{2}I + \mathcal{K}_\Omega^{k_0})[U - u_{\mathrm{meas}}]} & \text{on } \partial\Omega\ ,
\end{cases}
\tag{8.19}
$$

where u_{meas} is the boundary pressure in the presence of the inclusion. The function w is obtained by backpropagating the Neumann data

$$
(-\frac{1}{2}I + (\mathcal{K}_\Omega^{k_0})^*)(-\frac{1}{2}I + \mathcal{K}_\Omega^{k_0})[U - u_{\mathrm{meas}}]
$$

inside the background medium (without any inclusion). Note that $\overline{(\mathcal{K}_\Omega^{k_0})^*} = (\mathcal{K}_\Omega^{-k_0})^*$.

The function w can be used to image the inclusion. It corresponds to backpropagating the discrepancy between the measured and the background solutions. However, we introduce here a functional that exploits better the coherence between the phases of the background and perturbed fields at the location of the inclusion. This functional turns out to be exactly the topological derivative imaging functional introduced in [20].

For a single measurement, we set

$$\mathcal{I}_{\mathrm{TD}}[U](z^S) = \Re\left\{\nabla U(z^S) \cdot M(\lambda', B')\nabla w(z^S) + k_0^2(\frac{\varepsilon'}{\varepsilon_0} - 1)|B'|U(z^S)w(z^S)\right\} . \tag{8.20}$$

The functional $\mathcal{I}_{\mathrm{TD}}[U](z^S)$ gives, at every search point $z^S \in \Omega$, the sensitivity of the misfit function

$$\mathcal{E}[U](z^S) := \frac{1}{2}\int_{\partial\Omega}\left|(-\frac{1}{2}I + \mathcal{K}_\Omega^{k_0})[u_{z^S} - u_{\mathrm{meas}}](x)\right|^2 d\sigma(x) ,$$

where u_{z^S} is the solution of (2.70) with the inclusion $D' = z^S + \delta'B'$. The location of the maximum of $z^S \mapsto \mathcal{I}_{\mathrm{TD}}[U](z^S)$ corresponds to the point at which the insertion of an inclusion centered at that point maximally decreases the misfit function. Using the Helmholtz-Kirchhoff identity (2.89) and the relation (3.28) between the Neumann function N_{k_0}, defined by (3.27), and fundamental solution Γ_{k_0}, we can show that the functional $\mathcal{I}_{\mathrm{TD}}$ attains its maximum at $z^S = z$; see [20]. It is also shown in [20] that the postprocessing of the data set by applying the integral operator $(-\frac{1}{2}I + \mathcal{K}_\Omega^{k_0})$ is essential in order to obtain an efficient topological derivative based imaging functional, both in terms of resolution and stability. By postprocessing the data, we ensure that the topological derivative based imaging functional attains its maximum at the true location of the inclusion.

For multiple measurements, $U_l, l = 1, \ldots, n$, the topological derivative based imaging functional is simply given by

$$\mathcal{I}_{\mathrm{TD}}(z^S, \omega) := \frac{1}{n}\sum_{l=1}^n \mathcal{I}_{\mathrm{TD}}[U_l](z^S) . \tag{8.21}$$

Let, for simplicity, $(\boldsymbol{\theta}_1, \ldots, \boldsymbol{\theta}_n)$ be n uniformly distributed directions over the unit sphere and consider U_l to be the plane wave

$$U_l(x) = e^{ik_0\boldsymbol{\theta}_l \cdot x}, \quad x \in \Omega, \quad l = 1, \ldots, n . \tag{8.22}$$

Let

$$r_{k_0}(z^S, z) := \int_{\partial\Omega}\Gamma_{k_0}(x, z^S)\overline{\Gamma_{k_0}}(x, z)\, d\sigma(x) , \tag{8.23}$$

$$R_{k_0}(z^S, z) := \int_{\partial\Omega}\nabla_z\Gamma_{k_0}(x, z^S)\nabla_z\overline{\Gamma_{k_0}}(x, z)^T\, d\sigma(x) . \tag{8.24}$$

Note that $R_{k_0}(z^S, z)$ is a $d \times d$ matrix. When $\mu_* = \mu_0$, it is proved in [20] that

$$\mathcal{I}_{\mathrm{TD}}[U](z^S) \approx \delta^d k_0{}^4 (\frac{\varepsilon'}{\varepsilon_0} - 1)(\frac{\varepsilon_*}{\varepsilon_0} - 1)|B'| \Re e \left\{ U(z^S) r_{k_0}(z^S, z) \overline{U}(z) \right\}, \quad (8.25)$$

where r_{k_0} is given by (8.23). Therefore, by computing the topological derivatives for the n plane waves (n sufficiently large), it follows from (8.15) together with

$$\int_{\partial\Omega} \overline{\Gamma_{k_0}}(x, z) \Gamma_{k_0}(x, z^S) \, d\sigma(x) \sim \frac{1}{k_0} \Im m \left\{ \Gamma_{k_0}(z^S, z) \right\}, \qquad d = 2, 3, \quad (8.26)$$

where $A \sim B$ means $A \approx CB$ for some constant C independent of k_0, that

$$\frac{1}{n} \sum_{l=1}^{n} \mathcal{I}_{\mathrm{TD}}[U_l](z^S) \sim k_0{}^{5-d} (\Im m \left\{ \Gamma_{k_0}(z^S, z) \right\})^2 .$$

Similarly, when $\varepsilon_* = \varepsilon_0$, by computing the topological derivatives for the n plane waves, $U_l, l = 1, \ldots, n$, given by (8.22), we obtain

$$\frac{1}{n} \sum_{l=1}^{n} \mathcal{I}_{\mathrm{TD}}[U_l](z^S) \approx$$
$$\delta^d k_0{}^2 \frac{1}{n} \sum_{l=1}^{n} \Re e \left\{ e^{ik_0\boldsymbol{\theta}_l \cdot (z^S - z)} \left[\boldsymbol{\theta}_l \cdot M(\lambda', B') R_{k_0}(z^S, z) M(\lambda, B) \boldsymbol{\theta}_l \right] \right\} .$$

Using B' the unit disk, the polarization tensor $M(\lambda', B') = C_d I$, where C_d is a constant, is proportional to the identity; see (3.8).

If, additionally, we assume that $M(\lambda, B)$ is approximately proportional to the identity, which occurs in particular when B is a disk or a ball, then by using

$$\int_{\partial\Omega} \nabla_z \Gamma_{k_0}(x, z^S) \nabla_z \overline{\Gamma_{k_0}}(x, z)^T \, d\sigma(x)$$
$$\sim k_0 \Im m \left\{ \Gamma_{k_0}(z^S, z) \right\} \left(\frac{z - z^S}{|z - z^S|} \right) \left(\frac{z - z^S}{|z - z^S|} \right)^T, \qquad (8.27)$$

we arrive at

$$\frac{1}{n} \sum_{l=1}^{n} \mathcal{I}_{\mathrm{TD}}[U_l](z^S) \sim k_0{}^{5-d} (\Im m \left\{ \Gamma_{k_0}(z^S, z) \right\})^2 . \qquad (8.28)$$

Therefore, $\mathcal{I}_{TD}$ attains its maximum at z. Moreover, the resolution for the location estimation is given by the diffraction limit. We refer the reader to [20] for a detailed stability analysis of $\mathcal{I}_{TD}$ with respect to both medium and measurement noises as well as its resolution. In the case of measurement noise, the SNR of $\mathcal{I}_{TD}$,

$$\mathrm{SNR}(\mathcal{I}_{TD}) = \frac{\mathbb{E}[\mathcal{I}_{TD}(z,\omega)]}{\mathrm{Var}(\mathcal{I}_{TD}(z,\omega))^{1/2}},$$

is equal to

$$\mathrm{SNR}(\mathcal{I}_{TD}) = \frac{\sqrt{2}\pi^{1-d/2}k_0^{(d+1)/2}|U(z)|(\frac{\varepsilon_\star}{\varepsilon_0}-1)|D|}{\sigma_{\mathrm{noise}}}.$$

In the case of medium noise, let us introduce the kernel

$$Q(z^S, z) := \Re e\left\{ U^{(0)}(z^S)\overline{U^{(0)}(z)}\int_{\partial\Omega} \Gamma_{k_0}(x, z^S)\overline{\Gamma_{k_0}(x,z)}\,d\sigma(x)\right\}.$$

We can express the topological derivative imaging functional as follows [20]:

$$\begin{aligned}
\mathcal{I}_{TD}[U^{(0)}](z^S) &\approx k_0^4(\frac{\varepsilon'}{\varepsilon_0}-1)|B'|\int_\Omega \nu_{\mathrm{noise}}(y)Q(z^S,y)\,dy \\
&+ k_0^4(\frac{\varepsilon'}{\varepsilon_0}-1)(\frac{\varepsilon_\star}{\varepsilon_0}-1)|B'||D|Q(z^S,z)e^{-\frac{k_0^2\mathrm{Var}(T)}{2}},
\end{aligned} \tag{8.29}$$

provided, once again, that the correlation length of the random process ν_{noise} is small and the amplitude of ν_{noise} is also small. Consequently, the topological derivative has the form of a peak centered at the location z of the inclusion (second term of the right-hand side of (8.29)) buried in a zero-mean Gaussian field or speckle pattern (first term of the right-hand side of (8.29)) that we can characterize statistically.

8.3 Direct Imaging for the Helmholtz Equation at Multiple Frequencies

Let $(\boldsymbol{\theta}_1, \ldots, \boldsymbol{\theta}_n)$ be n uniformly distributed directions over the unit sphere. We consider plane wave illuminations at multiple frequencies, $(\omega_j)_{j=1,\ldots,m}$, instead of a fixed frequency:

$$U_{lj}(x) := U(x, \boldsymbol{\theta}_l, \omega_j) = e^{ik_j\boldsymbol{\theta}_l \cdot x},$$

where $k_j := \sqrt{\varepsilon_0 \mu_0} \omega_j$, and record the perturbations due to the inclusion. In this case, we can construct the topological derivative imaging functional by summing over frequencies

$$\mathcal{I}_{\mathrm{TDF}}(z^S) := \frac{1}{m} \sum_{j=1}^{m} \mathcal{I}_{\mathrm{TD}}(z^S, \omega_j) . \tag{8.30}$$

Suppose for simplicity that $\mu_\star = \mu_0$. Then, (8.25) and (8.26) yield

$$\mathcal{I}_{\mathrm{TDF}}(z^S) \sim \int_{k_0} k_0^{5-d} \left(\Im\{ \Gamma_{k_0}(z^S, z) \} \right)^2 dk_0, \quad d = 2,3 ,$$

and hence, $\mathcal{I}_{\mathrm{TDF}}(z^S)$ has a large peak only at z. In the case where $\mu_\star \neq \mu_0$, we can use (8.27) to state the same behavior at z.

An alternative imaging functional when searching for an inclusion using multiple frequencies is the Reverse-Time migration imaging functional [50]:

$$\begin{aligned}
\mathcal{I}_{\mathrm{RMF}}(z^S) := &\frac{1}{nm} \sum_{l=1}^{n} \sum_{j=1}^{m} \overline{U}(z^S, \boldsymbol{\theta}_l, \omega_j) \\
&\times \int_{\partial\Omega} (-\frac{1}{2} I + \mathcal{K}_\Omega^{k_j})[u - U](x, \boldsymbol{\theta}_l, \omega_j) \overline{\Gamma_{k_j}}(x, z^S)\, d\sigma(x) .
\end{aligned} \tag{8.31}$$

In fact, when for instance $\mu_\star = \mu_0$,

$$\mathcal{I}_{\mathrm{RMF}}(z^S) \sim \frac{1}{nm} \sum_{l=1}^{m} \sum_{j=1}^{m} \omega_j^3 U(z, \boldsymbol{\theta}_l, \omega_j) \overline{U}(z^S, \boldsymbol{\theta}_l, \omega_j) \Im\{ \Gamma_{k_j}(z^S, z) \} ,$$

and therefore, it is approximately proportional to

$$\begin{aligned}
\int_{\mathcal{S}^{d-1}} \int_{k_0} & k_0^3 e^{ik_0\boldsymbol{\theta}\cdot(z^S-z)} \Im\{ \Gamma_{k_0}(z^S, z) \} dk_0 d\sigma(\boldsymbol{\theta}) \\
&\sim \int_{k_0} k_0^{5-d} \left(\Im\{ \Gamma_{k_0}(z^S, z) \} \right)^2 dk_0 ,
\end{aligned}$$

where $\mathcal{S}^{d-1}$ is the unit sphere and $d = 2,3$. Hence,

$$\mathcal{I}_{\mathrm{RMF}}(z^S) \sim \mathcal{I}_{\mathrm{TDF}}(z^S).$$

Finally, it is possible to use a backpropagation imaging functional:

$$\mathcal{I}_{\mathrm{BPF}}(z^S) := \frac{1}{m} \sum_{j=1}^{m} \mathcal{I}_{\mathrm{BP}}(z^S, \omega_j) ,$$

or a Kirchhoff imaging functional:

$$\mathcal{I}_{\mathrm{KMF}}(z^S) := \frac{1}{m}\sum_{j=1}^{m}\mathcal{I}_{\mathrm{KM}}(z^S,\omega_j)\;.$$

We can also use the matched field imaging functional:

$$\mathcal{I}_{\mathrm{MF}}(z^S) := \frac{1}{m}\sum_{j=1}^{m}|\mathcal{I}_{\mathrm{KM}}(z^S,\omega_j)|^2\;,$$

in which the phase coherence between the different frequency-dependent perturbations is not exploited. This makes sense when the different frequency-dependent perturbations are incoherent.

If the measurement noises $\nu_{\mathrm{noise}}(x,\omega_j), j = 1,\dots,m$, are independent and identically distributed, the multiple frequencies enhance the detection performance via a higher "effective" SNR.

If some correlation between frequency-dependent perturbations exist, for example because of a medium noise, then summing over frequencies an imaging functional is not appropriate. A single-frequency imaging functional at the frequency which maximizes the SNR may give a better reconstruction.

In the presence of a medium noise, a Coherent Interferometry (CINT) procedure may be appropriate. CINT consists of backpropagating the cross correlations of the recorded signals over appropriate space-time or space-frequency windows rather than the signals themselves. Here, we provide a CINT strategy in inclusion imaging.

Following [51, 52] a CINT-like algorithm is given by

$$\mathcal{I}_{\mathrm{CINT}}(z^S) = \int_{\mathcal{S}^{d-1}}\int_{\omega_1}\int_{\omega_2}\int_{\partial\Omega}\int_{\partial\Omega} e^{-\frac{|\omega_1-\omega_2|^2}{2\Omega_D^2}}\, e^{-\frac{|x_1-x_2|^2}{2X_D^2}}$$

$$(-\tfrac{1}{2}I + \mathcal{K}_{\Omega}^{k_1})[u - U](x_1,\boldsymbol{\theta},\omega_1)\overline{\Gamma_{k_1}}(x_1,z^S) \tag{8.32}$$
$$\overline{U}(z^S,\boldsymbol{\theta},\omega_1)(-\tfrac{1}{2}I + \mathcal{K}_{\Omega}^{k_2})[u - U](x_2,\boldsymbol{\theta},\omega_2)$$

$$\Gamma_{k_2}(x_2,z^S)U(z^S,\boldsymbol{\theta},\omega_2)d\sigma(x_1)d\sigma(x_2)d\omega_1 d\omega_2 d\sigma(\boldsymbol{\theta})\;,$$

where X_D and Ω_D are two cut-off parameters.

The purpose of the CINT-like imaging functional $\mathcal{I}_{\mathrm{CINT}}$ is to keep in (8.32) the pairs (x_1,ω_1) and (x_2,ω_2) for which the postprocessed data $(-\tfrac{1}{2}I + \mathcal{K}_{\Omega}^{k_1})[u - U](x_1,\omega_1)$ and $(-\tfrac{1}{2}I + \mathcal{K}_{\Omega}^{k_2})[u - U](x_1,\omega_1)$ are coherent, and to remove the pairs that do not bring information.

Depending on the parameters X_D,Ω_D, we get different trade-offs between resolution and stability. When X_D and Ω_D become small, $\mathcal{I}_{\mathrm{CINT}}$ presents better stability properties at the expense of a loss of resolution. In the limit $X_D \to \infty$, $\Omega_D \to \infty$, we get the square of the topological derivative functional $\mathcal{I}_{\mathrm{TDF}}$.

Bibliography and Discussion

The imaging techniques developed in this chapter could be seen as a regularizing method in comparison with iterative approaches; they reduce the set of admissible solution. Their robustness and accuracy are related to the fact that the number of unknowns is reduced and the imaging problem is sparse. The algorithms designed for the Helmholtz equation use the phase information on the measured wave in an essential way. They can not be used to locate the target from intensity-only measurements [63].

The reader can refer to [40, 120] for further details for the projection algorithm for the conductivity problem. The MUSIC algorithm was originally developed for source separation in signal theory [154]. The imaging functional introduced in (8.6) can be interpreted as of a topological derivative type. It extends the one developed for wave imaging to conductivity imaging. The MUSIC-type algorithm for locating small electromagnetic inclusions from the multi-static response matrix at a fixed frequency was developed in [26]. Kirchhoff migration and backpropagation-type algorithms for inclusion imaging have been investigated in [19, 23]. The topological derivative imaging functional, which is quite robust with respect to both measurement and medium noises, was introduced in [20]. The Helmholtz-Kirchhoff identity plays a key role in justifying its performance. The comparison of the performance of direct imaging functionals at a fixed or multiple frequencies as well as the stability and resolution analysis in the presence of medium and measurement noises were carried out in [19]. In [51, 52], CINT has been shown to achieve a good compromise between resolution and deblurring for imaging in noisy environments from multiple frequency measurements. A precise stability and resolution analysis for $\mathcal{I}_{\mathrm{CINT}}$ can be derived by exactly the same arguments as those in [5].

The Helmholtz-Kirchhoff identity shows that the sharper the behavior of the imaginary part of the Green function around the location of the inclusion is, the higher is the resolution. It would be quite challenging to explicitly see how this behavior depends on the heterogeneity of the surrounding medium. This would yield super-resolved imaging systems.

Taking into account the sparsity of the imaging problem, l_1 minimization-based imaging methods can be designed for locating small targets from boundary or MSR data. We refer the reader to [64] for studying such methods and comparing them with those described in this chapter.

Chapter 9
Detection and Imaging from MSR Measurements

The problem addressed in this chapter is to detect and localize point reflectors or small inclusions embedded in a medium from MSR measurements. We use random matrix theory tools and the results of Chap. 6 to study these problems in the presence of measurement noise. The measurement noise can be modeled by an additive complex Gaussian matrix with zero mean. We consider an SVD based detection test. By the Neyman-Pearson lemma we design the most powerful test for a given false alarm rate and provide the probability of detection of a point reflector hidden or not in noise. Then we build algorithms that estimate the number, the location, and the strength of points reflectors embedded in the medium. Using again the results in Chap. 6 we adopt these algorithms for small inclusion detection and localization.

9.1 Point Reflectors

Suppose for simplicity that the array of transmitters and receivers coincide and denote by $\{x_1, \ldots, x_N\}$ the set of sensors locations. From the Born approximation (6.12), the multistatic data set is modeled in the absence of noise by

$$A_{nm}^0(\omega) = \omega^2 \int_{\mathbb{R}^2} \Gamma_\omega(x_n, z) V_{\text{true}}(z) \Gamma_\omega(z, x_m) dz, \quad m, n = 1, \ldots, N . \quad (9.1)$$

We define, for a smooth compactly supported V,

$$[\mathcal{A}(\omega)V]_{nm} = \int_{\mathbb{R}^2} \Gamma_\omega(x_n, z) V(z) \Gamma_\omega(z, x_m) dz . \quad (9.2)$$

$\mathcal{A}(\omega)$ is the frequency-dependent, linear operator that maps the reflectivity function V to the array data.

H. Ammari et al., *Mathematical and Statistical Methods for Multistatic Imaging*, 189
Lecture Notes in Mathematics 2098, DOI 10.1007/978-3-319-02585-8_9,
© Springer International Publishing Switzerland 2013

9.1.1 Linearized Inversion

Let A^{meas} denote the measured MSR matrix. The least-squares inverse problem under the Born approximation consists in minimizing over the reflectivity functions V the misfit functional $J_{\mathrm{LS}}[V]$ where

$$J_{\mathrm{LS}}[V] := \int d\omega \sum_{n,m=1}^{N} \left| A_{nm}^{\mathrm{meas}}(\omega) - [\mathcal{A}(\omega)V]_{nm} \right|^2 .$$

The solution of the least-squares linearized inverse problem is

$$V_{\mathrm{LS}} = \left(\int \mathcal{A}^*(\omega)\mathcal{A}(\omega)d\omega \right)^{-1} \left(\int \mathcal{A}^*(\omega)A^{\mathrm{meas}}(\omega)d\omega \right) .$$

Here the adjoint operator $\mathcal{A}^*(\omega)$ is defined for $N \times N$ matrices $M = (M_{nm})$ by

$$\mathcal{A}^*(\omega)[M](y) = \sum_{n,m=1}^{N} \overline{\Gamma_\omega(y,x_n)\Gamma_\omega(x_m,y)}M_{nm} .$$

Remember that the complex conjugation in the frequency domain corresponds to the time-reversal operation in the time domain. This shows that the adjoint operator corresponds to the backpropagation of the array data both from the receiver point x_n and from the source point x_m to the test point y.

9.1.2 Point Reflectors: Kirchhoff Migration

The least-squares imaging functional is

$$\mathcal{I}_{\mathrm{LS}}(z^S) = \left[\left(\int \mathcal{A}^*(\omega)\mathcal{A}(\omega)d\omega \right)^{-1} \left(\int \mathcal{A}^*(\omega)A^{\mathrm{meas}}(\omega)d\omega \right) \right](z^S) . \qquad (9.3)$$

Motivated by the fact that we often have

$$\int \mathcal{A}^*(\omega)\mathcal{A}(\omega)d\omega \approx I ,$$

where I is the identity operator (in particular, this is a consequence of the Helmholtz-Kirchhoff identity when the array completely surrounds the region of interest), we can drop this term to get a simplified imaging functional.

The Reverse-Time migration imaging functional for the search point z^S is defined by

$$\mathcal{I}_{\mathrm{RT}}(z^S) := \int d\omega \left[\mathcal{A}^*(\omega)A^{\mathrm{meas}}(\omega)\right](z^S)$$

$$= \int d\omega \sum_{n,m=1}^{N} \overline{\Gamma_\omega(z^S, x_n)}\Gamma_\omega(x_m, z^S) A_{nm}^{\mathrm{meas}}(\omega) . \qquad (9.4)$$

The Kirchhoff migration (or travel time migration) is obtained as a simplification of the Reverse-Time migration imaging functional in which we replace $\Gamma_\omega(x, y)$ with $e^{i\omega T(x,y)}$, where $T(x, y) = |x - y|$ is the travel time from x to y (since $\varepsilon_0 = \mu_0 = 1$). Therefore the Kirchhoff migration imaging functional has the form:

$$\mathcal{I}_{\mathrm{KM}}(z^S) := \int d\omega \sum_{m,n=1}^{N} e^{-i\omega(T(x_n,z^S)+T(x_m,z^S))} A_{nm}^{\mathrm{meas}}(\omega) .$$

Imaging Point Reflectors at a Single Frequency

At a single frequency ω, reverse-time migration and Kirchhoff imaging functional can be defined by

$$\mathcal{I}_{\mathrm{RT}}(z^S, \omega) := \sum_{n,m=1}^{N} \overline{\Gamma_\omega(z^S, x_n)}\Gamma_\omega(x_m, z^S) A_{nm}^{\mathrm{meas}}(\omega) \qquad (9.5)$$

and

$$\mathcal{I}_{\mathrm{KM}}(z^S, \omega) := \sum_{m,n=1}^{N} e^{-i\omega(T(x_n,z^S)+T(x_m,z^S))} A_{nm}^{\mathrm{meas}}(\omega) . \qquad (9.6)$$

The locations of the maxima of $\mathcal{I}_{\mathrm{RT}}$ and $\mathcal{I}_{\mathrm{KM}}$ correspond to the point reflectors. Moreover, one can introduce for imaging at a single frequency a MUSIC-type imaging functional. Let $v(z^S, \omega)$ be the normalized vector of the fundamental solutions from the transmitter array to the search point z^S. Since the arrays of transmitters and receivers are assumed to be coincident the vector $v(z^S, \omega)$ is also the vector of the fundamental solutions from the receiver array to the search point z^S. Let $(v_{\mathrm{meas}}^{(l)}[\omega])_{l=1,...,N}$ be the singular vectors of $A^{\mathrm{meas}}[\omega]$. In the presence of r point reflectors, there are only r significant singular vectors of $A^{\mathrm{meas}}[\omega]$, i.e., r is the dimension of the image space of $A^{\mathrm{meas}}[\omega]$. The MUSIC-type imaging functional is defined by

$$\mathcal{I}_{MU}(z^S,\omega) := \left\| v(z^S,\omega) - \sum_{l=1}^{r} \langle v_{meas}^{(l)}[\omega], v(z^S,\omega) \rangle v_{meas}^{(l)}[\omega] \right\|^{-1} . \qquad (9.7)$$

The MUSIC-type imaging functional $\mathcal{I}_{MU}(z^S,\omega)$ has large peaks at the location of the point reflectors. As shown in [24], $\mathcal{I}_{MU}$ is nothing else than a weighted subspace migration algorithm. Introduce for the filter (complex) weights $w(z^S,\omega) = (w_l(z^S,\omega))_{l=1,\dots,N}$ the weighted subspace migration functional

$$\mathcal{I}_{SM}(z^S,\omega,w) := \sum_{l=1}^{N} w_l(z^S,\omega) \langle v(z^S,\omega), v_{meas}^{(l)}[\omega] \rangle^2 . \qquad (9.8)$$

Consider the weights

$$w_l^{(2)}(z^S,\omega) = \begin{cases} \exp\left(-i2\arg\langle v(z^S,\omega), v_{meas}^{(l)}[\omega]\rangle\right) & \text{for } l \leq r , \\ 0 & \text{elsewhere.} \end{cases}$$

We have the following connection of $\mathcal{I}_{SM}(x,\omega,w^{(2)})$ to the MUSIC algorithm:

$$\mathcal{I}_{MU}(z^S,\omega) = \left(1 - \sum_{l=1}^{r} |\langle v_{meas}^{(l)}[\omega], v(z^S,\omega)\rangle|^2\right)^{-1/2}$$

$$= \left(1 - \mathcal{I}_{SM}(z^S,\omega,w^{(2)})\right)^{-1/2} . \qquad (9.9)$$

Choosing now

$$w_l^{(1)}(z^S,\omega) = \sigma_{meas}^{(l)}[\omega] \quad \text{for } l = 1,\dots,N ,$$

where $\sigma_{meas}^{(l)}$ are the singular values of A^{meas} arranged in a decreasing sequence, we obtain that $\mathcal{I}_{SM}(x,\omega,w^{(1)})$ corresponds to Reverse-Time migration:

$$\mathcal{I}_{SM}(z^S,\omega,w^{(1)}) = \mathcal{I}_{RT}(z^S,\omega) = \overline{v(z^S,\omega)} \cdot A^{meas}[\omega]\overline{v(z^S,\omega)} . \qquad (9.10)$$

Assuming a zero-mean additive and uncorrelated measurement noise, it is proved in [24] that given the MSR matrix A^{meas}, the estimator

$$\hat{z} = \underset{z^S \in \Omega}{\arg\max} |\mathcal{I}_{RT}(z^S)|^2$$

is the best estimator among all the estimators obtained from the weighted subspace functionals (9.8) in the maximum likelihood sense to localize a reflector buried in the search domain Ω. Note that the maximum likelihood yields estimates for the unknown quantities (here the locations and the coefficients of reflectivity of the point reflectors) which maximize the probability of obtaining the observed set of data (here the MSR data). We also note that under the assumption of zero-mean Gaussian measurement noise maximum likelihood and minimum variance estimation yield the same exact results for the least-squares estimates; see [69].

9.1.3 Detection Test

The objective of this subsection is to design specific point reflector detection rules.

SVD Based Detection Test

Suppose that the MSR matrix A^{meas} consists of independent Gaussian noise coefficients with mean zero and variance $\sigma_{\text{noise}}^2/N$ (here we assume that the Hadamard acquisition scheme has been used as in Chap. 6). Let the ratio R of the first singular value $\sigma_1(A^{\text{meas}})$ over the normalized l^2-norm of the other singular values $(\sigma_j(A^{\text{meas}}))_{j=2,\dots,N}$ of the measured MSR marix A^{meas} be defined by

$$R := \frac{\sigma_1(A^{\text{meas}})}{\left(\frac{1}{N-1}\sum_{j=2}^{N}\sigma_j(A^{\text{meas}})^2\right)^{1/2}} . \tag{9.11}$$

Using Proposition 1.9 ((ii) and (iii)) and Slutsky's Theorem 1.3, we obtain the following result.

Proposition 9.1. *In the absence of any point reflector, the ratio R defined by (9.11) has the following statistical distribution*

$$R \overset{dist.}{=} 2 + \frac{1}{2^{2/3}N^{2/3}}Z_2 , \tag{9.12}$$

when N is large, where Z_2 is a random variable following a type 2 Tracy-Widom distribution.

This proposition describes the statistical distribution of the ratio (9.11) in the absence of a point reflector. As we will see, it allows us to compute explicitly the threshold of the likelihood-ratio test.

Now we turn to the case where the MSR matrix is obtained with a single point reflector in the presence of additive noise. Then,

$$A^{\text{meas}} = A^0 + \frac{1}{\sqrt{N}}W ,$$

where A^0 is the unperturbed MSR matrix (9.1) (corresponding to one point reflector) and the entries of the matrix W are independent complex Gaussian random variables with mean zero and variance σ_{noise}^2. The form of the measured matrix follows from the use of the Hadamard acquisition technique as explained in Chap. 6.

Let σ_0 be the nonzero singular value of A^0 and let $\sigma_1(A^{\text{meas}}) \geq \sigma_2(A^{\text{meas}}) \geq \cdots \geq \sigma_N(A^{\text{meas}})$ be the singular values of the measured MSR matrix A^{meas}. Using Theorem 6.2 we can describe the ratio of the maximal singular value over the normalized l^2-norm as follows.

Proposition 9.2. *Let us consider the symmetrized MSR matrix obtained in the presence of measurement noise with a point reflector. For $\sigma_{\text{noise}} < \sigma_0$, the ratio R defined by (9.11) has the following statistical distribution*

$$R \stackrel{dist.}{=} \frac{\sigma_0}{\sigma_{\text{noise}}} + \frac{\sigma_{\text{noise}}}{\sigma_0} + \frac{1}{\sqrt{2N}}\sqrt{1 - \sigma_{\text{noise}}^2 \sigma_0^{-2}}\, Z_0 , \qquad (9.13)$$

where Z_0 follows a Gaussian distribution with mean zero and variance one. For $\sigma_{\text{noise}} > \sigma_0$ we have (9.12).

This proposition describes the statistical distribution of the ratio (9.11) in the presence of a point reflector. It allows us to compute explicitly the power of the likelihood-ratio test which is the most powerful test for a given false alarm rate by the Neyman-Pearson lemma.

Statistical Test

As in the standard statistical hypothesis testing [70, 112], we postulate two hypotheses and derive a decision rule for deciding in between them based on the measured MSR matrix.

We define H_o the (null) hypothesis to be tested and H_a the (alternative) hypothesis:

- H_o: there is no point reflector,
- H_a: there is a point reflector.

We want to test H_o against H_a. Two types of independent errors can be made:

- Type I errors correspond to rejecting the null hypothesis H_o when it is correct (false alarm). Their probability is given by

$$\alpha := P[\text{accept } H_a | H_o \text{ true}] .$$

- Type II errors correspond to accepting H_o when it is false (missed detection) and have probability

$$\beta := P[\text{accept } H_o | H_a \text{ true}] .$$

The success of the test (probability of detection or detection power) is therefore given by $1 - \beta$.

Given the data the decision rule for accepting H_o or not can be derived from the Neyman-Pearson lemma which asserts that for a prescribed false alarm rate α the most powerful test corresponds to accepting H_a for the likelihood ratio of H_a to H_o exceeding a threshold value determined by α.

Neyman-Pearson Lemma: Let Y be the set of all possible data and let $f_0(y)$ and $f_1(y)$ be the probability densities of Y under the null and alternative hypotheses. The Neyman-Pearson lemma [70, p. 335] states that the most powerful test has a critical region defined by

$$\mathcal{Y}_\alpha := \left\{ y \in Y \mid \frac{f_1(y)}{f_0(y)} \geq \eta_\alpha \right\} , \tag{9.14}$$

for a threshold η_α satisfying

$$\int_{y \in \mathcal{Y}_\alpha} f_0(y) dy = \alpha . \tag{9.15}$$

Let the data be y. We reject H_o if the likelihood ratio $\frac{f_1(y)}{f_0(y)} > \eta_\alpha$ and accept H_o otherwise. The power of the (most powerful) test is

$$1 - \beta = \int_{y \in \mathcal{Y}_\alpha} f_1(y) dy . \tag{9.16}$$

Berens' Modeling

In [47] a framework for analyzing schemes for nondestructive inspection methods and testing for the presence of flaws was introduced. In this reliability analysis the probability of detection (POD) as a function of flaw size played a central role. In our notation the "flaw size" corresponds to the parameter ρ (given by (6.11)) and we are thus interested in designing reliability tests with a desirable performance in terms of the corresponding POD(ρ) function. In [47] a maximum likelihood approach was used for parameter estimation, and a log normal distribution was in particular postulated for the response variable's relation to point reflector strength. One parameter to be estimated is then the variance of the Gaussian residual. Our approach here is to introduce a physical model for the measurements, as we have described above, and then infer a corresponding "optimal" POD function that can be associated with the MSR matrix measurements. We describe the picture deriving from this approach below. It turns out that the resulting picture deviates somewhat from the one derived from Berens' modeling.

Consider the imaging of point reflectors from measurements of the MSR matrix at a single frequency ω in the presence of measurement noise, that is, we model with an additive Gaussian noise. Assuming availability of previous and/or multiple measurements we may assume that the variance of the entries of the MSR matrix (due to the measurement noise) is known and equal to $\sigma_{\text{noise}}^2/N$. In fact, we will see that we do not need to know the value σ_{noise}^2 in order to build the most powerful test with a prescribed false alarm rate.

In the absence of the point reflector (hypothesis H_o) the statistical distribution of the ratio R of the first singular value of the symmetrized MSR matrix over the normalized l^2-norm of the other singular values is of the form (9.12).

In the presence of a point reflector at position z and with coefficient of reflection ρ (hypothesis H_a), Proposition 9.2 shows that the ratio is of the form (9.13), with σ_0 given by (6.16):

$$\sigma_0(\rho, z) = \rho \left(\sum_{n=1}^{N} |\Gamma_\omega(z, x_n)|^2 \right) . \tag{9.17}$$

This result is correct as long as $\sigma_0 > \sigma_{\text{noise}}$. When $\sigma_0 < \sigma_{\text{noise}}$ we have (9.12).

If the data gives the ratio R, then we propose to use a test of the form $R > r$ for the alarm corresponding to the presence of a point reflector. By the Neyman-Pearson lemma the decision rule of accepting H_a if and only if $R > r_\alpha$ maximizes the probability of detection for a given false alarm probability α

$$\alpha = \mathbb{P}(R > r_\alpha | H_o) ,$$

with the threshold

$$r_\alpha = 2 + \frac{1}{2^{2/3} N^{2/3}} \Phi_{\text{TW2}}^{-1}(1 - \alpha) , \tag{9.18}$$

where $\Phi_{\text{TW2}}(x) = \int_{-\infty}^{x} p_{\text{TW2}}(y) dy$ is the cumulative distribution function of the Tracy-Widom distribution of type 2. The computation of the threshold is easy since it depends only on the number of sensors N and on the false alarm probability α. This test is therefore universal. Note that we should use a Tracy-Widom distribution table, and not a Gaussian table. We have, for instance, $\Phi_{\text{TW2}}^{-1}(0.99) \approx 0.48$.

The detection probability $1 - \beta$ is the probability to sound the alarm when there is a point reflector:

$$1 - \beta = \mathbb{P}(R > r_\alpha | H_a) .$$

For a given measurement array it depends on ρ and z through the value $\sigma_0(\rho, z)$ and also on the noise level σ_{noise}. Here we find that the detection probability is

$$\text{POD}(\rho, z) = 1 - \beta(\rho, z) = \Phi\left(\sqrt{N}\frac{\frac{\sigma_0}{\sigma_{\text{noise}}} + \frac{\sigma_{\text{noise}}}{\sigma_0} - r_\alpha}{\sqrt{1 - (\sigma_{\text{noise}}/\sigma_0)^2}}\right), \tag{9.19}$$

where $\Phi(x) = \int_{-\infty}^{x} \frac{1}{\sqrt{2\pi}} \exp(-y^2/2)dy$ is the cumulative distribution function of the normal distribution with mean zero and variance one. This result is valid as long as $\sigma_0 > \sigma_{\text{noise}}$. When $\sigma_0 < \sigma_{\text{noise}}$, so that the point reflector is "hidden in noise", then we have $1 - \beta = 1 - \Phi_{\text{TW2}}(\Phi_{\text{TW2}}^{-1}(1 - \alpha)) = \alpha$. Note that, as functions of the number of sensors N, the singular value σ_0 scales as N. This shows that the detection power increases with the number of sensors.

9.1.4   Localization and Reconstruction

In this subsection we consider the situation in which there are an unknown number r of point reflectors embedded in the medium. We would like to build algorithms that estimate the number r of point reflectors, estimate their locations z_j, and estimate their coefficients of reflection ρ_j (defined by (6.11)). In the first version of the algorithm we assume that the noise level σ_{noise} is known. The algorithm is then the following one, having the SVD of A^0 (6.15) in mind.

1. Compute the singular values $\sigma_j(A^{\text{meas}})$ of the measured MSR matrix A^{meas}.
2. Estimate the number of reflectors by

$$\hat{r} = \max\left\{j, \sigma_j(A^{\text{meas}}) > r_\alpha \sigma_{\text{noise}}\right\},$$

   where the threshold value r_α, given by (9.18), ensures that the false alarm rate (for the detection of a reflector) is α.
3. For each $j = 1, \ldots, \hat{r}$, estimate the positions z_j of the jth reflector by looking after the position $\hat{z}_j$ of the global maximum of the subspace imaging functional $\mathcal{I}_j(z)$ defined by

$$\mathcal{I}_j(z) = \left|w(z)^* w_j(A^{\text{meas}})\right|^2. \tag{9.20}$$

Here, $w_j(A^{\text{meas}})$ is the j-th left singular vector of the measured response matrix (i.e., the left singular vector associated to the j-th largest singular value) and $w(z)$ is the normalized vector of Green's functions defined by (6.13).

4. For each $j = 1, \ldots, \hat{r}$, estimate the amplitudes ρ_j of the j-th reflector by

$$\hat{\rho}_j = \left(\sum_{n=1}^{N} |\Gamma_\omega(z_j, x_n)|^2 \right)^{-1} \hat{\sigma}_j , \tag{9.21}$$

with $\hat{\sigma}_j$ being the estimator of $\sigma_j(A^0)$ defined by

$$\hat{\sigma}_j = \frac{\sigma_j(A^{\mathrm{meas}})}{2} + \left(\frac{\sigma_j(A^{\mathrm{meas}})^2}{4} - \sigma_{\mathrm{noise}}^2 \right)^{\frac{1}{2}} . \tag{9.22}$$

The form of the estimator $\hat{\sigma}_j$ comes from the inversion of relation (6.18). If we were using $\sigma_j(A^{\mathrm{meas}})$ as an estimator of $\sigma_j(A^0)$, then we would over-estimate the reflectivity coefficients of the reflectors.

Note that we do not need to compute all the singular values of the measured response matrix A^{meas}, only the singular values larger than $2\sigma_{\mathrm{noise}}$ need to be computed.

If the noise level is not known, the first two steps of the algorithm must be replaced by the following ones:

1. Set $j = 1$ and define $A^1 = A^{\mathrm{meas}}$.
2. (a) Compute the largest singular value $\sigma_1(A^j)$ (i.e., the spectral norm of A^j) and the associated singular vectors $v_1(A^j)$ and $w_1(A^j)$.
   (b) Compute the Frobenius norm $\|A^j\|_F$ and estimate the noise level by

$$\hat{\sigma}_{\mathrm{noise},j} = \left[\frac{\|A^j\|_F^2 - \sigma_1^2(A^j)}{N - 4j} \right]^{\frac{1}{2}} . \tag{9.23}$$

   (c) Compute the test

$$T_j = \begin{cases} 1 & \text{if } \sigma_1(A^j) > (2 + r_\alpha)\hat{\sigma}_{\mathrm{noise},j} , \\ 0 & \text{otherwise,} \end{cases} \tag{9.24}$$

   where the threshold value r_α is given by (9.18).
   (d) If $T_j = 1$ then define $A^{j+1} = A^j - \sigma_1(A^j)w_1(A^j)v_1(A^j)^*$, increase j by one, and go to (a).
   If $T_j = 0$ then set $\hat{r} = j - 1$ and $\hat{\sigma}_{\mathrm{noise}} = \hat{\sigma}_{\mathrm{noise},j-1}$ (if $j = 1$, then $\hat{\sigma}_{\mathrm{noise}} = \hat{\sigma}_{\mathrm{noise},0} = \|A^{\mathrm{meas}}\|_F/N^{\frac{1}{2}}$) and go to 3.

The sequence of singular values $\sigma_1(A^j)$, $j = 1, \ldots, \hat{r}$, is the list of the $\hat{r}$ largest singular values $\sigma_j(A^{\mathrm{meas}})$ of A^{meas}. Similarly the sequence of left singular vectors $w_1(A^j)$, $j = 1, \ldots, \hat{r}$, is the list of the left singular vectors $w_j(A^{\mathrm{meas}})$ associated to the $\hat{r}$ largest singular values of A^{meas}. In fact, it is not necessary to compute explicitly the Frobenius norm of A^j at each step in 2(a). We can compute the Frobenius norm of A and then use the relation

$$\|A^j\|_F^2 = \|A\|_F^2 - \sum_{l=1}^{j-1} \sigma_1^2(A^l) ,$$

or, equivalently, the recursive relation

$$\|A^1\|_F^2 = \|A^{\mathrm{meas}}\|_F^2, \qquad \|A^{j+1}\|_F^2 = \|A^j\|_F^2 - \sigma_1^2(A^j), \quad j \geq 1 .$$

This algorithm provides an estimator $\hat{\sigma}_{\mathrm{noise}}$ of the noise level σ_{noise} and an estimator $\hat{r}$ of the number r of significant singular values, that is, the number of reflectors. The steps 3 and 4 of the previous algorithm are then used for the localization and characterization of the reflectors, using the estimator $\hat{\sigma}_{\mathrm{noise}}$ for σ_{noise}.

An alternative algorithm to estimate the noise level is based on the minimization of the Kolmogorov-Smirnov distance between the empirical distribution of the (smallest) singular values of the perturbed matrix A and the theoretical deformed quarter-circle law [81]. This algorithm reduces significantly the bias but it is more computationally intensive. When N is very large, formula (9.23) is sufficient for the noise level estimation. When N is not very large, the algorithm presented in [81] should be used.

Instead of $\mathcal{I}_j(z)$ defined by (9.20), other subspace imaging functionals such as MUSIC or Kirchhoff-type algorithms can be used. The decomposition of the time-reversal operator (DORT) can also be used [65, 66, 72] for detecting and characterizing the reflectors.

9.2 Inclusions

We first apply Steps 1 and 2 of the algorithm described in Sect. 9.1.4 to estimate the number $\hat{r}$ of significant singular values of the perturbed matrix A^{meas}. We apply either the first version, in the case in which the noise level is known, or the second one, in the case in which it is unknown. We use in the following the same notation as in Sect. 9.1.4. The sequence of singular values $\sigma_1(A^j)$, $j = 1, \ldots, \hat{r}$, is the list of the $\hat{r}$ largest singular values $\sigma_j(A^{\mathrm{meas}})$ of A^{meas}. Similarly the sequence of left singular vectors $w_1(A^j)$, $j = 1, \ldots, \hat{r}$, is the list of the left singular vectors $w_j(A^{\mathrm{meas}})$ associated to the $\hat{r}$ largest singular values of A^{meas}.

Steps 1 and 2 are as in Sect. 9.1.4 and then the following steps are original compared to the previous algorithm. Indeed, in Sect. 9.1, the singular vectors are all of the same form. Here the singular vectors may have different forms depending on their nature (dielectric or magnetic).

3. For each $j = 1, \ldots, \hat{r}$, we consider the two functionals

$$\mathcal{I}_j(z) = |w(z)^* w_j(A^{\text{meas}})|^2, \qquad \mathcal{J}_j(z, \theta) = |U(z, \theta)^* w_j(A^{\text{meas}})|^2 ,$$

where $U(z, \theta)$ is given by (6.22). We find their maxima $\mathcal{I}_{j,\max}$ and $\mathcal{J}_{j,\max}$ (the optimization with respect to θ for $\mathcal{J}_j(z, \theta)$ can be carried out analytically). We estimate the theoretical angle between the unperturbed and perturbed vectors by

$$\hat{c}_j = 1 - \frac{\sigma_{\text{noise}}^2}{\hat{\sigma}_j^2} ,$$

with $\hat{\sigma}_j$ being the estimator of $\sigma_j(A^0)$ given by

$$\hat{\sigma}_j = \frac{\sigma_j(A^{\text{meas}})}{2} + \left(\frac{\sigma_j(A^{\text{meas}})^2}{4} - \sigma_{\text{noise}}^2 \right)^{\frac{1}{2}} . \qquad (9.25)$$

We decide the type (d for dielectric, m for magnetic) of the jth singular value as follows:

$$\mathcal{T}_j = \begin{cases} \text{d} & \text{if } |\mathcal{I}_{j,\max} - \hat{c}_j| \leq |\mathcal{J}_{j,\max} - \hat{c}_j| , \\ \text{m} & \text{if } |\mathcal{I}_{j,\max} - \hat{c}_j| > |\mathcal{J}_{j,\max} - \hat{c}_j| . \end{cases}$$

We also record the position $\hat{z}_j$ of the maximum of the imaging functional $\mathcal{I}_j(z)$ if $\mathcal{T}_j = \text{d}$ or the position and angle $(\hat{z}_j, \hat{\theta}_j)$ of the maximum of the imaging functional $\mathcal{J}_j(z, \theta)$ if $\mathcal{T}_j = \text{m}$.
4. We now cluster the results: we consider the set of positions $\hat{z}_j$ estimated in Step 3. We group the indices $\{1, \ldots, \hat{r}\}$ in subsets $(I_q)_{q=1,\ldots,\hat{R}}$ of up to three indices which contain the indices that correspond to positions close to each other (within one wavelength λ):

Set $I = \{2, \ldots, \hat{r}\}$, $j = 1$, $q = 1$, and $I_1 = \{1\}$.
While $I \neq \emptyset$, do

   – consider $\tilde{I} = \{l \in I$ such that $|\hat{z}_l - \hat{z}_j| < \lambda, \mathcal{T}_l$ is compatible$\}$.
   – if $\tilde{I} = \emptyset$, then increase q by one, set $j = \min(I)$, $I_q = \{j\}$, and remove j from I.
   – if $\tilde{I} \neq \emptyset$, then add $\min(\tilde{I})$ into I_q and remove it from I.

We say that the type $\mathcal{T}_l$ is not compatible if it is d in the case in which there is already one index with type d in I_q, or if it is m in the case in which there are already two or three indices with type m in I_q (note that this implies that $\mathcal{T}_l$ is never compatible as soon as $|I_q| = 3$)
   This procedure gives the decomposition:

$$\{1, \ldots, \hat{r}\} = \cup_{q=1}^{\hat{R}} I_q, \qquad I_q = \{j_1^{(q)}, \ldots, j_{n_q}^{(q)}\} ,$$

with $1 \leq n_q \leq 3$. The subset I_q contains n_q indices that correspond to positions close to each other and it does not contain two indices with type d or three indices with type m.

The estimators of the relevant quantities are the following ones:

The number of inclusions is estimated by $\hat{R}$.

The position of the q-th inclusion is estimated by the barycenter (Method 1):

$$\hat{Z}_q = \frac{1}{n_q} \sum_{l=1}^{n_q} \hat{z}_{j_l^{(q)}} \,.$$

An alternative estimator (Method 2) is obtained by an optimization method. It is given by

$$\hat{Z}_q = \underset{z}{\mathrm{argmax}} \sum_{l=1}^{n_q} \sigma_{j_l^{(q)}} (A^{\mathrm{meas}})^2 \Big[\delta_{T_{j_l^{(q)}}=\mathrm{d}} \big| w(z)^* w_{j_l^{(q)}} (A^{\mathrm{meas}}) \big|^2$$

$$+ \delta_{T_{j_l^{(q)}}=\mathrm{m}} \big| U(z, \hat{\theta}_{j_l^{(q)}})^* w_{j_l^{(q)}} (A^{\mathrm{meas}}) \big|^2 \Big] \,,$$

with δ being the counting measure. The first estimator can be used as a first guess for Method 2. More exactly the second estimator can be implemented in the form of an iterative algorithm (steepest descent) starting from this first guess.

The dielectric coefficients ε_q of the q-th inclusion cannot be estimated if there is no index with type d in I_q, otherwise it can be estimated by

$$\hat{\varepsilon}_q = \Big(\sum_{n=1}^{N} |\Gamma_\omega(\hat{Z}_q, x_n)|^2 \Big)^{-1} \hat{\sigma}_{j_l^{(q)}} \,, \tag{9.26}$$

with $\hat{\sigma}_j$ the estimator of $\sigma_j(A^0)$ given by (9.25) and $j_l^{(q)}$ is here the index with type d in I_q.

The angle and the magnetic coefficient α and β of the q-th inclusion:

– cannot be estimated if there is no index with type m in I_q,
– can be estimated by

$$\hat{\Theta}_q = \hat{\theta}_{j_l^{(q)}} \,,$$

$$\hat{\alpha}_q = \Big(\sum_{n=1}^{N} |a(\hat{\Theta}_q) \cdot \nabla \Gamma_\omega(\hat{Z}_q, x_n)|^2 \Big)^{-1} \hat{\sigma}_{j_l^{(q)}} \,,$$

if there is one index $j_l^{(q)}$ with type m in I_q (here $a(\theta) = (\cos\theta, \sin\theta)^T$),
  − can be estimated by

$$\hat{\Theta}_q = \underset{\theta}{\mathrm{argmax}}\left\{\sigma_{j_l^{(q)}}(A^{\mathrm{meas}})^2|U(\hat{Z}_q,\theta)^*w_{j_l^{(q)}}(A^{\mathrm{meas}})|^2\right.$$

$$\left. +\sigma_{j_{l'}^{(q)}}(A^{\mathrm{meas}})^2|U(\hat{Z}_q,\pi/2+\theta)^*w_{j_{l'}^{(q)}}(A^{\mathrm{meas}})|^2\right\},$$

$$\hat{\alpha}_q = \left(\sum_{n=1}^{N}|a(\hat{\Theta}_q)\cdot\nabla\Gamma_\omega(\hat{Z}_q,x_n)|^2\right)^{-1}\hat{\sigma}_{j_l^{(q)}},$$

$$\hat{\beta}_q = \left(\sum_{n=1}^{N}|a(\hat{\Theta}_q+\pi/2)\cdot\nabla\Gamma_\omega(\hat{Z}_q,x_n)|^2\right)^{-1}\hat{\sigma}_{j_{l'}^{(q)}},$$

if there are two indices $j_l^{(q)}, j_{l'}^{(q)}$ with type m in I_q.

Bibliography and Discussion

The results on point reflector and inclusion localizations are from [18]. The results on detection tests and Berens' modeling in the context of detecting point reflectors from MSR measurements are from [23]. They also apply to crack imaging where not only the location of the crack but also its orientation can be estimated [23].

Part V
Dictionary Matching and Tracking Algorithms

Chapter 10
Reconstruction of GPTs from MSR Measurements

This chapter aims to reconstruct GPTs from MSR measurements. We consider the effect of the presence of measurement noise in the MSR on the reconstruction of the GPTs of a small conductivity inclusion. Given a signal-to-noise ratio, we determine the statistical stability in the reconstruction of the GPTs, and show that such an inverse problem is exponentially unstable. This is the well-known ill-posedness of the inverse conductivity problem.

10.1 Least-Squares Formulation

The first step in the target identification procedure of a small conductivity inclusion is to reconstruct GPTs from the MSR matrix A, which has expression (7.14). Define the linear operator $L : \mathbb{R}^{2K \times 2K} \to \mathbb{R}^{N \times N}$ by

$$L(\mathbb{M}) := V \mathbb{M} V^T .\tag{10.1}$$

Here, $\mathbb{M}$ is defined by (7.13).

We reconstruct GPTs as the least-squares solution of the above linear system, i.e.,

$$\mathbb{M}^{\text{est}} = \underset{\mathbb{M}^{\text{test}} \perp \ker(L)}{\arg\min} \|A - L(\mathbb{M}^{\text{test}})\|_F ,\tag{10.2}$$

where $\ker(L)$ denotes the kernel of L and $\| \cdot \|_F$ denotes the Frobenius norm of matrices. In general we take N large enough so that $2K \leq N$. When V has full rank $2K$, L is rank preserving and $\ker(L)$ is trivial; in that case, the admissible set above can be replaced by $\mathbb{R}^{2K \times 2K}$ and

$$\mathbb{M}^{\text{est}} = (V^T V)^{-1} V^T A V (V^T V)^{-1} .$$

H. Ammari et al., *Mathematical and Statistical Methods for Multistatic Imaging*, 205
Lecture Notes in Mathematics 2098, DOI 10.1007/978-3-319-02585-8_10,
© Springer International Publishing Switzerland 2013

From the structure of the matrix V in (7.13) and the expression of the MSR matrix, we observe that the contribution of a GPT decays as its order grows. Consequently, one does not expect the inverse procedure to be stable for higher-order GPTs. The remainder of this chapter is devoted to such stability analysis.

10.2   Analytical Formula in the Circular Setting

To simplify the analysis, we assume that the receivers (sources) are evenly distributed along a circle of radius R centered at z. That is, $\theta_r = 2\pi r/N$, $r = 1, 2, \ldots, N$, and $R_r = R$. In this setting, we have $V = CD$, where C is an $N \times 2K$ matrix constructed from the block $C_{rm} = (\cos m\theta_r, \sin m\theta_r)$ and D is $2K \times 2K$ diagonal matrix:

$$
C = \begin{pmatrix} C_{11} & C_{12} & \cdots & C_{1K} \\ C_{21} & C_{22} & \cdots & C_{2K} \\ \cdots & \cdots & \ddots & \cdots \\ C_{N1} & C_{N2} & \cdots & C_{NK} \end{pmatrix} ; D = \frac{1}{2\pi} \begin{pmatrix} I_2/R & & & \\ & I_2/(2R^2) & & \\ & & \ddots & \\ & & & I_2/(KR^K) \end{pmatrix} .
$$

(10.3)

Here I_2 is the 2×2 identity matrix. We note that C and D account for the angular and radial coefficients in the expansion of MSR, respectively. The matrix C satisfies the following important property.

Proposition 10.1. *Suppose that $2K \leq N$ holds. Then*

$$
C^T C = \frac{N}{2} I_{2K} .
$$

(10.4)

Henceforth, we assume that the number of receivers is large enough so that $2K \leq N$. In this setting, the least-squares solution to (10.2) admits an analytical expression as follows.

Lemma 10.2. *In the above circular setting with sufficiently many receivers, i.e., $2K \leq N$, the least-squares estimation (10.2) is given by*

$$
\mathbb{M}^{\text{est}} = (\frac{2}{N})^2 D^{-1} C^T A C D^{-1} .
$$

(10.5)

Proof. Firstly, (10.4) implies that V has full rank, so $\ker(L) = \{0\}$. Moreover,

$$
(V^T V)^{-1} = \frac{2}{N} D^{-2} .
$$

Hence,

$$\mathbb{M}^{\text{est}} = (\frac{2}{N})^2 D^{-2} D C^T A C D D^{-2} \, ,$$

which yields (10.5).

10.3 Measurement Noise and Stability Analysis

We develop a stability analysis for the least-squares reconstruction of GPTs from the MSR matrix, in the circular setting.

Counting some additive measurement noise, we modify the expression of MSR to

$$A = C D \mathbb{M} D C^T + E + \sigma_{\text{noise}} W \, . \tag{10.6}$$

Here, E is the truncation error due to the finite order K in expansion (7.3), W is an $N \times N$ real-valued random matrix with independent and identically Gaussian entries with mean zero and unit variance, and σ_{noise} is a small positive number modeling the standard deviation of the noise.

Recall that the unknown $\mathbb{M}$ consists of GPTs of order up to K of the relative domain $\delta B = D - z$, where δ denotes the typical length scale of the domain D. The receivers and sources are located along a circle of radius R centered at z. Let $\eta = \delta/R$ be the ratio between the two scales, and assume that it is smaller than one. Due to the scaling property of GPT, the entries of the GPT block $\mathbb{M}_{mn}(\delta B)$ is $\delta^{m+n} \mathbb{M}_{mn}(B)$. Consequently, the size of $V \mathbb{M} V^T$ itself is of order η^2, which is the order of the first term in the expansion (7.12). The truncation error E is of order η^{K+2}. It is due to the scaling property of GPTs and it corresponds to the highest order GPT used in the truncated expansion.

According to the above analysis, we assume that the size of the noise satisfies

$$N \eta^{K+2} \ll \sigma_{\text{noise}} \ll \eta^2 \, . \tag{10.7}$$

This is the regime where the measurement noise is much smaller than the signal but much larger than the truncation error. The presence of N in (10.7) will become clear later; see Remark 10.4. We define the signal-to-noise ratio (SNR) to be

$$\text{SNR} = \frac{\eta^2}{\sigma_{\text{noise}}} .$$

We will investigate the error made by the least-squares estimation of the GPT matrix, in particular the manner of its growth with respect to the order of the GPTs. Given a SNR and a tolerance number τ_0, we can define the resolving order m_0 to be

$$m_0 = \min\left\{1 \le m \le K \; : \; \sqrt{\frac{\mathbb{E}\|\mathbb{M}_{mm}^{\text{est}} - \mathbb{M}_{mm}\|_F^2}{\|\mathbb{M}_{mm}\|_F^2}} \le \tau_0\right\}. \tag{10.8}$$

We are interested in the growth of m_0 with respect to SNR.

We have used the notation $\mathbb{M}_{mn}$, $m, n = 1, \ldots, K$, to denote the building block of the GPT matrix $\mathbb{M}$ in (7.13). In the following, we also use the notation $(\mathbb{M})_{jk}$, $j, k = 1, \ldots, 2K$, to denote the real-valued entries of the GPT matrix.

The following result can be derived using the same lines as in Proposition 4.10.

Theorem 10.3. *Assume that the condition of Lemma 10.2 holds; assume also that the additive noise is in the regime (10.7), Then for j, k so that $(\mathbb{M})_{jk}$ is nonzero, the relative error in its reconstructed GPT satisfies*

$$\sqrt{\frac{\mathbb{E}|(\mathbb{M}^{\text{est}})_{jk} - (\mathbb{M})_{jk}|^2}{|(\mathbb{M})_{jk}|^2}} \approx \frac{\sigma_{\text{noise}}}{N}\eta^{-\lceil j/2\rceil - \lceil k/2\rceil}\left[\frac{j}{2}\right]\left[\frac{k}{2}\right]. \tag{10.9}$$

Here, the symbol $\lceil l \rceil$ is the smallest natural number larger than or equal to l. For vanishing $(\mathbb{M})_{jk}$, the error $\sqrt{\mathbb{E}|(\mathbb{M}^{\text{est}})_{jk} - (\mathbb{M})_{jk}|^2}$ can be bounded by the right-hand side above with η replaced by R^{-1}. In particular, the resolving order m_0 satisfies

$$(m_0\eta^{1-m_0})^2 \approx \tau_0 \text{SNR}, \tag{10.10}$$

where τ_0 is the tolerance number. Estimate (10.10) shows that under the assumption (10.7)

$$m_0 \approx -\frac{\log \text{SNR}}{2\log\eta},$$

and so, in order to achieve high resolving order we need exponentially high SNR.

Proof. From the analytical formula of the least-squares reconstruction (10.5) and the expression of V (10.6), we see that for each fixed $j, k = 1, \ldots, 2K$,

$$(\mathbb{M}^{\text{est}} - \mathbb{M})_{jk} = \frac{2^2\sigma_{\text{noise}}}{N^2}(D^{-1}C^T WCD^{-1})_{jk} + \frac{2^2}{N^2}(D^{-1}C^T ECD^{-1})_{jk}.$$

Let us denote these two terms by $\mathcal{I}_{jk1}$ and $\mathcal{I}_{jk2}$ respectively. For the first term, define $\tilde{W}$ to be $(\sqrt{2/N}C)^T W (\sqrt{2/N}C)$, which is an $N \times N$ random matrix. Due to the orthogonality (10.4), $\tilde{W}$ remains to have mean zero Gaussian entries with unit variance. Because D is diagonal, we have for each $j, k = 1, \ldots, 2K$,

$$
\begin{aligned}
\mathbb{E}(\mathcal{I}_{jk1})^2 &= \frac{2^2 \sigma_{\text{noise}}^2}{N^2} (D_{jj})^{-2} \mathbb{E}|\tilde{W}_{jk}|^2 (D_{kk})^{-2} \\
&= \frac{2^6 \pi^4 \sigma_{\text{noise}}^2}{N^2} R^{2(\lceil j/2 \rceil + \lceil k/2 \rceil)} \left\lceil \frac{j}{2} \right\rceil^2 \left\lceil \frac{k}{2} \right\rceil^2 .
\end{aligned}
$$

Note that $\lceil j/2 \rceil \lceil k/2 \rceil$ is the order of GPT element $(\mathbb{M})_{jk}$; see (7.13). It is known from the scaling property of the GPTs that

$$
(\mathbb{M})_{jk}(\delta B) = \delta^{\lceil j/2 \rceil + \lceil k/2 \rceil} (\mathbb{M})_{jk}(B) .
$$

When this term is nonzero, it is of order $\delta^{\lceil j/2 \rceil + \lceil k/2 \rceil}$. This fact and the above control of $\mathcal{I}_{jk1}$ show that $\sqrt{\mathbb{E}|\mathcal{I}_{jk1}|^2 / |(\mathbb{M})_{jk}|^2}$ satisfies the estimate in (10.9).

For the second term, since E is symmetric, it has the decomposition $E = P^T \mathcal{E} P$, where P is an $N \times N$ orthonormal matrix, and $\mathcal{E}$ is an $N \times N$ diagonal matrix consisting of eigenvalues of E. Then $(\sqrt{2/N}C)^T E (\sqrt{2/N}C)$ can be written as $Q^T \mathcal{E} Q$ where $Q = \sqrt{2/N} PC$ is an $N \times 2K$ matrix satisfying $Q^T Q = I_{2K}$. The calculation for $\mathcal{I}_{jk1}$ can therefore be applied to yield

$$
(\mathcal{I}_{jk2})^2 = \frac{2^6 \pi^4}{N^2} R^{2(\lceil j/2 \rceil + \lceil k/2 \rceil)} \left\lceil \frac{j}{2} \right\rceil^2 \left\lceil \frac{k}{2} \right\rceil^2 \left(\sum_{l=1}^N \mathcal{E}_{ll} Q_{jl}^T Q_{lk} \right)^2 .
$$

Since E is of order η^{K+2} as shown in (4.46), the sum is of order $N\eta^{K+2}$. Therefore, we have

$$
\sqrt{\mathbb{E}|\mathcal{I}_{jk2}|^2} \leq C \eta^{K+2 - \lceil j/2 \rceil - \lceil k/2 \rceil} \lceil \frac{j}{2} \rceil \lceil \frac{k}{2} \rceil .
$$

Since we assumed that (10.7) holds, this error is dominated by the one due to the noise. Hence, (10.9) is proved.

For diagonal blocks $\mathbb{M}_{mm}$, their Frobenius norms do not vanish and (10.8) is well defined. In particular, (10.9) applied to the case $j, k = 2m - 1, 2m$, shows that the relative error made in the block $\mathbb{M}_{mm}$ is of order $\sigma_{\text{noise}} m^2 \eta^{-2m}$. Using the definition of SNR, we verify (10.10).

Remark 10.4. If E has only several (of order one) nonzero eigenvalues, then the preceding calculation shows that $(\mathcal{I}_{jk2})^2 \leq C \eta^{2(K+2)}$ and condition (10.7) can be replaced with $\eta^{K+2} \ll \sigma_{\text{noise}} \ll \eta^2$.

Bibliography and Discussion

The results of this chapter are from [8]. They were extended to the electromagnetic case in [38]. The FDPTs can be extracted from the multistatic data in a similar way. It would be challenging to develop a stability analysis for the least-squares reconstruction of GPTs or FDPTs from multistatic data in the limited-view setting. The ill-posedness of the inverse conductivity problem was shown in this chapter. Nevertheless, in the circular full-view case, it was proved in [9] that the reconstruction problem of GPTs from MSR measurements has the remarkable property that low order GPTs are not affected by the error caused by the instability of higher-orders in the presence of measurement noise.

Chapter 11
Target Identification and Tracking

In this chapter we first recall the notion of contracted GPTs. Then we show that the CGPTs have some nice properties, such as simple rotation and translation formulas, simple relation with shape symmetry, etc. More importantly, we derive new invariants for the CGPTs. Based on those invariants, we develop a dictionary matching algorithm. We suppose that the unknown shape of the target is an exact copy of some element from the dictionary, up to a rigid transform and dilatation. Using the invariants, we identify the target in the dictionary with a low computational cost. We also apply the Extended Kalman Filter to track both the location and the orientation of a mobile target from MSR data.

11.1 Complex CGPTs Under Rigid Motions and Scaling

As we will see later, a complex combination of CGPTs is most convenient when we consider the transforms of CGPTs under dilatation and rigid motions, i.e., shift and rotation. Therefore, for a double index mn, with $m, n = 1, 2, \ldots$, we make use of the complex combination of CGPTs given by (7.10), where the CGPTs, $M_{mn}^{cc}, M_{mn}^{ss}, M_{mn}^{cs}$, and M_{mn}^{sc}, are defined by (7.6)–(7.9).

Then, from (4.2), we observe that

$$\mathbb{N}_{mn}^{(1)}(\lambda, D) = \int_{\partial D} P_n(y)(\lambda I - \mathcal{K}_D^*)^{-1}[\langle \nu, \nabla P_m \rangle](y) \, d\sigma(y) \,,$$

$$\mathbb{N}_{mn}^{(2)}(\lambda, D) = \int_{\partial D} P_n(y)(\lambda I - \mathcal{K}_D^*)^{-1}[\langle \nu, \nabla \overline{P_m} \rangle](y) \, d\sigma(y) \,,$$

H. Ammari et al., *Mathematical and Statistical Methods for Multistatic Imaging*, Lecture Notes in Mathematics 2098, DOI 10.1007/978-3-319-02585-8_11, © Springer International Publishing Switzerland 2013

where P_n and P_m are defined by (7.4). In order to simplify the notation, we drop λ in the following and write simply $N_{mn}^{(1)}(D), N_{mn}^{(2)}(D)$.

We consider the translation, the rotation and the dilatation of the domain D by introducing the following notation:

(i) Shift: $T_z D = \{x + z, x \in D\}$, for $z \in \mathbb{R}^2$;
(ii) Rotation: $R_\theta D = \{e^{i\theta}x, x \in D\}$, for $\theta \in [0, 2\pi)$;
(iii) Scaling: $sD = \{sx, x \in D\}$, for $s > 0$.

The following properties for the complex CGPTs hold. They are much simpler than those associated with the GPTs, which are derived in Sect. 4.2.

Proposition 11.1. *For all integers m, n, and geometric parameters θ, s, and z, the following holds:*

$$N_{mn}^{(1)}(R_\theta D) = e^{i(m+n)\theta} N_{mn}^{(1)}(D), \quad N_{mn}^{(2)}(R_\theta D) = e^{i(n-m)\theta} N_{mn}^{(2)}(D) ,$$
(11.1)

$$N_{mn}^{(1)}(sD) = s^{m+n} N_{mn}^{(1)}(D), \quad N_{mn}^{(2)}(sD) = s^{m+n} N_{mn}^{(2)}(D) ,$$
(11.2)

$$N_{mn}^{(1)}(T_z D) = \sum_{l=1}^{m} \sum_{k=1}^{n} C_{ml}^z N_{lk}^{(1)}(D) C_{nk}^z, \quad N_{mn}^{(2)}(T_z D) = \sum_{l=1}^{m} \sum_{k=1}^{n} \overline{C_{ml}^z} N_{lk}^{(2)}(D) C_{nk}^z ,$$
(11.3)

where C^z is a lower triangle matrix with the m, n-th entry given by

$$C_{mn}^z = \binom{m}{n} z^{m-n} ,$$
(11.4)

and $\overline{C^z}$ denotes its conjugate. Here, we identify $z = (z_1, z_2)$ with $z = z_1 + iz_2$.

An ingredient that we will need in the proof is the following chain rule between the gradient of a function and its push forward under transformation. In fact, for any diffeomorphism Ψ from $\mathbb{R}^2$ to $\mathbb{R}^2$ and any scalar-valued differentiable map f on $\mathbb{R}^2$, we have

$$d(f \circ \Psi)\big|_x (h) = \left(df\big|_{\Psi(x)} \circ d\Psi\big|_x \right)(h) ,$$
(11.5)

for any tangent vector $h \in \mathbb{R}^2$, with $d\Psi$ being the differential of Ψ.

Proof (of Proposition 11.1). We will follow proofs of similar relations that can be found in Chap. 4. Let us first show (11.1) for the rotated domain $D_\theta := R_\theta D$. For a function $\phi(y), y \in \partial D$, we define a function $\phi^\theta(y_\theta), y_\theta := R_\theta y \in \partial D_\theta$ by

$$\phi^\theta(y_\theta) = \phi \circ R_{-\theta}(y_\theta) = \phi(y) .$$

It is proved in (4.10) that $\lambda I - \mathcal{K}_D^*$ is invariant under the rotation map, that is,

$$(\lambda I - \mathcal{K}_{D_\theta}^*)[\phi^\theta](y_\theta) = (\lambda I - \mathcal{K}_D^*)[\phi](y) . \tag{11.6}$$

We also check that $P_m(R_\theta y) = e^{im\theta} P_m(y)$.

We will focus on the relation for $\mathbb{N}_{mn}^{(1)}$, the other one can be proved in the same way. By definition, we have

$$\mathbb{N}_{mn}^{(1)}(D) = \int_{\partial D} P_n(y) \phi_{D,m}(y) d\sigma(y) ,$$

$$\mathbb{N}_{mn}^{(1)}(D_\theta) = \int_{\partial D_\theta} P_n(y_\theta) \phi_{D_\theta,m}(y_\theta) d\sigma(y_\theta) , \tag{11.7}$$

where

$$\phi_{D,m}(y) = (\lambda I - \mathcal{K}_D^*)^{-1}[\langle \nu, \nabla P_m \rangle](y) ,$$

$$\phi_{D_\theta,m}(y_\theta) = (\lambda I - \mathcal{K}_{D_\theta}^*)^{-1}[\langle \nu, \nabla P_m \rangle](y_\theta) .$$

Note that the last function differs from $\phi_{D,m}^\theta$. By the change of variables $y_\theta = R_\theta y$ in the first expression of (11.7), we obtain

$$\mathbb{N}_{mn}^{(1)}(D) = \int_{\partial D_\theta} P_n(R_{-\theta} y_\theta) \phi_{D,m}(R_{-\theta} y_\theta) d\sigma(y_\theta)$$

$$= e^{-in\theta} \int_{\partial D_\theta} P_n(y_\theta) \phi_{D,m}^\theta(y_\theta) d\sigma(y_\theta) .$$

From (11.6), we have

$$(\lambda I - \mathcal{K}_{D_\theta}^*)[\phi_{D,m}^\theta](y_\theta) = (\lambda I - \mathcal{K}_D^*)[\phi_{D,m}](y)$$

$$= \langle \nu_y, \nabla P_m(y) \rangle .$$

Moreover, $P_m(y) = e^{-im\theta} P_m(y_\theta)$ so that, by applying the chain rule (11.5) with $f = P_m$, $T = R_\theta$, $x = y$ and $h = \nu_y$, we can conclude that

$$\langle \nu_y, \nabla P_m(y) \rangle = e^{-im\theta} \langle R_\theta \nu_y, \nabla P_m(y_\theta) \rangle$$

$$= e^{-im\theta} \langle \nu_{y_\theta}, \nabla P_m(y_\theta) \rangle .$$

Therefore, $\phi_{D,m}^{\theta} = e^{-im\theta} \phi_{D_\theta,m}$, and we conclude that

$$\mathbb{N}_{mn}^{(1)}(D_\theta) = e^{i(m+n)\theta} \mathbb{N}_{mn}^{(1)}(D) .$$

The second identity in (11.1) results from the same computation as above (the minus sign comes from the conjugate in the definition of $\mathbb{N}^{(2)}$), and the two equations in (11.2) are proved in the same way, replacing the transformed function ϕ^θ by

$$\phi^s(sy) = \phi(y) .$$

Thus, only (11.3) remains. Since the difference between these two comes from the conjugation, we will focus only on the first identity in (11.3). The strategy will be once again the following: for a function $\phi(y), y \in \partial D$, we define a function $\phi^z(y_z), y_z = y + z \in \partial D_z$, with $D_z := T_z D$, by

$$\phi^z(y_z) = \phi \circ T_{-z}(y_z) = \phi(y) ,$$

which also verifies an invariance relation similar to (11.6)

$$(\lambda I - \mathcal{K}_{D_z}^*)[\phi^z](y_z) = (\lambda I - \mathcal{K}_D^*)[\phi](y) . \qquad (11.8)$$

Moreover, for every integer $q \in \mathbb{N}$ one has the following

$$P_q(y_z) = (y+z)^q = \sum_{r=0}^{q} \binom{q}{r} y^r z^{q-r} . \qquad (11.9)$$

Equations (11.7) become

$$\mathbb{N}_{mn}^{(1)}(D) = \int_{\partial D} P_n(y) \phi_{D,m}(y) d\sigma(y) ,$$

$$\mathbb{N}_{mn}^{(1)}(D_z) = \int_{\partial D_z} P_n(y_z) \phi_{D_z,m}(y_z) d\sigma(y_z) ,$$

where

$$\phi_{D,m}(y) = (\lambda I - \mathcal{K}_D^*)^{-1} [\langle \nu, \nabla P_m \rangle](y) ,$$

$$\phi_{D_z,m}(y_z) = (\lambda I - \mathcal{K}_{D_z}^*)^{-1} [\langle \nu, \nabla P_m \rangle](y_z) .$$

Thus, combining (11.8) and (11.9) leads us to

$$(\lambda I - \mathcal{K}_{D_z}^*)[\phi_{D_z,m}](y_z) = \langle \nu_{y_z}, \nabla P_m(y_z) \rangle$$

$$= \langle \nu_y, \sum_{l=1}^{m} \binom{m}{l} z^{m-l} \nabla P_l(y) \rangle$$

$$= \sum_{l=1}^{m} \binom{m}{l} z^{m-l} (\lambda I - \mathcal{K}_D^*)[\phi_{D,l}](y)$$

$$= \sum_{l=1}^{m} \binom{m}{l} z^{m-l} (\lambda I - \mathcal{K}_{D_z}^*)[\phi_{D,l}^z](y_z) ,$$

so that we have

$$\phi_{D_z,m}(y) = \sum_{l=1}^{m} \binom{m}{l} z^{m-l} \phi_{D,l}^z(y_z) .$$

Hence, returning to the definition of $\mathbb{N}_{mn}^{(1)}(D_z)$ with the substitution $y_z \leftrightarrow y$, we obtain

$$\mathbb{N}_{mn}^{(1)}(D_z) = \sum_{l=1}^{m} \binom{m}{l} z^{m-l} \int_{\partial D_z} P_n(y_z) \phi_{D,l}^z(y_z) d\sigma(y_z) ,$$

$$= \sum_{l=1}^{m} \sum_{k=1}^{n} \binom{m}{l} \binom{n}{k} z^{m-l} z^{n-k} \mathbb{N}_{lk}^{(1)}(D) ,$$

which is the desired result. Note that the index k begins with $k = 1$ because $\int_{\partial D_z} \phi_{D,l}^z = 0$. This completes the proof.

11.1.1 Some Properties of the Complex CGPTs

We define the complex CGPT matrices by $\mathbb{N}^{(1)} := (\mathbb{N}_{mn}^{(1)})_{m,n}$ and $\mathbb{N}^{(2)} := (\mathbb{N}_{mn}^{(2)})_{m,n}$. We set $w = se^{i\theta}$ and introduce the diagonal matrix G^w with the m-th diagonal entry given by $w^m = s^m e^{im\theta}$. Proposition 11.1 implies immediately that

$$\mathbb{N}^{(1)}(T_z s R_\theta D) = C^z G^w \mathbb{N}^{(1)}(D) G^w (C^z)^T , \tag{11.10}$$

$$\mathbb{N}^{(2)}(T_z s R_\theta D) = \overline{C^z G^w} \mathbb{N}^{(2)}(D) G^w (C^z)^T , \tag{11.11}$$

where C^z is defined by (11.4). Relations (11.10) and (11.11) still hold for the truncated CGPTs of finite order, due to the triangular shape of the matrix C^z.

Using the symmetry of the CGPTs [31, Theorem 4.11] and the positivity of the GPTs as proved in [31], we easily establish the following result.

Proposition 11.2. *The complex CGPT matrix* $\mathbb{N}^{(1)}$ *is symmetric:* $(\mathbb{N}^{(1)})^T = \mathbb{N}^{(1)}$, *and* $\mathbb{N}^{(2)}$ *is Hermitian:* $(\overline{\mathbb{N}^{(2)}})^T = \mathbb{N}^{(2)}$. *Consequently, the diagonal elements of* $\mathbb{N}^{(2)}$ *are strictly positive if* $\lambda > 0$ *and strictly negative if* $\lambda < 0$.

Furthermore, the CGPTs of rotation invariant shapes have special structures:

Proposition 11.3. *Suppose that* D *is invariant under rotation of angle* $2\pi/p$ *for some integer* $p \geq 2$, *i.e.,* $R_{2\pi/p}D = D$, *then*

$$\mathbb{N}^{(1)}_{mn}(D) = 0 \text{ if } p \text{ does not divide } (m+n), \tag{11.12}$$

$$\mathbb{N}^{(2)}_{mn}(D) = 0 \text{ if } p \text{ does not divide } (m-n). \tag{11.13}$$

Proof. Suppose that p does not divide $(m+n)$, and define $r := 2\pi(n+m)/p \mod 2\pi$. Then by the rotation symmetry of D and the symmetry property of the CGPTs, we have

$$\mathbb{N}^{(1)}_{mn}(D) = \mathbb{N}^{(1)}_{mn}(R_{2\pi/p}D) = e^{i(m+n)2\pi/p}\mathbb{N}^{(1)}_{mn}(D) = e^{ir}\mathbb{N}^{(1)}_{mn}(D).$$

Since $r < 2\pi$ and $r \neq 0$, we conclude that $\mathbb{N}^{(1)}_{mn}(D) = 0$. The proof of (11.13) is similar.

11.2   Shape Identification by the CGPTs

We call a *dictionary* $\mathcal{D}$ a collection of standard shapes, which are centered at the origin and with characteristic sizes of order 1. Given the CGPTs of an unknown shape D, and assuming that D is obtained from a certain element $B \in \mathcal{D}$ by applying some unknown rotation θ, scaling s and translation z, i.e., $D = T_z s R_\theta B$, our objective is to recognize B from $\mathcal{D}$. For doing so, one may proceed by first reconstructing the shape D using its CGPTs through some optimization procedures as proposed in [37], and then match the reconstructed shape with $\mathcal{D}$. However, such a method may be time-consuming and the recognition efficiency depends on the shape reconstruction algorithm.

We propose in Sects. 11.2.1 and 11.2.2 two shape identification algorithms using the CGPTs. The first one matches the CGPTs of data with that of the dictionary element by estimating the transform parameters, while the second one is based on a transform invariant shape descriptor obtained from the CGPTs. The second approach is computationally more efficient. Both of them operate directly in the data domain which consists of CGPTs and avoid the need for reconstructing the shape D. The heart of our approach is some basic algebraic equations between the CGPTs of D and B that can be deduced easily from (11.10) and (11.11). Particularly, the first four equations read:

$$N_{11}^{(1)}(D) = w^2 N_{11}^{(1)}(B) \,, \tag{11.14}$$

$$N_{12}^{(1)}(D) = 2N_{11}^{(1)}(D)z + w^3 N_{12}^{(1)}(B) \,, \tag{11.15}$$

$$N_{11}^{(2)}(D) = s^2 N_{11}^{(2)}(B) \,, \tag{11.16}$$

$$N_{12}^{(2)}(D) = 2N_{11}^{(2)}(D)z + s^2 w N_{12}^{(2)}(B) \,, \tag{11.17}$$

where $w = se^{i\theta}$.

11.2.1 CGPTs Matching

Determination of Transform Parameters

Suppose that the complex CGPT matrices $N^{(1)}(B), N^{(2)}(B)$ of the true shape B are given. Then, from (11.16), we obtain that

$$s = \sqrt{N_{11}^{(2)}(D)/N_{11}^{(2)}(B)} \,. \tag{11.18}$$

Case 1: Rotational Symmetric Shape.

If the shape B has rotational symmetry, i.e., $R_{2\pi/p}B = B$ for some $p \geq 2$, then from Proposition 11.3 we have $N_{12}^{(2)}(B) = 0$ and the translation parameter z is uniquely determined from (11.17) by

$$z = \frac{N_{12}^{(2)}(D)}{2N_{11}^{(2)}(D)} \,. \tag{11.19}$$

On the contrary, the rotation parameter θ (or $e^{i\theta}$) can only be determined up to a multiple of $2\pi/p$, from CGPTs of order $\lceil p/2 \rceil$ at least. Although explicit expressions of $e^{ip\theta}$ can be deduced from (11.14)–(11.17) (or higher-order equations if necessary), we propose to recover $e^{ip\theta}$ by solving the least-squares problem:

$$\min_{\theta} \left(\|N^{(1)}(T_z s R_\theta B) - N^{(1)}(D)\|_F^2 + \|N^{(2)}(T_z s R_\theta B) - N^{(2)}(D)\|_F^2 \right) \,. \tag{11.20}$$

Here, s and z are given by (11.18) and (11.19) respectively, and $\mathbb{N}^{(1)}(D)$ and $\mathbb{N}^{(2)}(D)$ are the truncated complex CGPTs matrices of dimension $\lceil p/2 \rceil \times \lceil p/2 \rceil$.

Case 2: Non Rotational Symmetric Shape.

Consider a non rotational symmetric shape B which satisfies the assumption:

$$\mathbb{N}^{(1)}_{11}(B) \neq 0 \quad \text{and} \quad \det \begin{pmatrix} \mathbb{N}^{(1)}_{11}(B) & \mathbb{N}^{(2)}_{11}(B) \\ \mathbb{N}^{(1)}_{12}(B) & \mathbb{N}^{(2)}_{12}(B) \end{pmatrix} \neq 0 \,. \tag{11.21}$$

From (11.15) and (11.17), it follows that we can uniquely determine the translation z and the rotation parameter $w = e^{i\theta}$ from CGPTs of orders one and two by solving the following linear system:

$$\mathbb{N}^{(1)}_{12}(D)/\mathbb{N}^{(1)}_{11}(D) = 2z + w\mathbb{N}^{(1)}_{12}(B)/\mathbb{N}^{(1)}_{11}(B) \,,$$

$$\mathbb{N}^{(2)}_{12}(D)/\mathbb{N}^{(2)}_{11}(D) = 2z + w\mathbb{N}^{(2)}_{12}(B)/\mathbb{N}^{(2)}_{11}(B) \,. \tag{11.22}$$

Debiasing by Least-Squares Solutions

In practice (for both the rotational symmetric and non rotational symmetric cases), the values of the parameters z, s and θ provided by the analytical formulas and numerical procedures above may be inexact, due to the noise in the data and the ill-conditioned character of the linear system (11.22). Let z^*, s^*, θ^* be the true transform parameters, which can be considered as perturbations around the estimations z, s, θ obtained above:

$$z^* = z + \delta_z, \quad s^* = s\delta_s, \quad \text{and } \theta^* = \theta + \delta_\theta \,, \tag{11.23}$$

for δ_z, δ_θ small and δ_s close to 1. To find these perturbations, we solve a nonlinear least-squares problem:

$$\min_{z', s', \theta'} \left(\|\mathbb{N}^{(1)}(T_{z'}s'R_{\theta'}B) - \mathbb{N}^{(1)}(D)\|^2_F + \|\mathbb{N}^{(2)}(T_{z'}s'R_{\theta'}B) - \mathbb{N}^{(2)}(D)\|^2_F \right) \,, \tag{11.24}$$

with (z, s, θ) as an initial guess. Here, the order of the CGPTs in (11.24) is taken to be 2 in the non rotational case and $\max(2, \lceil p/2 \rceil)$ in the rotational symmetric case. Thanks to the relations (11.10) and (11.11), one can calculate explicitly the derivatives of the objective function, therefore can solve (11.24) by means of standard gradient-based optimization methods.

Algorithm 11.1 Shape identification based on CGPT matching

Input: the first k-th order CGPTs $\mathbb{N}^{(1)}(D), \mathbb{N}^{(2)}(D)$ of an unknown shape D

for $B_n \in \mathcal{D}$ **do**

    1. Estimate z, s, and θ using the procedures described in Sect. 11.2.1;

    2. $\tilde{D} \leftarrow R_{-\theta} s^{-1} T_{-z} D$, and calculate $\mathbb{N}^{(1)}(\tilde{D})$ and $\mathbb{N}^{(2)}(\tilde{D})$;

    3. $E^{(1)} \leftarrow \mathbb{N}^{(1)}(B_n) - \mathbb{N}^{(1)}(\tilde{D})$, and $E^{(2)} \leftarrow \mathbb{N}^{(2)}(B_n) - \mathbb{N}^{(2)}(\tilde{D})$;

    4. $e_n \leftarrow (\|E^{(1)}\|_F^2 + \|E^{(2)}\|_F^2)^{1/2} / (\|\mathbb{N}^{(1)}(B_n)\|_F^2 + \|\mathbb{N}^{(2)}(B_n)\|_F^2)^{1/2}$;

    5. $n \leftarrow n + 1$;

end for

Output: the true dictionary element $n^* \leftarrow \operatorname{argmin}_n e_n$.

First Algorithm for Shape Identification

For each dictionary element, we determine the transform parameters as above, then measure the similarity of the complex CGPT matrices using the Frobenius norm, and choose the most similar element as the identified shape. Intuitively, the true dictionary element will give the correct transform parameters and hence the most similar CGPTs. This procedure is described in Algorithm 11.1.

11.2.2 Transform Invariant Shape Descriptors

From (11.16) and (11.17) we deduce the following identity:

$$\frac{\mathbb{N}_{12}^{(2)}(D)}{2\mathbb{N}_{11}^{(2)}(D)} = z + se^{i\theta} \frac{\mathbb{N}_{12}^{(2)}(B)}{2\mathbb{N}_{11}^{(2)}(B)} , \qquad (11.25)$$

which is well defined since $\mathbb{N}_{11}^{(2)} \neq 0$ thanks to the Proposition 11.2. Identity (11.25) shows a very simple relationship between $\frac{\mathbb{N}_{12}^{(2)}(B)}{2\mathbb{N}_{11}^{(2)}(B)}$ and $\frac{\mathbb{N}_{12}^{(2)}(D)}{2\mathbb{N}_{11}^{(2)}(D)}$ for $D = T_z s R_\theta B$.

Let $u = \frac{\mathbb{N}_{12}^{(2)}(D)}{2\mathbb{N}_{11}^{(2)}(D)}$. We first define the following quantities which are translation invariant:

$$\mathcal{J}^{(1)}(D) = \mathbb{N}^{(1)}(T_{-u}D) = C^{-u}\mathbb{N}^{(1)}(D)(C^{-u})^T , \qquad (11.26)$$

$$\mathcal{J}^{(2)}(D) = \mathbb{N}^{(2)}(T_{-u}D) = \overline{C^{-u}}\mathbb{N}^{(2)}(D)(C^{-u})^T , \qquad (11.27)$$

with the matrix C^{-u} being the same as in Proposition 11.1. From $\mathcal{J}^{(1)}(D) = (\mathcal{J}_{mm}^{(1)}(D))_{m,n}$ and $\mathcal{J}^{(2)}(D) = (\mathcal{J}_{mm}^{(2)}(D))_{m,n}$, we define, for any indices m, n, the scaling invariant quantities:

Algorithm 11.2 Shape identification based on transform invariant descriptors

---

Input: the first k-th order shape descriptors $\mathcal{I}^{(1)}(D), \mathcal{I}^{(2)}(D)$ of an unknown shape D
for $B_n \in \mathcal{D}$ **do**
    1. $e_n \leftarrow \left(\|\mathcal{I}^{(1)}(B_n) - \mathcal{I}^{(1)}(D)\|_F^2 + \|\mathcal{I}^{(2)}(B_n) - \mathcal{I}^{(2)}(D)\|_F^2 \right)^{1/2}$;
    2. $n \leftarrow n + 1$;
end for
Output: the true dictionary element $n^* \leftarrow \operatorname{argmin}_n e_n$.

$$
\mathcal{S}_{mn}^{(1)}(D) = \frac{\mathcal{J}_{mn}^{(1)}(D)}{\left(\mathcal{J}_{mm}^{(2)}(D) \mathcal{J}_{nn}^{(2)}(D) \right)^{1/2}} \ , \quad \mathcal{S}_{mn}^{(2)}(D) = \frac{\mathcal{J}_{mn}^{(2)}(D)}{\left(\mathcal{J}_{mm}^{(2)}(D) \mathcal{J}_{nn}^{(2)}(D) \right)^{1/2}} \ .
$$

$$(11.28)$$

Finally, we introduce the CGPT-based shape descriptors $\mathcal{I}^{(1)} = (\mathcal{I}_{mn}^{(1)})_{m,n}$ and $\mathcal{I}^{(2)} = (\mathcal{I}_{mn}^{(2)})_{m,n}$:

$$
\mathcal{I}_{mn}^{(1)}(D) = |\mathcal{S}_{mn}^{(1)}(D)| \quad \text{and} \quad \mathcal{I}_{mn}^{(2)}(D) = |\mathcal{S}_{mn}^{(2)}(D)| \ , \tag{11.29}
$$

where $|\cdot|$ denotes the modulus of a complex number. Constructed in this way, $\mathcal{I}^{(1)}$ and $\mathcal{I}^{(2)}$ are clearly invariant under translation, rotation, and scaling.

It is worth emphasizing the symmetry property, $\mathcal{I}_{mn}^{(1)} = \mathcal{I}_{nm}^{(1)}, \mathcal{I}_{mn}^{(2)} = \mathcal{I}_{nm}^{(2)}$, and the fact that $\mathcal{I}_{mm}^{(2)} = 1$ for any m.

Second Algorithm for Shape Identification

Thanks to the transform invariance of the new shape descriptors, there is no need now for calculating the transform parameters, and the similarity between a dictionary element and the unknown shape can be directly measured from $\mathcal{I}^{(1)}$ and $\mathcal{I}^{(2)}$. As in Algorithm 11.1, we use the Frobenius norm as the distance between two shape descriptors and compare with all the elements of the dictionary. We propose a simplified method for shape identification, as described in Algorithm 11.2.

11.3   Target Tracking

In this section we apply an Extended Kalman Filter to track both the location and the orientation of a mobile target from multistatic response measurements. As shown in Sect. 1.10.2, the Extended Kalman Filter (EKF) is a generalization of the Kalman Filter (KF) to nonlinear dynamical systems.

It is robust with respect to noise and computationally inexpensive, therefore is well suited for real-time applications such as tracking. One should have in mind that, in real applications, one would like to localize the target and reconstruct its orientation directly from the MSR data without reconstructing the GPTs.

11.3.1 Location and Orientation Tracking of a Mobile Target

We denote by $z_t = (x_t, y_t)^T \in \mathbb{R}^2$ the position and $\theta_t \in [0, 2\pi)$ the orientation of a target D_t at the instant t, such that the shape of the target D_t is given by:

$$D_t = z_t + R_{\theta_t} B , \tag{11.30}$$

where R_{θ_t} is the rotation by θ_t. We assume that the CGPTs of B have been reconstructed and the shape has been correctly identified from a dictionary, so that the CGPT matrix $\mathbb{M} := \mathbb{M}(B)$ of order $K \geq 2$ is available. We use the same notation as in the previous chapter. Then we have the MSR matrix:

$$A_t = L(\mathbb{M}_t) + E_t + W_t , \tag{11.31}$$

where $\mathbb{M}_t$ is the CGPT of D_t, E_t is the truncation error, and W_t the measurement noise of time t. In the case of circular configuration with coincident arrays of sources and receivers, the linear operator L takes the form:

$$L(\mathbb{M}_t) = CD\mathbb{M}_t DC^T . \tag{11.32}$$

The objective of *tracking* is to estimate the target location z_t and orientation θ_t from the MSR data stream A_t. Before developing a CGPT-based tracking algorithm, we establish a simple relation between $\mathbb{M}_t$ and $\mathbb{M}$.

Time Relationship Between CGPTs

Let $u = (1, i)^T$. The complex CGPT $\mathbb{N}^{(1)}, \mathbb{N}^{(2)}$ are defined by

$$\mathbb{N}^{(1)}_{mn} = (M^{cc}_{mn} - M^{ss}_{mn}) + i(M^{cs}_{mn} + M^{sc}_{mn}) = u^T M_{mn} u ,$$

$$\mathbb{N}^{(2)}_{mn} = (M^{cc}_{mn} + M^{ss}_{mn}) + i(M^{cs}_{mn} - M^{sc}_{mn}) = u^* M_{mn} u .$$

Therefore, we have

$$\mathbb{N}^{(1)} = U^T \mathbb{M} U, \quad \text{and} \quad \mathbb{N}^{(2)} = U^* \mathbb{M} U , \tag{11.33}$$

where the matrix U of dimension $2K \times K$ is defined by

$$U = \begin{pmatrix} u & 0 & \dots & 0 \\ 0 & u & \dots & 0 \\ \vdots & & \ddots & \vdots \\ 0 & \dots & 0 & u \end{pmatrix} . \tag{11.34}$$

To recover the CGPT $\mathbb{M}_{mn}$ from the complex CGPTs $\mathbb{N}^{(1)}, \mathbb{N}^{(2)}$, we simply use the relations

$$
\begin{aligned}
M^{cc}_{mn} &= \frac{1}{2}\Re e(\mathbb{N}^{(1)}_{mn} + \mathbb{N}^{(2)}_{mn}), \quad M^{cs}_{mn} = \frac{1}{2}\Im m(\mathbb{N}^{(1)}_{mn} + \mathbb{N}^{(2)}_{mn}) , \\
M^{sc}_{mn} &= \frac{1}{2}\Im m(\mathbb{N}^{(1)}_{mn} - \mathbb{N}^{(2)}_{mn}), \quad M^{ss}_{mn} = \frac{1}{2}\Re e(\mathbb{N}^{(2)}_{mn} - \mathbb{N}^{(1)}_{mn}) .
\end{aligned}
\tag{11.35}
$$

For two targets D_t and B satisfying (11.30), the following relationships between their complex CGPTs hold:

$$\mathbb{N}^{(1)}(D_t) = F_t^T \mathbb{N}^{(1)}(B) F_t , \tag{11.36a}$$

$$\mathbb{N}^{(2)}(D_t) = F_t^* \mathbb{N}^{(2)}(B) F_t , \tag{11.36b}$$

where F_t is a upper triangle matrix with the (m, n)-th entry given by

$$(F_t)_{mn} = \binom{n}{m}(x_t + iy_t)^{n-m} e^{im\theta_t} . \tag{11.37}$$

Linear Operator T_t

Now one can find explicitly a linear operator T_t which depends only on z_t, θ_t, such that $\mathbb{M}_t = T_t(\mathbb{M})$, and the equation (11.31) becomes:

$$A_t = L(T_t(\mathbb{M})) + E_t + W_t . \tag{11.38}$$

For doing so, we set $J_t := U F_t$, where U is given by (11.34). Then, a straightforward computation using (11.33), (11.35), and (11.36) shows that

$$
\begin{aligned}
M^{cc}(D_t) &= \Re e J_t^T \mathbb{M} \Re e J_t, \quad M^{cs}(D_t) = \Re e J_t^T \mathbb{M} \Im m J_t , \\
M^{sc}(D_t) &= \Im m J_t^T \mathbb{M} \Re e J_t, \quad M^{ss}(D_t) = \Im m J_t^T \mathbb{M} \Im m J_t ,
\end{aligned}
\tag{11.39}
$$

where $M^{cc}(D_t), M^{cs}(D_t), M^{sc}(D_t), M^{ss}(D_t)$ are the CGPTs. By interlacing all these four terms we get the operator T_t:

$$T_t(\mathbb{M}) = \Re eU(\Re eJ_t^T\mathbb{M}\Re eJ_t)\Re eU^T + \Re eU(\Re eJ_t^T\mathbb{M}\Im mJ_t)\Im mU^T +$$

$$\Im mU(\Im mJ_t^T\mathbb{M}\Re eJ_t)\Re eU^T + \Im mU(\Im mJ_t^T\mathbb{M}\Im mJ_t)\Im mU^T = \mathbb{M}_t . \tag{11.40}$$

Tracking by CGPTs

A naive way to track the location z_t and the orientation θ_t is as follows. At each time t we first reconstruct $\mathbb{M}_t$ to get the complex CGPTs

$$\mathbb{N}_{1,1}^{(1)}(D_t), \mathbb{N}_{1,2}^{(1)}(D_t), \mathbb{N}_{1,1}^{(2)}(D_t), \mathbb{N}_{1,2}^{(2)}(D_t) .$$

Then we find the relative movement $\Delta z_t = z_t - z_{t-1}$ and $\Delta\theta_t = \theta_t - \theta_{t-1}$ by solving a linear system:

$$\mathbb{N}_{12}^{(1)}(D_t)/\mathbb{N}_{11}^{(1)}(D_t) = 2(\Re\Delta z_t + i\Im\Delta z_t) + e^{i\Delta\theta_t}\mathbb{N}_{12}^{(1)}(D_{t-1})/\mathbb{N}_{11}^{(1)}(D_{t-1}),$$

$$\mathbb{N}_{12}^{(2)}(D_t)/\mathbb{N}_{11}^{(2)}(D_t) = 2(\Re\Delta z_t + i\Im\Delta z_t) + e^{i\Delta\theta_t}\mathbb{N}_{12}^{(2)}(D_{t-1})/\mathbb{N}_{11}^{(2)}(D_{t-1}) . \tag{11.41}$$

The estimated path is then $z_t = \sum_{s=1}^t \Delta z_s + z_0$, and $\theta_t = \sum_{s=1}^t \Delta\theta_s + \theta_0$. However, such an algorithm has no practical interest. In fact, the error in the estimated path (z_t, θ_t) will propagate over time, since the noise presented in data is not properly taken into account here. In the following we apply the Extended Kalman Filter to the system (11.38) which takes advantage of the operator T_t and handles correctly the noise.

11.3.2   Tracking by the Extended Kalman Filter

In the next we establish first the *system state* and the *observation* equations, then linearize the observation equation and apply the EKF algorithm.

System State Observation Equations

We assume that the position of the target is subject to an external driving force that has the form of a white noise. In other words the velocity $(V(\tau))_{\tau\in\mathbb{R}^+}$ of the target is given in terms of a two-dimensional Brownian motion $(W_a(\tau))_{\tau \in \mathbb{R}^+}$ and its position $(Z(\tau))_{\tau \in \mathbb{R}^+}$ is given in terms of the integral of this Brownian motion:

$$V(\tau) = V_0 + \sigma_a W_a(\tau), \qquad Z(\tau) = Z_0 + \int_0^\tau V(s)ds .$$

The orientation $(\Theta(\tau))_{\tau \in \mathbb{R}^+}$ of the target is subject to random fluctuations and its angular velocity is given in terms of an independent white noise, so that the orientation is given in terms of a one-dimensional Brownian motion $(W_\theta(\tau))_{\tau \in \mathbb{R}^+}$:

$$\Theta(\tau) = \Theta_0 + \sigma_\theta W_\theta(\tau) .$$

We observe the target at discrete times $t\Delta\tau$, $t \in \mathbb{N}$, with time step $\Delta\tau$. We denote $z_t = Z(t\Delta\tau)$, $v_t = V(t\Delta\tau)$, and $\theta_t = \Theta(t\Delta\tau)$. These functions obey the recursive relations

$$
\begin{aligned}
v_t &= v_{t-1} + a_t, & a_t &= \sigma_a\big(W_a(t\Delta\tau) - W_a((t-1)\Delta\tau)\big) , \\
z_t &= z_{t-1} + v_{t-1}\Delta\tau + b_t , & b_t &= \sigma_a \int_{(t-1)\Delta\tau}^{t\Delta\tau} W_a(s) - W_a((t-1)\Delta\tau)ds , \\
\theta_t &= \theta_{t-1} + c_t, & c_t &= \sigma_\theta\big(W_\theta(t\Delta\tau) - W_\theta((t-1)\Delta\tau)\big) .
\end{aligned}
$$

$$(11.42)$$

Since the increments of the Brownian motions are independent from each other, the vectors $(U_t)_{t\geq 1}$ given by

$$U_t = \begin{pmatrix} a_t \\ b_t \\ c_t \end{pmatrix}$$

are independent and identically distributed with the multivariate normal distribution with mean zero and covariance matrix Σ given by

$$\Sigma = \Delta\tau \begin{pmatrix} \sigma_a^2 I_2 & \frac{\sigma_a^2}{2}\Delta\tau I_2 & 0 \\ \frac{\sigma_a^2}{2}\Delta\tau I_2 & \frac{\sigma_a^2}{3}\Delta\tau^2 I_2 & 0 \\ 0 & 0 & \sigma_\theta^2 \end{pmatrix} . \qquad (11.43)$$

The evolution of the state vector

$$X_t = \begin{pmatrix} v_t \\ z_t \\ \theta_t \end{pmatrix}$$

takes the form

$$X_t = FX_{t-1} + U_t, \qquad F = \begin{pmatrix} I_2 & 0 & 0 \\ \Delta\tau I_2 & I_2 & 0 \\ 0 & 0 & 1 \end{pmatrix} . \tag{11.44}$$

The observation made at time t is the MSR matrix given by (11.38), where the system state X_t is implicitly included in the operator T_t. For the sake of simplicity, we suppose that the truncation error E_t is small compared to the measurement noise so that it can be dropped in (11.38), and that the Gaussian white noise W_t of different time are mutually independent. We emphasize that the velocity vector v_t of the target does not contribute to (11.38), which can be seen from (11.30). To highlight the dependence on z_t, θ_t, we introduce a function h which is nonlinear in z_t, θ_t, and takes $\mathbb{M}$ as a parameter, such that

$$h(X_t; \mathbb{M}) = h(z_t, \theta_t; \mathbb{M}) = L(T_t(\mathbb{M})) . \tag{11.45}$$

Then together with (11.44) we get the following *system state* and *observation* equations:

$$X_t = FX_{t-1} + U_t , \tag{11.46a}$$

$$V_t = h(X_t; \mathbb{M}) + W_t . \tag{11.46b}$$

Note that (11.46a) is linear, so in order to apply EKF on (11.46), we only need to linearize (11.46b), or in other words, to calculate the partial derivatives of h with respect to x_t, y_t, θ_t.

Linearization of the Observation Equation

Clearly, the operator L contains only the information concerning the acquisition system and does not depend on x_t, y_t, θ_t. So, by (11.45), we have

$$\partial_{x_t} h = L(\partial_{x_t} T_t(\mathbb{M})) . \tag{11.47}$$

Moreover, the calculation for $\partial_{x_t} T_t$ is straightforward using (11.40). We have

$$\partial_{x_t} T_t(\mathbb{M}) = \Re e U \partial_{x_t} (\Re e J_t^T \mathbb{M} \Re e J_t) \Re e U^T + \Re e U \partial_{x_t} (\Re e J_t^T \mathbb{M} \Im m J_t) \Im m U^T +$$
$$\Im m U \partial_{x_t} (\Im m J_t^T \mathbb{M} \Re e J_t) \Re e U^T + \Im m U \partial_{x_t} (\Im m J_t^T \mathbb{M} \Im m J_t) \Im m U^T , \tag{11.48}$$

where the derivatives are found by the product rule:

$$\partial_{x_t}(\Re eJ_t^T \mathbb{M}\Re eJ_t) = \Re e(\partial_{x_t} J_t^T)\mathbb{M}\Re eJ_t + \Re eJ_t^T \mathbb{M}\Re e(\partial_{x_t} J_t) ,$$

$$\partial_{x_t}(\Re eJ_t^T \mathbb{M}\Im mJ_t) = \Re e(\partial_{x_t} J_t^T)\mathbb{M}\Im mJ_t + \Re eJ_t^T \mathbb{M}\Im m(\partial_{x_t} J_t) ,$$

$$\partial_{x_t}(\Im mJ_t^T \mathbb{M}\Re eJ_t) = \Im m(\partial_{x_t} J_t^T)\mathbb{M}\Re eJ_t + \Im mJ_t^T \mathbb{M}\Re e(\partial_{x_t} J_t) ,$$

$$\partial_{x_t}(\Im mJ_t^T \mathbb{M}\Im mJ_t) = \Im m(\partial_{x_t} J_t^T)\mathbb{M}\Im mJ_t + \Im mJ_t^T \mathbb{M}\Im m(\partial_{x_t} J_t) ,$$

and $\partial_{x_t} J_t = U\partial_{x_t} F_t$. The (m, n)-th entry of the matrix $\partial_{x_t} F_t$ is given by

$$(\partial_{x_t} F_t)_{m,n} = \binom{n}{m}(n - m)z_t^{n-m-1}e^{im\theta_t} . \tag{11.49}$$

The derivatives $\partial_{y_t} T_t(\mathbb{M})$ and $\partial_{\theta_t} T_t(\mathbb{M})$ are calculated in the same way.

Bibliography and Discussion

The results of this chapter on target identification are from [8]. They provide an efficient approach for real-time target identification using dictionary matching. They show that GPT-based representations are appropriate and natural tools for multistatic imaging. They can be generalized to electromagnetic wave propagation as well. As shown in this chapter, they can be used for tracking a mobile target from multistatic data. The results of this chapter on the location and orientation tracking are from [9]. An analysis of the ill-posed character of both the location and orientation tracking in the case of limited-view data was carried out in [9]. In [11], transformation formulas for the GPTs under rigid motions and scaling in three dimensions are given. Moreover, invariants under those transformations, which can be used as shape descriptors for dictionary matching in three dimensions, are constructed.

In [41] a shape identification and classification algorithm in echolocation is proposed. The approach is based on first extracting scattering coefficients from the reflected waves and then matching them with precomputed ones associated with a dictionary of targets. The construction of such frequency-dependent shape descriptors is based on the properties of the scattering coefficients described in Chap. 5 and some new invariants.

Part VI
Imaging of Extended Targets

Chapter 12
Time-Reversal and Diffraction Tomography for Inverse Source Problems

The aim of this chapter is to provide classical techniques for solving inverse extended source problems. From time-domain or broadband measurements, time-reversal techniques yield direct reconstruction of the source. In the frequency domain, from measurements at a single frequency or bandlimited measurements, diffraction tomography can be used to reconstruct within the Born approximation a low-pass version of the electromagnetic target.

The main idea of time-reversal is to take advantage of the reversibility of the wave equation in order to back-propagate signals to the sources that emitted them. In the context of inverse source problems, one measures the emitted wave on a closed surface surrounding the source, and retransmits it through the background medium in a time-reversed chronology. Then the perturbation will travel back to the location of the source.

In diffraction tomography, we compute the Fourier transform of the reflectivity function of the weakly scattering target from the Fourier transform of the measured scattered data. The computation is based on the Weyl representation (2.63) of cylindrical and spherical waves.

12.1 Time-Reversal Techniques

Consider the wave equation in the free space $\mathbb{R}^d$, $d = 2$ or 3,

$$\begin{cases} \dfrac{\partial^2 u}{\partial t^2}(x,t) - \Delta u(x,t) = \dfrac{d\delta_0}{dt}(t)f(x), & (x,t) \in \mathbb{R}^d \times \mathbb{R}, \\[2mm] u(x,t) = 0 \quad \text{and} \quad \dfrac{\partial u(x,t)}{\partial t} = 0, & t < 0, \end{cases} \tag{12.1}$$

where δ_0 is the Dirac mass at 0 and the source f is real-valued, smooth and has a smooth compact support.

H. Ammari et al., *Mathematical and Statistical Methods for Multistatic Imaging*, Lecture Notes in Mathematics 2098, DOI 10.1007/978-3-319-02585-8_12, © Springer International Publishing Switzerland 2013

Let Ω be a smooth bounded domain in $\mathbb{R}^d$ containing the support of f. Let $g(y, t)$ be defined as $g(y, t) := u(y, t)$ for all $y \in \partial\Omega$ and $t \in [0, T]$, where, in three dimensions, T is supposed to be sufficiently large such that

$$u(x, t) = \frac{\partial u(x, t)}{\partial t} = 0$$

for $t \geq T$ and $x \in \Omega$. It is easy to see that g is smooth.

Our aim in this section is to reconstruct an approximation of the source f from g on $\partial\Omega \times [0, T]$.

We introduce the time-dependent Green function

$$\Gamma(x, y, s, t) = \frac{1}{2\pi} \int_{\mathbb{R}} \Gamma_\omega(x, y) \exp(-i\omega(t - s)) d\omega , \qquad (12.2)$$

where Γ_ω is the outgoing fundamental solution to the Helmholtz equation $(\Delta + \omega^2)$ in $\mathbb{R}^d$:

$$(\Delta + \omega^2)\Gamma_\omega(x, y) = \delta_x(y) \quad \text{in } \mathbb{R}^d ,$$

subject to the outgoing radiation condition. Here δ_x is the Dirac mass at x. It is the solution of the free space wave equation

$$\begin{cases} \dfrac{\partial^2 \Gamma}{\partial t^2}(x, y, s, t) - \Delta_y \Gamma(x, y, s, t) = -\delta_x(y)\delta_s(t), & (y, t) \in \mathbb{R}^d \times \mathbb{R} , \\ \Gamma(x, y, s, t) = \dfrac{\partial \Gamma}{\partial t}(x, y, s, t) = 0, & y \in \mathbb{R}^d, \quad t < s , \end{cases}$$

where δ_x and δ_s are the Dirac masses at x and at s. Note that $U_y(x, t)$ introduced in (2.105) is nothing else than $\Gamma(x, y, 0, t)$. Moreover,

$$g(y, t) = -\int_\Omega \frac{\partial \Gamma}{\partial t}(z, y, 0, t)f(z) \, dz, \quad y \in \partial\Omega, \quad t \in [0, T] . \qquad (12.3)$$

12.1.1 Ideal Time-Reversal Imaging Technique

We introduce the solution v of the following wave problem

$$\begin{cases} \dfrac{\partial^2 v}{\partial t^2}(x, t) - \Delta_x v(x, t) = 0, & (x, t) \in \Omega \times [0, T] , \\ v(x, 0) = \dfrac{\partial v}{\partial t}(x, 0) = 0, & x \in \Omega , \\ v(x, t) = \dfrac{1}{2} g(x, T - t), & (x, t) \in \partial\Omega \times [0, T] . \end{cases}$$

The time-reversal imaging functional $\mathcal{I}_{\mathrm{TR}}^{(1)}(x)$ is defined by

$$\mathcal{I}_{\mathrm{TR}}^{(1)}(x) = v(x, T), \quad x \in \Omega .$$

In order to make $\mathcal{I}_{\mathrm{TR}}^{(1)}(x)$ explicit, we introduce the causal Dirichlet Green function (also called retarded Dirichlet Green function) $G^{\mathrm{c}}(x, y, s, t)$ defined as the solution of the following wave equation

$$\begin{cases} \dfrac{\partial^2 G^{\mathrm{c}}}{\partial t^2}(x, y, s, t) - \Delta_y G^{\mathrm{c}}(x, y, s, t) = -\delta_x(y)\delta_s(t), & (y, t) \in \Omega \times \mathbb{R}, \\[2mm] G^{\mathrm{c}}(x, y, s, t) = 0, \quad \dfrac{\partial G^{\mathrm{c}}}{\partial t}(x, y, s, t) = 0, & y \in \Omega, \quad t < s, \\[2mm] G^{\mathrm{c}}(x, y, s, t) = 0, & (y, t) \in \partial\Omega \times \mathbb{R}. \end{cases}$$

We also introduce the anticausal Dirichlet Green function (also called advanced Dirichlet Green function) $G^{\mathrm{a}}(x, y, s, t)$ defined as the solution of the following wave equation

$$\begin{cases} \dfrac{\partial^2 G^{\mathrm{a}}}{\partial t^2}(x, y, s, t) - \Delta_y G^{\mathrm{a}}(x, y, s, t) = -\delta_x(y)\delta_s(t), & (y, t) \in \Omega \times \mathbb{R}, \\[2mm] G^{\mathrm{a}}(x, y, s, t) = 0, \quad \dfrac{\partial G^{\mathrm{a}}}{\partial t}(x, y, s, t) = 0, & y \in \Omega, \quad t > s, \\[2mm] G^{\mathrm{a}}(x, y, s, t) = 0, & (y, t) \in \partial\Omega \times \mathbb{R}. \end{cases}$$

Using the time-reversibility of the wave equation, the anticausal Green function is given in terms of the causal Green function by

$$G^{\mathrm{a}}(x, y, s, t) = G^{\mathrm{c}}(x, y, s, 2s - t) = G^{\mathrm{c}}(x, y, t, s). \tag{12.4}$$

Finally we introduce the symmetric Dirichlet Green function:

$$G(x, y, s, t) = \frac{1}{2}\Big(G^{\mathrm{a}}(x, y, s, t)\chi(]-\infty, s[)(t) + G^{\mathrm{c}}(x, y, s, t)\chi(]s, \infty[)(t) \Big) ,$$

where $\chi(]-\infty, s[)$ and $\chi(]s, \infty[)$ are the characteristic functions of the intervals $]-\infty, s[$ and $]s, \infty[$, respectively. The function G contains both the causal and anticausal Green functions and it is a solution of

$$\begin{cases} \dfrac{\partial^2 G}{\partial t^2}(x, y, s, t) - \Delta_y G(x, y, s, t) = -\delta_x(y)\delta_s(t), & (y, t) \in \Omega \times \mathbb{R}, \\[2mm] G(x, y, s, t) = 0, & (y, t) \in \partial\Omega \times \mathbb{R}. \end{cases}$$

We can then express the time-reversal imaging functional $\mathcal{I}_{\mathrm{TR}}^{(1)}$ as

$$\mathcal{I}_{\mathrm{TR}}^{(1)}(x) = v(x,T) = \frac{1}{2}\int_0^T \int_{\partial\Omega} \frac{\partial G^{\mathrm{c}}(x,y,s,T)}{\partial\nu_y} g(y,T-s)d\sigma(y)ds \ ,$$

where $\partial/\partial\nu_y$ denotes the outward normal derivative at $y \in \partial\Omega$. The relation (12.4) yields

$$\mathcal{I}_{\mathrm{TR}}^{(1)}(x) = \frac{1}{2}\int_0^T \int_{\partial\Omega} \frac{\partial G^{\mathrm{a}}(x,y,T,s)}{\partial\nu_y} g(y,T-s)d\sigma(y)ds$$

$$= \int_0^T \int_{\partial\Omega} \frac{\partial G(x,y,T,s)}{\partial\nu_y} g(y,T-s)d\sigma(y)ds$$

$$= \int_0^T \int_{\partial\Omega} \frac{\partial G(x,y,0,t)}{\partial\nu_y} g(y,t)d\sigma(y)dt \ ,$$

because G is even. Since $g(x,t) = 0$ for $t \geq T$ and for $t \leq 0$,

$$\mathcal{I}_{\mathrm{TR}}^{(1)}(x) = \int_{\mathbb{R}} \int_{\partial\Omega} \frac{\partial G(x,y,0,t)}{\partial\nu_y} g(y,t)d\sigma(y)dt \ . \tag{12.5}$$

In identity (12.5), the dependence of the time-reversal functional $\mathcal{I}_{\mathrm{TR}}^{(1)}$ on the boundary data g is explicitly shown. Moreover, it follows from (12.3) that

$$\mathcal{I}_{\mathrm{TR}}^{(1)}(x) = -\int_{\Omega} f(z) \int_{\mathbb{R}} \int_{\partial\Omega} \frac{\partial G(x,y,0,t)}{\partial\nu_y} \frac{\partial\Gamma}{\partial t}(z,y,0,t)d\sigma(y)dt\,dz \ . \tag{12.6}$$

The reason why we have chosen to express the functional $\mathcal{I}_{\mathrm{TR}}^{(1)}$ in terms of the symmetric Green function G rather than in terms of the causal Green function G^{c} will be clear later.

Now we prove that $\mathcal{I}_{\mathrm{TR}}^{(1)}(x)$ gives a perfect image of $f(x)$. For doing so, we denote

$$G_\omega(x,y) = \int_{\mathbb{R}} G(x,y,0,t)e^{i\omega t}\,dt \ .$$

Then from (12.2), (12.6), and Parseval's relation (1.1), we have

$$\mathcal{I}_{\mathrm{TR}}^{(1)}(x) = -\frac{i}{2\pi}\int_{\Omega} f(z) \int_{\mathbb{R}} \omega \int_{\partial\Omega} \frac{\partial G_\omega}{\partial\nu_y}(x,y)\overline{\Gamma_\omega}(z,y)d\sigma(y)\,d\omega\,dz \ .$$

Moreover, by integrating by parts over Ω we get

$$\int_{\partial\Omega} \frac{\partial G_\omega}{\partial\nu_y}(x,y)\overline{\Gamma_\omega}(z,y)d\sigma(y) = \overline{\Gamma_\omega}(z,x) - G_\omega(x,z) = \overline{\Gamma_\omega}(x,z) - G_\omega(x,z) \ ,$$

and recalling that G_ω is real-valued because $t \to G(x, y, 0, t)$ is real and even, we have

$$\Im m \int_{\partial\Omega} \frac{\partial G_\omega}{\partial \nu_y}(x, y)\overline{\Gamma_\omega}(z, y)d\sigma(y) = \Im m\left\{\overline{\Gamma_\omega}(x, z)\right\} = -\Im m\left\{\Gamma_\omega(x, z)\right\} .$$

Therefore,

$$\mathcal{I}_{\mathrm{TR}}^{(1)}(x) = -\frac{1}{2\pi}\int_\Omega f(z)\int_{\mathbb{R}}\omega\Im m\left\{\Gamma_\omega(x, z)\right\} d\omega \, dz . \tag{12.7}$$

Recall that

$$\frac{1}{\pi}\int_{\mathbb{R}}\omega\Im m\left\{\Gamma_\omega(x, z)\right\} d\omega = -\delta_z(x) , \tag{12.8}$$

which follows from (1.35) or equivalently, from using

$$\lim_{t\to 0^+}\partial\Gamma/\partial t(x, z, 0, 0) = -\delta_z(x)$$

and (12.2).

Finally, from (12.7) it follows that

$$2\mathcal{I}_{\mathrm{TR}}^{(1)}(x) = f(x) .$$

Note that the choice of the symmetric Dirichlet Green function is justified by the fact that its Fourier transform is real-valued. The ideal time-reversal imaging functional $\mathcal{I}_{\mathrm{TR}}^{(1)}$ yields a perfect image of f. However, we need to compute the symmetric Dirichlet Green function G associated with the domain Ω. In the general case, it may be difficult to find an explicit expression for G. In the next section we introduce a modified time-reversal imaging functional where G is replaced with the free space fundamental solution Γ and show that the modified functional yields a good approximation of the source term f.

12.1.2 A Modified Time-Reversal Imaging Technique

In this section, we present a modified approach to the time-reversal concept using "free boundary conditions". For $s \in [0, T]$ we introduce the function v_s defined as the solution to the wave problem

$$\begin{cases} \dfrac{\partial^2 v_s}{\partial t^2}(x,t) - \Delta_x v_s(x,t) = \dfrac{d\delta_s}{dt}(t)g(x,T-s)\delta_{\partial\Omega}(x), & (x,t) \in \mathbb{R}^d \times \mathbb{R}, \\[3mm] v_s(x,t) = 0, \quad \dfrac{\partial v_s}{\partial t}(x,t) = 0, & x \in \mathbb{R}^d, \quad t < s. \end{cases}$$

Here, $\delta_{\partial\Omega}$ is the surface Dirac measure on $\partial\Omega$ and g is the measured data. We define a modified time-reversal imaging functional by

$$\mathcal{I}_{\mathrm{TR}}^{(2)}(x) = \int_0^T v_s(x,T)\,ds, \quad x \in \Omega. \tag{12.9}$$

Note that

$$v_s(x,t) = -\int_{\partial\Omega} \frac{\partial\Gamma}{\partial t}(x,y,s,t)g(y,T-s)\,d\sigma(y).$$

Consequently, the functional $\mathcal{I}_{\mathrm{TR}}^{(2)}$ can be expressed in terms of the free-space fundamental solution Γ as follows:

$$\begin{aligned} \mathcal{I}_{\mathrm{TR}}^{(2)}(x) &= -\int_0^T \int_{\partial\Omega} \frac{\partial\Gamma}{\partial t}(x,y,s,T)g(y,T-s)\,d\sigma(y)\,ds, \\ &= -\int_{\mathbb{R}} \int_{\partial\Omega} \frac{\partial\Gamma}{\partial t}(x,y,0,t)g(y,t)\,d\sigma(y)\,dt \quad x \in \Omega. \end{aligned} \tag{12.10}$$

Note that $\mathcal{I}_{\mathrm{TR}}^{(2)}$ is not exactly equivalent to $\mathcal{I}_{\mathrm{TR}}^{(1)}$ but is an approximation. Indeed, denoting by

$$g_\omega(y) = \int g(y,s)e^{i\omega s}\,ds$$

the Fourier transform of $g(s,y)$, we have from (12.2) and (12.3)

$$g_\omega(y) = i\omega \int_\Omega \Gamma_\omega(z,y)f(z)\,dz.$$

Parseval's relation and (12.2) give

$$\begin{aligned} \mathcal{I}_{\mathrm{TR}}^{(2)}(x) &= -\int_0^T \int_{\partial\Omega} \frac{\partial\Gamma}{\partial t}(x,y,0,t)g(y,t)\,d\sigma(y)\,dt \\ &= \frac{1}{2\pi} \int_{\mathbb{R}} \int_{\partial\Omega} i\omega\Gamma_\omega(x,y)\overline{g}_\omega(y)\,d\sigma(y)\,d\omega, \\ &= \frac{1}{2\pi} \int_{\mathbb{R}^d} f(z) \int_{\mathbb{R}} \int_{\partial\Omega} \omega^2 \Gamma_\omega(x,y)\overline{\Gamma}_\omega(z,y)\,d\sigma(y)\,d\omega\,dz. \end{aligned}$$

Using the Helmholtz-Kirchhoff identity (2.91)

$$\int_{\partial\Omega} \Gamma_\omega(x,y)\overline{\Gamma}_\omega(z,y)d\sigma(y) \approx -\frac{1}{\omega}Im\left\{\Gamma_\omega(x,z)\right\} ,$$

which is valid when Ω is a sphere with a large radius in $\mathbb{R}^d$, we find

$$2\mathcal{I}_{\mathrm{TR}}^{(2)}(x) \approx -\frac{1}{\pi}\int_{\mathbb{R}^d} f(z)\int_{\mathbb{R}} \omega\Im m\left\{\Gamma_\omega(x,z)\right\}d\omega\,dz .$$

Using (12.8), we finally obtain that

$$2\mathcal{I}_{\mathrm{TR}}^{(2)}(x) \approx f(x) ,$$

which yields

$$\mathcal{I}_{\mathrm{TR}}^{(2)}(x) \approx \mathcal{I}_{\mathrm{TR}}^{(1)}(x) .$$

Note that, from (12.3), the operator $\mathcal{T} : f \rightarrow g$ can be expressed in the form

$$\mathcal{T}(f)(y,t) = g(y,t) = -\int_{\mathbb{R}^d} \frac{\partial\Gamma}{\partial t}(x,y,0,t)f(x)\,dx, \quad (y,t) \in \partial\Omega \times [0,T] .$$

Then its adjoint $\mathcal{T}^*$ satisfies

$$\mathcal{T}^*(g)(x) = -\int_{\mathbb{R}}\int_{\partial\Omega} \frac{\partial\Gamma}{\partial t}(x,y,0,t)g(y,t)d\sigma(y)dt ,$$

which can be seen from (12.10) to be the time-reversal functional $\mathcal{I}_{\mathrm{TR}}^{(2)}$.

12.2 Diffraction Tomography Within the Born Approximation

12.2.1 Born Approximation

Suppose that the object to be imaged is irradiated by a plane wave $e^{i\omega x\cdot\theta}$, with the wavelength $\lambda := 2\pi/\omega$, travelling in the direction of the unit vector θ. Consider the Helmholtz equation

$$\Delta u + \omega^2(1+V)u = 0 \quad \text{in } \mathbb{R}^d ,$$

subject to the Sommerfeld radiation condition on the scattered wave

$$u^{(s)}(x) := u(x) - e^{i\omega x \cdot \theta}$$

at infinity, where the object is given by the reflectivity function V, which vanishes outside the object. The total field u is measured outside the object for many irradiation directions θ. From all these measurements, the function V has to be determined.

The scattered field $u^{(s)}$ satisfies the Sommerfeld radiation condition and the Helmholtz equation

$$\Delta u^{(s)}(x) + \omega^2 u^{(s)}(x) = -\omega^2 (e^{i\omega x \cdot \theta} + u^{(s)}(x)) V(x), \quad x \in \mathbb{R}^d . \qquad (12.11)$$

Now we consider the Born approximation for weakly scattering extended target. We refer to (6.10) for point reflectors. We assume that the function V is supported in $|x| < \rho$ and $|V| \ll 1$. Then we can neglect $u^{(s)}$ on the right-hand side of (12.11), obtaining

$$\Delta u^{(s)}(x) + \omega^2 u^{(s)}(x) \approx -\omega^2 e^{i\omega x \cdot \theta} V(x), \quad x \in \mathbb{R}^d .$$

This equation can be solved for $u^{(s)}$ with the help of the outgoing fundamental solution Γ_ω for the Helmholtz operator $\Delta + \omega^2$. We have

$$u^{(s)}(x) \approx -\omega^2 \int_{|y|<\rho} \Gamma_\omega(x,y) e^{i\omega\theta \cdot y} V(y) \, dy, \quad x \in \mathbb{R}^d . \qquad (12.12)$$

12.2.2   Diffraction Tomography Algorithm

In diffraction tomography, one computes for weakly scattering objects the Fourier transform of the reflectivity function from the Fourier transform of the measured scattered data.

To present the basics of diffraction tomography, we first recall that the fundamental solution Γ_ω has the plane wave decomposition (2.63). Substituting (2.63) into the Born approximation (12.12) for the scattered field $u^{(s)}$ yields

$$u^{(s)}(x) \approx i\omega^2 c_d \int_{\mathbb{R}^{d-1}} \int_{|y|<\rho} \frac{V(y)}{\beta(\alpha)} e^{i(\beta(\alpha)|x_d - y_d| + \alpha \cdot (\tilde{x} - \tilde{y}))} e^{i\omega\theta \cdot y} \, dy \, d\alpha ,$$

$$(12.13)$$

where $y = (\tilde{y}, y_d)$.

Suppose for simplicity that $d = 2$, $\theta = (0, 1)$ and $u^{(s)}$ is measured on the line $x_2 = l$, where l is greater than any y_2-coordinate within the object. Then (12.13) may be rewritten as

$$u^{(s)}(x_1, l) \approx \frac{i\omega^2}{4\pi} \int_{-\infty}^{+\infty} d\alpha \int_{|y|<\rho} \frac{V(y)}{\beta(\alpha)} e^{i(\beta(\alpha)(l-y_2)+\alpha(x_1-y_1))} e^{i\omega y_2} \, dy \, .$$

Recognizing part of the inner integral as the two-dimensional Fourier transform of the reflectivity function V evaluated at $(\alpha, \beta(\alpha) - \omega)$ we find

$$u^{(s)}(x_1, l) \approx \frac{i\omega^2}{2} \int_{-\infty}^{+\infty} \frac{1}{\beta(\alpha)} e^{i(\beta(\alpha)l+\alpha x_1)} \mathcal{F} V(\alpha, \beta(\alpha) - \omega) \, d\alpha \, ,$$

where

$$\mathcal{F}(V)(\alpha, \beta) = \frac{1}{2\pi} \int V(y) e^{-i\alpha y_1 - i\beta y_2} \, dy \, .$$

Taking the one-dimensional Fourier transform $\mathcal{F}_1$ of $u^{(s)}(x_1, l)$,

$$\mathcal{F}_1(u^{(s)})(\alpha, l) = \frac{1}{\sqrt{2\pi}} \int u^{(s)}(y_1, l) e^{-i\alpha y_1} \, dy_1 \, ,$$

we obtain

$$\mathcal{F}_1(u^{(s)})(\alpha, l) \approx i\omega^2 \sqrt{\frac{\pi}{2}} \frac{1}{\sqrt{\omega^2 - \alpha^2}} e^{i\sqrt{\omega^2-\alpha^2}\,l} \mathcal{F}(V)(\alpha, \sqrt{\omega^2 - \alpha^2} - \omega) \, ,$$

for $|\alpha| < \omega$.

This expression relates the two-dimensional Fourier transform of the reflectivity function to the one-dimensional Fourier transform of the scattered field at the receiver line $x_2 = l$.

The factor

$$i\omega^2 \sqrt{\frac{\pi}{2}} \frac{1}{\sqrt{\omega^2 - \alpha^2}} e^{i\sqrt{\omega^2-\alpha^2}\,l}$$

is a simple function of α for a fixed receiver line and operating frequency ω. As α varies from $-\omega$ to ω, the coordinates $(\alpha, \sqrt{\omega^2 - \alpha^2} - \omega)$ in the Fourier transform of V trace out a semicircular arc. The endpoints of this semicircular arc are at the distance $\sqrt{2}\,\omega$ from the origin in the Fourier domain. Therefore, if the object is illuminated from many different θ-directions, we can fill up a disk of diameter $\sqrt{2}\,\omega$ in the Fourier domain and then approximately reconstruct the reflectivity function $V(x)$ by direct Fourier inversion. The reconstructed object is a low-pass filtered version of the original one.

Bibliography and Discussion

The physics literature on time-reversal is quite rich. One refers, for instance, to [76] and the references therein. Many interesting mathematical works have dealt with different aspects of time-reversal phenomena: see, for instance, [44, 45] for time-reversal in the time-domain, [65, 66, 73, 128] for time-reversal in the frequency domain, [53, 79] for time-reversal in random media, and [4, 6, 104] for time-reversal in attenuating media. The results provided in this chapter on time-reversal techniques are from [4].

The reader is referred to Devaney [72] for diffraction tomography. We also refer to Zhao [162] for the analysis of the response matrix for an extended target within the Born approximation. Direct imaging approaches within the Born approximation have been designed in [98, 99].

Chapter 13
Imaging Small Shape Deformations of an Extended Target from MSR Measurements

Let D be a bounded smooth domain in $\mathbb{R}^d, d = 2, 3$. We assume that D is known. We consider a perturbation, denoted by D^h, of the domain D given by

$$\partial D^h := \left\{ x + h(x)\nu(x), x \in \partial D \right\}, \qquad (13.1)$$

where ν is the outward unit normal to ∂D and h is a $\mathcal{C}^1$ function on ∂D.

We suppose that D^h represents an inclusion of electromagnetic parameters $\varepsilon_\star$ and $\mu_\star$. The electromagnetic parameters of the background are denoted by μ_0 and ε_0.

Suppose for simplicity that the array of transmitters and receivers coincide and denote by $\{y_1, \ldots, y_N\}$ the set of locations of sensors, which are away from the inclusion D^h. We measure the MSR matrix A^{meas} as before: A^{meas}_{nm} is the time-harmonic signal recorded at y_n in the presence of the inclusion D^h when y_m emits a unit-amplitude signal, from which the direct wave $\Gamma_{k_0}(y_n, y_m)$ is removed, for $n, m = 1, \ldots, N$.

Our aim in this chapter is to reconstruct the perturbations h from the MSR matrix A^{meas}. As for small volume inclusions, we present direct imaging algorithms and analyze their resolution and stability. Our algorithms are based on an asymptotic expansion for the perturbations in the data due to small shape deformations. A concept equivalent to the polarization tensor for small volume targets is introduced.

13.1 Asymptotic Expansion of the MSR Matrix

The purpose of this section is to investigate the asymptotic behavior of the MSR matrix as $\|h\|_{\mathcal{C}^1}$ tends to 0.

H. Ammari et al., *Mathematical and Statistical Methods for Multistatic Imaging*, 239
Lecture Notes in Mathematics 2098, DOI 10.1007/978-3-319-02585-8_13,
© Springer International Publishing Switzerland 2013

Let $a \otimes b := (a_i b_j)_{i,j=1}^d$ denote the tensor product between vectors in $\mathbb{R}^d$. Introduce the polarization tensor in the two-dimensional case as follows:

$$M[\frac{\mu_0}{\mu_\star}](x) = \left(\frac{\mu_0}{\mu_\star} - 1\right)\left(\frac{\mu_0}{\mu_\star}\nu(x) \otimes \nu(x) + \tau(x) \otimes \tau(x)\right), \qquad x \in \partial D ,$$

where $\tau(x)$ is the unit tangential vector to ∂D at x.

In the three-dimensional case, we define the polarization tensor by

$$M[\frac{\mu_0}{\mu_\star}](x) = \left(\frac{\mu_0}{\mu_\star} - 1\right)\left(\frac{\mu_0}{\mu_\star}\nu(x) \otimes \nu(x) + \sum_{k=1}^{2} \tau_k(x) \otimes \tau_k(x)\right), \qquad x \in \partial D ,$$

where $\tau_1(x)$ and $\tau_2(x)$ are two orthogonal unit tangential vectors to ∂D at x. The polarization tensors are diagonal matrices in the normal, tangential basis associated with ∂D.

Let $u[D](x,y)$ be the solution to (3.34). The following result holds.

Theorem 13.1. *The following asymptotic formula holds as* $\|h\|_{\mathcal{C}^1} \to 0$:

$$u[D^h](x,y) - u[D](x,y) = \int_{\partial D} h(z)\left[\nabla_z u[D](x,z) \cdot M[\frac{\mu_0}{\mu_\star}](z)\nabla_z u[D](z,y)\right.$$

$$+k_0^2(\frac{\varepsilon_\star}{\varepsilon_0} - 1)u[D](x,z)u[D](z,y)\Big] d\sigma(z)$$

$$+o(\|h\|_{\mathcal{C}^1}) ,$$

$$(13.2)$$

for x *away from* D.

In the presence of an inclusion D, the MSR matrix is

$$A_{nm}[D] = -u[D](y_n, y_m) + \Gamma_{k_0}(y_n, y_m), \qquad n, m = 1, \ldots, N . \qquad (13.3)$$

From Theorem 13.1, we have (using the reciprocity property $u(x,y) = u(y,x)$)

$$A_{nm}[D^h] - A_{nm}[D] = -\int_{\partial D} h(x)B_{nm}[D](x)\, d\sigma(x) + o(\|h\|_{\mathcal{C}^1}) , \qquad (13.4)$$

where $B[D](x) = (B_{nm}[D](x))_{n,m=1}^{N}$ is the matrix defined by

$$B_{nm}[D](x) := \nabla_x u[D](x, y_n) \cdot M[\frac{\mu_0}{\mu_\star}](x)\nabla_x u[D](x, y_m)$$

$$+k_0^2(\frac{\varepsilon_\star}{\varepsilon_0} - 1)u[D](x, y_n)u[D](x, y_m) . \qquad (13.5)$$

Note that B depends not only on D but also on ω and on the contrasts $\mu_0/\mu_\star$ and $\varepsilon_0/\varepsilon_\star$. Note also that the matrix B corresponds to the backpropagation of the array data both from the receiver point x_n and from the source point x_m to the point x on the boundary ∂D.

13.2   Direct Reconstructions

The matrix $A[D]$ is symmetric by definition, but the measured matrix A^{meas} may not be symmetric due to an additive noise. We symmetrize the measured matrix by the transform $A \to (A + A^T)/2$. The symmetrization of the MSR matrix reduces the variance of its off-diagonal entries by a factor of 2 in the case of additive noise:

$$\mathbb{E}\left[\left|\frac{W_{mn} + W_{nm}}{2}\right|^2\right] = \frac{1}{4}\left(\mathbb{E}[|W_{mn}|^2] + \mathbb{E}[|W_{nm}|^2]\right) = \frac{1}{2}\mathbb{E}[|W_{mn}|^2] \text{ if } m \neq n .$$

The first functional that is used for reconstructing the function $h(x)$ is the one given by

$$\mathcal{J}_1[h] := \frac{1}{2}\|A[D^h] - A^{\text{meas}}\|_F^2, \tag{13.6}$$

with $\|\cdot\|_F$ being the Frobenius norm. Using (13.4) and neglecting the remainder $o(\|h\|_{\mathcal{C}^1})$, we have

$$\mathcal{J}_1[h] = \frac{1}{2}\sum_{n,m=1}^{N}\left|\int_{\partial D} B_{nm}[D](x)h(x)\,d\sigma(x) - A_{nm}[D] + A_{nm}^{\text{meas}}\right|^2 .$$

The function h solving this least-squares minimization problem is the solution to

$$\mathcal{C}_1[h] = \mathcal{D}_1 \quad \text{on } \partial D , \tag{13.7}$$

where $\mathcal{C}_1$ is the integral operator with the kernel

$$\mathcal{C}_1(x, x') = \Re \sum_{n,m=1}^{N} B_{nm}[D](x)\overline{B_{nm}[D](x')} ,$$

and

$$\mathcal{D}_1(x) = \Re \sum_{n,m=1}^{N} (A_{nm}[D] - A_{nm}^{\text{meas}})\overline{B_{nm}[D](x)} .$$

The solution to (13.7) will be made explicit in Sect. 13.4 in the case in which D is a disk. In the general case, it may be difficult to find an explicit expression for the inverse of $\mathcal{C}_1$ and then one implements iterative methods or an approximate direct solution.

A simple but approximate direct solution to the minimization problem can be obtained as follows. Note that the Fréchet derivative of $\mathcal{J}_1$ in the direction h is given by

$$d\mathcal{J}_1(h) = -\int_{\partial D} \mathcal{D}_1(x)h(x)d\sigma(x) .$$

One looks at the expansion of $\mathcal{J}_1[h]$ for small h:

$$\mathcal{J}_1[h] = \mathcal{J}_1[0] - \int_{\partial D} \mathcal{D}_1(x)h(x)d\sigma(x) + O(\|h\|_{\mathcal{C}^1}^2) .$$

One takes a solution pointing into the direction of $\mathcal{D}_1(x)$:

$$h(x) = \frac{1}{X_1}\mathcal{D}_1(x) . \tag{13.8}$$

One chooses the positive real number X_1 by substituting (13.8) into $\mathcal{J}_1[h]$ and by equating to zero, which gives a simple quadratic equation for X_1:

$$X_1^2 \sum_{n,m=1}^{N} \left|A_{nm}^{\mathrm{meas}} - A_{nm}[D]\right|^2 - 2X_1 \int_{\partial D} \mathcal{D}_1[D](x)^2\, d\sigma(x)$$

$$+ \sum_{n,m=1}^{N} \left|\int_{\partial D} B_{nm}[D](x)\mathcal{D}_1[D](x)\, d\sigma(x)\right|^2 = 0 .$$

A second algorithm for reconstructing the perturbations h is as follows. Let $\sigma_{\mathrm{meas}}^{(l)}$, $l = 1, \ldots, L$, be the L first singular values of A^{meas} counted according to multiplicity and $v_{\mathrm{meas}}^{(l)}$ be the singular vector associated with $\sigma_{\mathrm{meas}}^{(l)}$, so that $(v_{\mathrm{meas}}^{(l)})_{l=1}^{L}$ is a basis of the image space of A^{meas}. The choice of L will be discussed in the next section and results from a trade-off between stability and resolution. In order to reconstruct the function $h(x)$ one minimizes over $h \in \mathcal{C}^1$ the functional defined by

$$\mathcal{J}_2[h] := \frac{1}{2}\sum_{l=1}^{L} \left\|(A[D^h] - A^{\mathrm{meas}})v_{\mathrm{meas}}^{(l)}\right\|^2 . \tag{13.9}$$

Here, the MSR discrepancy is minimized with respect to the signal space generated by the first L singular vectors of A^{meas}. The least-squares solution to $\min_h \mathcal{J}_2[h]$ is given by

$$h(x) = \mathcal{C}_2^{-1} \left\{ \Re \sum_{l=1}^{L} \left\langle (A[D] - A^{\text{meas}})v_{\text{meas}}^{(l)}, B[D](\cdot)v_{\text{meas}}^{(l)} \right\rangle \right\}(x), \qquad (13.10)$$

where $\langle \, , \, \rangle$ denotes the Hermitian product and $\mathcal{C}_2$ is the operator with the kernel:

$$\mathcal{C}_2(x, x') = \Re \sum_{l=1}^{L} \left\langle B[D](x)v_{\text{meas}}^{(l)}, B[D](x')v_{\text{meas}}^{(l)} \right\rangle. \qquad (13.11)$$

The expression (13.10) will be made explicit in Sect. 13.4 in the case in which D is a disk. An approximate solution is

$$h(x) = \frac{1}{X_2} \Re \sum_{l=1}^{L} \left\langle (A[D] - A^{\text{meas}})v_{\text{meas}}^{(l)}, B[D](x)v_{\text{meas}}^{(l)} \right\rangle, \qquad (13.12)$$

where the positive real number X_2 is chosen so that $\mathcal{J}_2[h] = 0$ for h given by (13.12).

Let $\sigma^{(l')}[D]$ be the singular values of $A[D]$ arranged in a descending order and counted according to their multiplicities. Let $v^{(l')}[D]$, $l' = 1, \ldots, L'$, be the first L' singular vectors associated with $\sigma^{(l')}[D_j]$. A third algorithm is to minimize over $h \in \mathcal{C}^1$ the functional given by

$$\mathcal{J}_3[h] := \frac{1}{2} \sum_{l'=1}^{L'} \sum_{l=1}^{L} \left| \left\langle (A[D^h] - A_{\text{meas}})v_{\text{meas}}^{(l)}, v^{(l')}[D] \right\rangle \right|^2, \qquad (13.13)$$

where $\langle \, , \, \rangle$ denotes the Hermitian product. Here, the MSR discrepancy of the search direction is minimized in the direction of the signal space. The least-squares solution to $\min_h \mathcal{J}_3[h]$ is given by

$$h(x) = \mathcal{C}_3^{-1} \left\{ \Re \sum_{l=1}^{L} \sum_{l'=1}^{L'} \left\langle (A[D] - A^{\text{meas}})v_{\text{meas}}^{(l)}, v^{(l')}[D] \right\rangle \right.$$
$$\left. \times \overline{\left\langle B[D](\cdot)v_{\text{meas}}^{(l)}, v^{(l')}[D] \right\rangle} \right\}(x), \qquad (13.14)$$

where $\mathcal{C}_3$ is the operator with the kernel:

$$C_3(x, x') = \Re \sum_{l,l'=1}^{L} \left\langle B[D](x)v_{\text{meas}}^{(l)}, \, v^{(l')}[D] \right\rangle \overline{\left\langle B[D](x')v_{\text{meas}}^{(l)}, \, v^{(l')}[D] \right\rangle}.$$

$$(13.15)$$

The expression (13.14) will be made explicit in Sect. 13.4 in the case in which D is a disk. An approximate solution is

$$h(x) = \frac{1}{X_3} \Re \sum_{l=1}^{L} \left\langle (A[D] - A^{\text{meas}})v_{\text{meas}}^{(l)}, \, v^{(l')}[D] \right\rangle$$

$$\times \overline{\left\langle B[D](x)v_{\text{meas}}^{(l)} \, v^{(l')}[D] \right\rangle},$$

$$(13.16)$$

where the positive real number X_3 is chosen so that $\mathcal{J}_3[h] = 0$ for h given by (13.16).

13.3   The Born Approximation

In this section we assume $\mu = \mu_0$ and address the case where the Born approximation is valid. We consider a circular geometry of array of sources (which is the same as the array of receivers) with target in center. This concentric configuration allows us to do a detailed resolution and stability analysis in the presence of an additive measurement noise.

13.3.1   Asymptotic Formulation of the Response Matrix

Let D_{true} be the true inclusion. If we set $\mu_\star = \mu_0$ and assume that $\varepsilon_\star$ is not far from ε_0, then we can use the Born approximation

$$u[D_{\text{true}}](x, y_n) \approx \Gamma_{k_0}(x, y_n), \quad \forall \, 1 \le n \le N \quad \text{and} \quad x \in D_{\text{true}}.$$

Therefore, we have

$$A_{nm}^{\text{meas}}[D_{\text{true}}] = -u[D_{\text{true}}](x_n, y_m) + \Gamma_{k_0}(x_n, y_m)$$

$$\approx k_0^2 (\frac{\varepsilon_\star}{\varepsilon_0} - 1) \int_{D_{\text{true}}} \Gamma_{k_0}(x, y_n)\Gamma_{k_0}(x, y_m) \, dx.$$

If we define the matrix

$$B[\omega](x) := \left(\Gamma_{k_0}(x, y_n) \Gamma_{k_0}(x, y_m) \right)_{n,m=1}^{N} \quad \text{for } x \in D_{\text{true}},$$

then, one can write

$$A^{\text{meas}} \approx k_0^2 \left(\frac{\varepsilon_\star}{\varepsilon_0} - 1 \right) \int_{D_{\text{true}}} B[\omega](x) \, dx \, .$$

Note that in this case $B[\omega]$ does not depend on D_{true}.

Below we assume $d = 2$. If ω is large, then

$$B[\omega](x) \approx \frac{i}{8\pi k_0} \left(\frac{e^{ik_0(|x-y_n|+|x-y_m|)}}{\sqrt{|x - y_n||x - y_m|}} \right)_{n,m=1}^{N} .$$

Assuming further that the distance L_F between the array and the origin of the target is much larger than $\text{diam}(D_{\text{true}})$ yields

$$B[\omega](x) \approx \frac{i}{8\pi k_0 L_F} \left(e^{ik_0(|x-y_n|+|x-y_m|)} \right)_{n,m=1}^{N} .$$

In polar coordinates, let the points of the array be as follows:

$$y_n = L_F(\cos\theta_n, \sin\theta_n) \, ,$$

and let the domain D_{true} be of the form

$$D_{\text{true}} = \{ y = (r\cos\theta, r\sin\theta) \, , \, 0 \le r \le R(\theta) \, , \, 0 \le \theta \le 2\pi \} \, . \tag{13.17}$$

Using the Taylor series expansion

$$|y_n - x| = |y_n| - \frac{y_n \cdot x}{|y_n|} + O\left(\frac{|x|^2}{|y_n|} \right) \, , \tag{13.18}$$

we find that, in polar coordinates $x = (r\cos\theta, r\sin\theta)$,

$$A_{mn}^{\text{meas}}[D_{\text{true}}] = e^{2ik_0 L_F} \int_0^{2\pi} d\theta \int_0^{R(\theta)} r \, dr \, e^{-ik_0 r[\cos(\theta-\theta_m)+\cos(\theta-\theta_n)]} \, , \tag{13.19}$$

up to a multiplicative constant. In view of (13.18), (13.19) is valid if $k_0 \text{diam}^2(D)$ is much smaller than the distance L_F from the origin of the target D to the array. This is the so-called Fraunhofer regime.

Note that the phase factor $e^{2ik_0 L_F}$ in the response matrix (13.19) does not modify the singular values and it only modifies the singular vectors by a phase term independent of the singular value itself. In the following this factor is removed.

13.3.2 The Unperturbed Domain

We assume in this subsection that the domain $D_{\text{true}} := D_0$, a disk with radius r_0. In the continuum approximation (the number of array elements $N \to +\infty$) the response matrix is proportional to the operator whose kernel is

$$\mathcal{A}[D_0](\theta_1, \theta_2) = \frac{1}{\pi r_0^2} \int_0^{2\pi} d\theta \int_0^{r_0} r dr e^{-ik_0 r[\cos(\theta - \theta_1) + \cos(\theta - \theta_2)]} .$$

The kernel can be written as (see formulas (1.13) and (1.12))

$$\mathcal{A}[D_0](\theta_1, \theta_2) = a(\theta_1 - \theta_2) \text{ with } a(\theta) = 2 \frac{J_1\big(2k_0 r_0 \cos(\frac{\theta}{2})\big)}{2k_0 r_0 \cos(\frac{\theta}{2})} .$$

From (1.16) the function $a(\theta)$ can be expanded in Fourier series as

$$a(\theta) = \sum_{n=-\infty}^{\infty} \hat{a}_n e^{in\theta} \text{ with } \hat{a}_n = (-1)^n \big(J_n^2 - J_{n-1}J_{n+1}\big)(k_0 r_0) ,$$

where J_n is the Bessel function of the first kind of order n, which shows that the singular values of $\mathcal{A}[D_0]$ are $(\sqrt{2\pi}|\hat{a}_p|)_{p \in \mathbb{N}}$, each of which (except $\sqrt{2\pi}|\hat{a}_0|$) is of multiplicity two. The associated singular vectors are $(\psi^{(p,\pm)}(\theta))_{\theta \in [0,2\pi]} = (e^{\pm ip\theta}/\sqrt{2\pi})_{\theta \in [0,2\pi]}$. Moreover, in the asymptotic framework $k_0 r_0 \gg 1$, there are about $2k_0 r_0$ significant singular values. More exactly, using (1.21), we find that, for $k_0 r_0 \gg 1$ and $n \in [-k_0 r_0, k_0 r_0]$:

$$|\hat{a}_n| \approx \frac{2}{\pi k_0 r_0} \sqrt{1 - \Big(\frac{n}{k_0 r_0}\Big)^2} .$$

For $k_0 r_0 \gg 1$ and $|n| > k_0 r_0$ the singular values are exponentially small.

13.3.3 The Perturbed Domain

We assume in this subsection that the domain D_{true} is a deformed disk (around the perfect disk D_0 with radius r_0). In polar coordinates $x = (r \cos \theta, r \sin \theta)$ the domain D_{true} is given by (13.17) with

$$R(\theta) = r_0 + h_{\text{true}}(\theta), \qquad h_{\text{true}}(\theta) = \sum_{p=-\infty}^{\infty} \hat{h}_{\text{true},p} e^{ip\theta} . \tag{13.20}$$

We address the regime in which $k_0 \|h_{\text{true}}\|_\infty \ll 1 < k_0 r_0$. In the continuum approximation the perturbation of the response matrix is the operator with the kernel

$$\mathcal{H}[D_{\text{true}}](\theta_1, \theta_2) := \mathcal{A}[D_{\text{true}}](\theta_1, \theta_2) - \mathcal{A}[D_0](\theta_1, \theta_2)$$

$$\approx \frac{1}{\pi r_0} \int_0^{2\pi} d\theta \, h_{\text{true}}(\theta) e^{-ik_0 r_0 [\cos(\theta - \theta_1) + \cos(\theta - \theta_2)]} .$$

The results of the previous subsection indicate that we should represent the perturbation of the response matrix in the Fourier domain, since the singular vectors of the unperturbed response matrix are the Fourier modes. The Fourier coefficients of the kernel of the operator $\mathcal{H}[D_{\text{true}}]$ are defined by

$$\hat{\mathcal{H}}_{jl}[D_{\text{true}}] = \frac{1}{(2\pi)^2} \int \mathcal{H}[D_{\text{true}}](\theta_1, \theta_2) e^{-ij\theta_1 - il\theta_2} d\theta_1 d\theta_2 .$$

Using (1.13) they are given by

$$\hat{\mathcal{H}}_{jl}[D_{\text{true}}] = \frac{2}{r_0} \hat{h}_{\text{true},j+l} J_j(k_0 r_0) J_l(k_0 r_0) i^{-j-l} . \tag{13.21}$$

13.4 Resolution and Stability Analysis of the Imaging Functionals

Assuming measurement noise, we perform a resolution and stability analysis of the proposed reconstruction formulas. We assume that the receiver-transmitter array covers in a dense manner a closed surface surrounding the inclusion D.

We assume that the domain is the deformed disk D_{true} given by (13.20) and that the response matrix is corrupted by an additive Gaussian white noise W or equivalently in the continuum approximation, the kernel of the operator is given by

$$\mathcal{A}^{\text{meas}}(\theta_1, \theta_2) = \mathcal{A}[D_{\text{true}}](\theta_1, \theta_2) + W(\theta_1, \theta_2) .$$

In this section we assume that $W(\theta_1, \theta_2)$ is a white noise, with mean zero and covariance function

$$\mathbb{E}\big[W(\theta_1, \theta_2) W(\theta_1', \theta_2')\big] = (2\pi)^2 \sigma_{\text{noise}}^2 \delta(\theta_1 - \theta_1') \delta(\theta_2 - \theta_2') .$$

The purpose of the imaging process is to estimate the function $h_{\text{true}}(\theta)$ that characterizes D_{true}. The results of the previous subsection indicate that we should look for the Fourier coefficients $(\hat{h}_{\text{true},p})_{p \in \mathbb{Z}}$ that characterize the boundary of the domain D_{true}.

13.4.1 First Reconstruction Formula

The first imaging functional defined in (13.6) is

$$\mathcal{J}_1[D] = \frac{1}{2} \left\| \mathcal{A}[D](\cdot,\cdot) - \mathcal{A}^{\text{meas}}(\cdot,\cdot) \right\|_F^2$$

$$= \frac{1}{2} \left\| \mathcal{H}[D](\cdot,\cdot) - \mathcal{H}[D_{\text{true}}](\cdot,\cdot) - W(\cdot,\cdot) \right\|_F^2 .$$

The domain D is characterized by r_0 and the function $(h(\theta))_{\theta \in [0,2\pi)}$. Using Parseval's formula and (13.21) the first imaging functional can be written as

$$\mathcal{J}_1[D] = 2\pi^2 \sum_{l',l=-\infty}^{\infty} \left| (\mathcal{Q}\hat{h})_{l'l} - (\mathcal{Q}\hat{h}_{\text{true}})_{l'l} - \hat{W}_{l'l} \right|^2 ,$$

where $\hat{W}_{l'l}$ are the Fourier coefficients of $W(\cdot,\cdot)$:

$$\hat{W}_{l'l} = \frac{1}{(2\pi)^2} \iint W(\theta_1,\theta_2) e^{-il'\theta_1 - il\theta_2} d\theta_1 d\theta_2 ,$$

and

$$(\mathcal{Q}\hat{h})_{l'l} = \frac{2}{r_0} \hat{h}_{l'+l} J_{l'}(k_0 r_0) J_l(k_0 r_0) i^{-l'-l}, \qquad l',l \in \mathbb{Z} . \tag{13.22}$$

Note that $\mathbb{E}[\hat{W}_{l'l}] = 0$ and

$$\mathbb{E}\left[\overline{\hat{W}_{l'l}} \hat{W}_{j'j} \right] = \sigma_{\text{noise}}^2 \delta_{l'j'} \delta_{lj} .$$

Here the least-squares solution can be written explicitly because the operator is $\mathcal{Q}^*\mathcal{Q}$ is diagonal and can be inverted easily. We find that the least-squares solution to the minimization of the first imaging functional is

$$\left((\mathcal{Q}^*\mathcal{Q})^{-1} \mathcal{Q}^* \hat{W} \right)_p = \frac{r_0 \sum_{l=-\infty}^{\infty} J_l(k_0 r_0) J_{p-l}(k_0 r_0) i^p \hat{W}_{l,p-l}}{2 \sum_{l=-\infty}^{\infty} J_l^2(k_0 r_0) J_{p-l}^2(k_0 r_0)}, \qquad p \in \mathbb{Z} .$$

Therefore, given the measured kernel $\mathcal{A}^{\text{meas}}$, the least-square estimation $(\hat{h}_p^{\text{est}})_{p\in\mathbb{Z}}$ of the Fourier coefficients of the shape $h^{\text{true}}(\theta)$ of the domain D_{true} is

$$(\hat{h}_p^{\text{est}})_{p\in\mathbb{Z}} = (\mathcal{Q}^*\mathcal{Q})^{-1}\mathcal{Q}^*\big((\hat{\mathcal{A}}_{l'l}^{\text{meas}} - \hat{\mathcal{A}}_{l'l}[D_0])_{l',l\in\mathbb{Z}}\big) , \qquad (13.23)$$

which gives, for all $p \in \mathbb{Z}$,

$$\hat{h}_p^{\text{est}} = \hat{h}_p^{\text{true}} + \frac{r_0 \sum_{l=-\infty}^{\infty} J_l(k_0 r_0) J_{p-l}(k_0 r_0) i^p \hat{W}_{l,p-l}}{2 \sum_{l=-\infty}^{\infty} J_l^2(k_0 r_0) J_{p-l}^2(k_0 r_0)} . \qquad (13.24)$$

It shows that the estimation is unbiased, i.e.,

$$\mathbb{E}[\hat{h}_p^{\text{est}}] = \hat{h}_p^{\text{true}} ,$$

and has the variance

$$\text{Var}(\hat{h}_p^{\text{est}}) = \frac{r_0^2 \sigma_{\text{noise}}^2}{4 \sum_{l=-\infty}^{\infty} J_l^2(k_0 r_0) J_{p-l}^2(k_0 r_0)} ,$$

where $\sigma_{\text{noise}}^2 = \mathbb{E}(|\hat{W}_{l'l}|^2)$ (independent on l', l for a white noise).

Now, from Neumann's formula (1.17), we have for any $l \in \mathbb{Z}$:

$$J_l(k_0 r_0) J_{p-l}(k_0 r_0) = \frac{1}{2\pi} \int_0^{2\pi} J_p(2k_0 r_0 \cos\theta) \cos\big((2l-p)\theta\big) d\theta .$$

Using Parseval's formula (1.1) gives

$$\sum_{l=-\infty}^{\infty} J_l^2(k_0 r_0) J_{p-l}^2(k_0 r_0) = \frac{1}{2\pi} \int_0^{2\pi} J_p^2(2k_0 r_0 \cos\theta) d\theta .$$

It follows from (1.21) that in the asymptotic framework when $k_0 r_0 \gg 1$ and p is smaller than $2k_0 r_0$,

$$\sum_{l=-\infty}^{\infty} J_l^2(k_0 r_0) J_{p-l}^2(k_0 r_0) \approx \frac{1}{\pi^2 k_0 r_0}\Big[\log k_0 r_0 + 5\log 2 + \gamma$$
$$-2\big(1 + \frac{1}{3} + \cdots + \frac{1}{2p-1}\big) + O\big(\frac{1}{(k_0 r_0)^{1/2}}\big)\Big] ,$$

where γ is the Euler constant, while when p is larger than $2k_0 r_0$ the sum is exponentially close to zero. So, if $p \leq 2k_0 r_0$ (and $k_0 r_0 \gg 1$), then

$$\text{Var}(\hat{h}_p^{\text{est}}) \approx \frac{\pi^2 k_0 r_0^3 \sigma_{\text{noise}}^2}{4 \log(k_0 r_0)} .$$

We can therefore conclude that, in the presence of a small additive noise:

(i) the estimation of $\hat{h}_{\text{true},p}$ is possible for $p < 2k_0r_0$ with the accuracy (standard deviation) of the order of $(\sigma_{\text{noise}}r_0/2)\pi(k_0r_0)^{1/2}/\log^{1/2}(k_0r_0)$, and impossible for $p > 2k_0r_0$;

(ii) the coefficient $\hat{h}_p$ corresponds to a feature at the surface of the unperturbed disk D_0 whose characteristic length scale is $2\pi r_0/p$, and therefore the limitation $p < 2k_0r_0$ corresponds to a length scale larger than half a wavelength, which is the classical diffraction limit.

13.4.2   The Second Reconstruction Formula

The second imaging functional defined in (13.9) is

$$
\mathcal{J}_2[D] = \frac{1}{2}\sum_l \left\| \left(\mathcal{A}[D](\cdot,\cdot) - \mathcal{A}^{\text{meas}}(\cdot,\cdot)\right)v_{\text{meas}}^{(l)}(\cdot) \right\|_2^2
$$

$$
= \frac{1}{2}\sum_l \left\| \left(\mathcal{H}[D](\cdot,\cdot) - \mathcal{H}[D_{\text{true}}](\cdot,\cdot) - W(\cdot,\cdot)\right)v_{\text{meas}}^{(l)}(\cdot) \right\|_2^2 ,
$$

where $\sigma_{\text{meas}}^{(l)}$ and $v_{\text{meas}}^{(l)}$ are the l-th singular value and singular vector of $\mathcal{A}^{\text{meas}}$. If D_{true} is a small deformation of the disk D_0 and the additive white noise is small, then the difference between the singular vectors of $\mathcal{A}^{\text{meas}}$ and those of $\mathcal{A}[D_0]$ is small and therefore, after relabelling the vectors obtained in Sect. 13.3.2 and up to an error that is of higher-order, we have

$$
\mathcal{J}_2[D] = \frac{1}{2}\sum_{l=-L}^{L} \left\| \left(\mathcal{H}[D](\cdot,\cdot) - \mathcal{H}[D_{\text{true}}](\cdot,\cdot) - W(\cdot,\cdot)\right)\psi^{(l)}(\cdot) \right\|_2^2 ,
$$

where $\psi^{(l)}(\theta) = e^{il\theta}/\sqrt{2\pi}$. We have

$$
\left(\mathcal{H}[D]\psi^{(l)}\right)(\theta) = \frac{2\sqrt{2\pi}}{r_0}\sum_{p=-\infty}^{+\infty} \hat{h}_p J_{l+p}(k_0r_0)J_l(k_0r_0)i^{-2l-p}e^{i(l+p)\theta} .
$$

Using Parseval's formula, we get

$$
\mathcal{J}_2[D] = 2\pi^2 \sum_{l'=-\infty}^{\infty}\sum_{l=-L}^{L} \left| (\mathcal{Q}\hat{h})_{l'l} - (\mathcal{Q}\hat{h}_{\text{true}})_{l'l} - \hat{W}_{l'l} \right|^2 ,
$$

where $\mathcal{Q}$ is defined by (13.22). The least-squares inverse is given by

$$\left((\mathcal{Q}^*\mathcal{Q})^{-1}\mathcal{Q}^*\hat{W}\right)_p = \frac{r_0 \sum_{l=-L}^{L} J_l(k_0 r_0) J_{p+l}(k_0 r_0) i^{p+2l} \hat{W}_{p+l,-l}}{2 \sum_{l=-L}^{L} J_l^2(k_0 r_0) J_{p+l}^2(k_0 r_0)}, \qquad p \in \mathbb{Z}\,.$$

Therefore the least-square estimation $(\hat{h}_{\mathrm{est},p})_{p\in\mathbb{Z}}$ of the Fourier coefficients of the shape $h_{\mathrm{true}}(\theta)$ of the domain D_{true} is

$$(\hat{h}_p^{\mathrm{est}})_{p\in\mathbb{Z}} = ((\mathcal{Q}^*\mathcal{Q})^{-1}\mathcal{Q}^*((\hat{A}_{l'l}^{\mathrm{meas}} - \hat{A}_{l'l}[D_0])_{l'\in\mathbb{Z},l=-L,\dots,L})\,. \tag{13.25}$$

This gives, for all $p \in \mathbb{Z}$, that

$$\hat{h}_p^{\mathrm{est}} = \hat{h}_p^{\mathrm{true}} + \frac{r_0 \sum_{l=-L-p}^{L-p} J_l(k_0 r_0) J_{p-l}(k_0 r_0) i^p \hat{W}_{l,p+l}}{2 \sum_{l=-L}^{L} J_l^2(k_0 r_0) J_{p-l}^2(k_0 r_0)}, \tag{13.26}$$

which implies that the estimation is unbiased with the variance

$$\mathrm{Var}(\hat{h}_{\mathrm{est},p}) = \frac{\sigma_{\mathrm{noise}}^2 r_0^2}{4 \sum_{l=-L}^{L} J_l^2(k_0 r_0) J_{p-l}^2(k_0 r_0)}\,.$$

This result shows that the second functional is more sensitive to an additive white noise than the first one for small L, while they are equivalent when $L > k_0 r_0$.

Note that these conclusions hold because of the continuum approximation. However, in the discrete case when there is only a finite number of transmitters and receivers, Method 2 is better than Method 1 since the information in the noise space is filtered out [21].

13.4.3   The Third Reconstruction Functional

The third imaging functional defined in (13.13) is

$$\mathcal{J}_3[D] = \frac{1}{2} \sum_{l'} \sum_{l} \left| \left\langle (A[D] - A^{\mathrm{meas}})) v_{\mathrm{meas}}^{(l)}, v^{(l')}[D_0] \right\rangle \right|^2.$$

To leading order in the amplitude of the noise and the deformation of the domain, we have after relabelling the vectors

$$\mathcal{J}_3[D] = \frac{1}{2} \sum_{l=-L}^{L} \sum_{l'=-L'}^{L'} \left| \left\langle (\mathcal{H}[D] - \mathcal{H}[D_{\mathrm{true}}] - W) \psi^{(l)}, \psi^{(l')} \right\rangle \right|^2,$$

where $\psi^{(l)}$ is as before. Using Parseval's formula (1.1), we get

$$
\mathcal{J}_3[D] = 2\pi^2 \sum_{l'=-L'}^{L'} \sum_{l=-L}^{L} \left| (\mathcal{Q}'\hat{h})_{l'l} - (\mathcal{Q}'\hat{h}^{\mathrm{true}})_{l'l} - \hat{W}_{l',-l} \right|^2 ,
$$

where

$$
(\mathcal{Q}'\hat{h})_{l'l} = \frac{2}{r_0} \hat{h}_{l'-l} J_{l'}(k_0 r_0) J_l(k_0 r_0) i^{-l'-l} ,
$$

for $l' = -L', \ldots, L'$ and $l = -L, \ldots, L$. Note that $\mathcal{Q}'\hat{h}$ is a function of $(\hat{h}_p)_{p=-L-L',\ldots,L+L'}$ only. The least-square estimation $(\hat{h}_{\mathrm{est},p})_{p=-L-L',\ldots,L+L'}$ of the first Fourier coefficients of the shape $h^{\mathrm{true}}(\theta)$ of the domain D_{true} is

$$
(\hat{h}_{\mathrm{est},p})_{p=-L-L',\ldots,L+L'}
$$
$$
= ((\mathcal{Q}')^*(\mathcal{Q}'))^{-1}(\mathcal{Q}')^* \left((\hat{A}_{l'l}^{\mathrm{meas}} - \hat{A}_{l'l}[D_0])_{l'=-L',\ldots,L',l=-L,\ldots,L} \right) .
$$
$$(13.27)$$

This gives, for all $p = -L - L', \ldots, L + L'$:

$$
\hat{h}_{\mathrm{est},p} = \hat{h}_{\mathrm{true},p} + \frac{r_0 \sum_{l=-L\vee-L'+p}^{L\wedge L'+p} J_l(k_0 r_0) J_{p-l}(k_0 r_0) i^p \hat{W}_{l,p-l}}{2 \sum_{l=-L\vee-L'+p}^{L\wedge L'+p} J_l^2(k_0 r_0) J_{p-l}^2(k_0 r_0)} , \qquad (13.28)
$$

which implies that the estimation is unbiased with the variance

$$
\mathrm{Var}(\hat{h}_{\mathrm{est},p}) = \frac{\sigma_{\mathrm{noise}}^2 r_0^2}{4 \sum_{l=-L\vee-L'+p}^{L\wedge L'+p} J_l^2(k_0 r_0) J_{p-l}^2(k_0 r_0)} .
$$

This result shows that it is possible to reconstruct the Fourier coefficients $\hat{h}_p$ up to $p = (L + L') \wedge 2k_0 r_0$ using the third functional. Here $a \wedge b$ and $a \vee b$ respectively denotes the minimum and the maximum between a and b.

Bibliography and Discussion

The results of this chapter are from [21]. In order to carry out a detailed resolution and stability analysis, we have chosen a particular geometry. It is possible to carry out such detailed analysis in different geometries. For instance, if the array has finite-aperture and covers only an angular cone with angular width $\theta_a \ll 2\pi$, then the analysis is similar to the one carried out in Sect. 4.6 and leads to the conclusion that it is possible to reconstruct the shape of the inclusion with a resolution of the order of the Rayleigh resolution formula $(2\pi/\theta_a)(\pi/k_0)$.

Chapter 14
Nonlinear Optimization Algorithms

In this chapter we consider the nonlinear optimization problem for reconstructing the shape of an extended target from multistatic data. Because of the nonlinearity of the problem, iterative algorithms have to be introduced. For doing so, we consider the function $\mathcal{J}[D]$ defined as in (13.6):

$$\mathcal{J}[D] = \frac{1}{2}\|A[D] - A^{\mathrm{meas}}\|_F^2 \,,$$

where A^{meas} is the measured MSR in the presence of the inclusion of unknown shape and $A[D]$ is the MSR (13.3) that would be recorded if the inclusion were D. The goal is to minimize $\mathcal{J}[D]$ (or a regularized version of it) over the shape of the domain D. The method to obtain the first guess is discussed in Sect. 14.3. An iterative algorithm that starts from a first guess and builds a minimizing sequence is proposed in Sect. 14.1. A detailed resolution analysis in the case of a linear array is presented in Sect. 14.2.

14.1   Optimal Control

In this section we propose an iterative algorithm that starts from a first guess and builds a minimizing sequence D_j. The iteration step from D_j to D_{j+1} is carried out by performing a small modification of D_j of the form $D_{j+1} = D_j^{h_j}$ where D^h is defined as in (13.1) and $h_j(x)$ is a function to be identified.

We define the shape derivative of the cost functional $\mathcal{J}[D]$ by

$$(d_{\mathcal{S}}\mathcal{J}[D], h) = \lim_{\delta \to 0} \frac{\mathcal{J}[D^{\delta h}] - \mathcal{J}[D]}{\delta} \,,$$

where $D^{\delta h}$ is defined as in (13.1) with h replaced by δh:

H. Ammari et al., *Mathematical and Statistical Methods for Multistatic Imaging*, Lecture Notes in Mathematics 2098, DOI 10.1007/978-3-319-02585-8_14, © Springer International Publishing Switzerland 2013

$$D^{\delta h} = \{x + \delta h(x)\nu(x), x \in \partial D\}.$$

From (13.4) it follows that the shape derivative of the cost functional $\mathcal{J}$ is given by

$$(d_{\mathcal{S}}\mathcal{J}[D], h) = -\Re e \sum_{n,m=1}^{N} \left[\left(A_{nm}[D] - A_{nm}^{\mathrm{meas}}\right) \int_{\partial D} h(x)\overline{B_{nm}[D](x)} \, d\sigma(x) \right],$$

where the matrix B is defined by (13.5). Motivated by the form of the representation of the shape derivative, we look for the changes h in the vector space spanned by $\{\psi_p\}$ defined by

$$\{\psi_p\} = \{\Re e\,(B_{nm}[D])\}_{n,m=1}^{N} \cup \{\Im m\,(B_{nm}[D])\}_{n,m=1}^{N}.$$

We modify at each iteration step j the boundary ∂D_j to obtain $\partial D_{j+1} = \partial D_j^{h_j}$ by applying the gradient descent method, where

$$\partial D_j^{h_j} := \{x + h_j(x)\nu(x), x \in \partial D_j\}.$$

For doing so, we choose h_j as follows:

$$h_j(x) = -\frac{\mathcal{J}[D_j]}{\sum_p |(d_{\mathcal{S}}\mathcal{J}[D_j], \psi_p)|^2} \sum_p (d_{\mathcal{S}}\mathcal{J}[D_j], \psi_p)\,\psi_p. \tag{14.1}$$

In the case where $\mathcal{J}[D_{j+1}] \geq \mathcal{J}[D_j]$ we choose the descent step length by Armijo's rule [140]. We replace h_j by $h_j/2^s$ where s is the smallest integer such that

$$\mathcal{J}[D_j^{h_j/2^s}] < \mathcal{J}[D_j]; \quad \partial D_j^{h_j/2^s} := \{x + h_j(x)/2^s\,\nu(x), x \in \partial D_j\}.$$

If we have measurements of the MSR matrix at multiple frequencies $(\omega_p)_{p=1,\ldots,P}$ then the change in the step j is given by

$$h_j(x) = \frac{1}{P} \sum_{p=1}^{P} h_j[\omega_p](x). \tag{14.2}$$

14.2 Resolution Analysis in the High-Frequency Regime

In this section we make use of the high-frequency asymptotic expansion (2.103) of the MSR matrix A in order to connect the eigenvalues and the eigenvectors of the MSR matrix to the target shape and give a detailed resolution analysis in a simplified configuration. In order to make explicit calculations we assume that the array of transmitters and receivers is linear.

14.2.1 The Unperturbed Domain

Let us consider the situation in which the array is linear and densely samples the line $\{(y,0),\, y \in]-\alpha/2, \alpha/2[\}$ while the illuminated boundary ∂D_0 of the target is the line

$$\partial D_0 = \{(x, -L_F),\, x \in]-\beta/2, \beta/2[\} \,.$$

Assuming that the distance L_F from the array to the target is much larger than the diameter α of the array and the diameter β of the target, the response matrix is proportional to

$$A_{nm} = \int_{\partial D_0} e^{ik_0[|x-y_m|+|x-y_n|]} d\sigma(x) \,.$$

Using the Taylor series expansion (13.18), we find that, in the Fraunhofer regime $k_0\beta^2/L_F \ll 1$, the response matrix is

$$A_{nm} = \beta e^{2ik_0 L_F + ik_0 \frac{y_m^2+y_n^2}{2L_F}} \operatorname{sinc}\left(\frac{k_0\beta}{2L_F}(y_n + y_m)\right) ,$$

where sinc is the function $\operatorname{sinc}(x) = \sin(x)/x$. Note that the phase factor in the response matrix does not modify the singular values and it only modifies the singular vectors by a phase term independent of the singular value itself. In the following this factor is removed. Therefore, in the continuum approximation (writing $y_m = \alpha y/2$), the response matrix can be replaced by the operator (from $L^2(]-1,1[)$ to $L^2(]-1,1[)$):

$$A[D_0] = \frac{\pi\beta}{C} \mathcal{R}\mathcal{S} \,,$$

where $\mathcal{R}$ is the involution operator

$$\mathcal{R}[f](x) = f(-x) \,,$$

$\mathcal{S}$ is the sinc operator whose kernel is given by

$$\mathcal{S}(x, y) = \frac{\sin[C(x - y)]}{\pi(x - y)}, \qquad x, y \in]-1, 1[\,,$$

and $C = (k_0\beta\alpha)/(4L_F)$.

The singular values $(\sigma^{(l)})_{l\geq 1}$ and singular vectors $(\psi^{(l)})_{l\geq 1}$ of the sinc operator $\mathcal{S}$ are described in Sect. 1.3.2. In particular the singular vectors are the prolate spheroidal functions which are either odd or even functions, so that $(\sigma^{(l)}, \psi^{(l)})_{l\geq 1}$ are also the singular values and vectors of $\mathcal{A}$. We consider the situation $C \gg 1$. According to [150], the important facts in this regime are:

(i) there are about $[2C/\pi]$ significant singular values; more exactly, the first $[2C/\pi]$ singular values are close to one while the following ones are close to zero. The Fourier transforms of the significant singular vectors are concentrated in $]-C, C[$;

(ii) the first singular vectors are concentrated around the center of the interval $]-1, 1[$, and they contain only low-frequencies; more exactly, the first singular vectors are approximately concentrated on an interval with length of the order of $1/\sqrt{C}$ centered at 0, and their Fourier transforms are approximately concentrated on an interval with length of the order of $\sqrt{C}$;

(iii) the last significant singular vectors (i.e., those with indices close to $[2C/\pi]$) are concentrated at the edges of the interval $]-1, 1[$ and their Fourier transforms are approximately concentrated on $]-C, C[$.

14.2.2 The Perturbed Domain

Here we consider the case when the illuminated boundary ∂D of the target is the perturbed curve

$$\partial D = \left\{(x, -L_F + h(x)), \, x \in\,] -\beta/2, \beta/2[\right\} .$$

Denoting $\tilde{h}(y) = h(\beta y/2)$, $y \in\,]-1, 1[$, the response matrix in the continuum approximation can be replaced by the operator

$$\mathcal{A}[D] = \mathcal{A}[D_0] - ik_0\beta R\mathcal{H}[D] ,$$

where the kernel of the operator $\mathcal{H}[D]$ is given by

$$\mathcal{H}[D](x, y) = \int_{-1}^{1} \tilde{h}(z)e^{iCz(x-y)}dz, \quad x, y \in\,]-1, 1[.$$

By expanding the function $\tilde{h}(y)$ over the image basis of the unperturbed operator $\mathcal{A}[D_0]$ (the prolate spheroidal functions $(\psi^{(p)})_p$),

$$\tilde{h}(y) = \sum_{p=1}^{\infty} \tilde{h}_p \psi^{(p)}(y), \quad y \in]-1,1[\,,$$

we find using (1.41) that

$$\langle \psi^{(l')}, \mathcal{H}[D]\psi^{(l)} \rangle = (\mathcal{Q}\tilde{h})_{l'l} = \sum_{p=1}^{\infty} \mathcal{Q}_{l'lp}\tilde{h}_p \,,$$

with

$$\mathcal{Q}_{l'lp} = 2\pi i^{l-l'} \frac{\sqrt{\sigma^{(l)}\sigma^{(l')}}}{C} \int_{-1}^{1} \psi^{(l')}(y)\psi^{(l)}(y)\psi^{(p)}(y)dy \,.$$

Note that $\mathcal{Q}_{l'lp}$ is not vanishing as long as l', l, p are smaller than $[2C/\pi]$. If the response matrix corresponding to the true domain D_{true} is corrupted by an additive Gaussian white noise, then the imaging functional has the following form to leading order in the perturbation (up to a multiplicative constant):

$$\mathcal{J}[D] = \frac{1}{2} \sum_{l=1}^{\infty} \sum_{l'=1}^{\infty} \left| (\mathcal{Q}\tilde{h})_{l'l} - (\mathcal{Q}\tilde{h}^{\mathrm{true}})_{l'l} - \hat{W}_{l'l} \right|^2 \,,$$

where $\hat{W}_{l'l}$ are independent complex Gaussian random variables.

The minimization problem is solved by applying the operator $(\mathcal{Q}^*\mathcal{Q})^{-1}\mathcal{Q}^*$ to the data

$$\left(\langle \psi^{(l')}, \frac{i}{k_0\beta} \mathcal{R}(\mathcal{A}^{\mathrm{meas}} - \mathcal{A}[D_0])\psi^{(l)} \rangle \right)_{l',l} \,.$$

This gives an unbiased estimator of $(\tilde{h}_p^{\mathrm{true}})_p$.

Note that we have using (1.38) that

$$(\mathcal{Q}^*\mathcal{Q})_{p'p} = \frac{4\pi^2}{C^2} \int_{-1}^{1} \int_{-1}^{1} \mathcal{S}(x,y)^2 \psi^{(p')}(x)\psi^{(p)}(y) \, dxdy \,,$$

which is close to the identity operator (up to a factor $4\pi/C$) when restricted to

$$p, p' \le [2C/\pi].$$

Therefore, we come to the following conclusions:

(i) we can reconstruct the coefficients $\tilde{h}_p^{\mathrm{true}}$ up to $p \le (L + L') \wedge [2C/\pi]$;

(ii) the first coefficients $\tilde{h}_p$ (those which are estimated with the highest accuracy) correspond to low-frequency information about the central part of the boundary ∂D;

(iii) the coefficients $\tilde{h}_p$ for p close to $[2C/\pi]$ correspond to high-frequency information about the edges of the boundary ∂D. This implies that, if we want a sharp detection of the edges of the boundary, then we should choose a weighted cost function of the form (13.13) that enhances the contributions of the singular vectors in the plunge region of the singular values;

(iv) the coefficients $\tilde{h}_p$, $p = 1, \ldots, [2C/\pi]$, correspond to features whose minimal wavenumber is $C/(\beta/2) = k_0\alpha/(2L_F)$, which corresponds to a length scale of $2\lambda_0 L_F/\alpha$. This is the classical Rayleigh resolution formula.

14.3 Construction of an Initial Guess

In this section we develop a weighted subspace migration imaging functional for constructing a good initial guess. The idea behind this is to use the high-frequency asymptotic analysis of the MSR matrix. We show the optimality in the presence of an additive noise of the proposed method for choosing the prior guess. Optimality is to be understood in the sense that the location of the maximum of the proposed imaging functional is exactly the maximum likelihood estimator of a sampling of the inclusion shape.

14.3.1 Measurements at a Single Frequency

We first construct an initial guess from measurements of the response matrix at a single frequency ω. Let us introduce the normalized vector field

$$g(x, \omega) = \left(\frac{\exp(ik_0|x - y_n|)}{\sqrt{N}} \right)_{n=1,\ldots,N}. \tag{14.3}$$

The vector $g(x, \omega)$ is the vector of phases of the fundamental solutions from the receiver (and transmitter) array to the point x.

A good initial guess would be obtained using a weighted subspace migration:

$$\mathcal{I}_{\mathrm{SM}}(x, \omega, w) = \overline{g(x, \omega)}^T \sum_{l=1}^N w_l(x, \omega) v_{\mathrm{meas}}^{(l)}[\omega] (v_{\mathrm{meas}}^{(l)})^T [\omega] \overline{g(x, \omega)}$$

$$= \sum_{l=1}^{N} w_l(x,\omega) \langle g(x,\omega), v^{(l)}_{\mathrm{meas}}[\omega] \rangle^2 , \tag{14.4}$$

where $(v^{(l)}_{\mathrm{meas}}[\omega])_{l=1,\ldots,N}$ are the singular vectors of $A^{\mathrm{meas}}[\omega]$ and $w(x,\omega) = (w_l(x,\omega))_{l=1,\ldots,N}$ are filter (complex) weights.

As in Sect. 9.1.2, consider in particular the weights:

$$w^{(1)}_l(x,\omega) = \sigma^{(l)}_{\mathrm{meas}}[\omega] \quad \text{for } l = 1, \ldots, N ,$$

and

$$w^{(2)}_l(x,\omega) = \begin{cases} \exp\left(-i2\arg\langle g(x,\omega), v^{(l)}_{\mathrm{meas}}[\omega]\rangle\right) & \text{for } l \le L , \\ 0 & \text{elsewhere}, \end{cases}$$

where L is the number of the nonzero singular values (i.e., the dimension of the image space of $A^{\mathrm{meas}}[\omega]$). Then $\mathcal{I}_{\mathrm{SM}}(x,\omega,w^{(1)})$ corresponds to Kirchhoff migration:

$$\mathcal{I}_{\mathrm{SM}}(x,\omega,w^{(1)}) = \mathcal{I}_{\mathrm{KM}}(x,\omega) := \overline{g(x,\omega)}^T A^{\mathrm{meas}}[\omega]\overline{g(x,\omega)} . \tag{14.5}$$

An initial guess for the boundary of the target is obtained as the set of points x where $\mathcal{I}_{\mathrm{SM}}(x,\omega,w^{(1)})$ is approximately 1. Moreover, we have the following connection of $\mathcal{I}_{\mathrm{SM}}(x,\omega,w^{(2)})$ to the MUSIC algorithm:

$$\mathcal{I}_{\mathrm{MU}}(x,\omega) = \left\| g(x,\omega) - \sum_{l=1}^{L} \langle v^{(l)}_{\mathrm{meas}}[\omega], g(x,\omega)\rangle v^{(l)}_{\mathrm{meas}}[\omega] \right\|^{-1/2}$$

$$= \left(1 - \sum_{l=1}^{L} |\langle v^{(l)}_{\mathrm{meas}}[\omega], g(x,\omega)\rangle|^2\right)^{-1/2}$$

$$= \left(1 - \mathcal{I}_{\mathrm{SM}}(x,\omega,w^{(2)})\right)^{-1/2} . \tag{14.6}$$

The MUSIC-type imaging functional $\mathcal{I}_{\mathrm{MU}}(x,\omega)$ has large peaks at the boundary of the target.

The next subsection will make it clear that an appropriate weighted subspace migration is optimal to find an initial guess in the presence of additive noise.

14.3.2 Optimality

We present here a particular context in which a weighted subspace migration imaging functional gives the "optimal" approach to choosing the prior guess for the target support D, or rather the illuminated part of its boundary. This generalizes the results of Chap. 9 obtained for a point target to the case of an extended target.

For simplicity, we drop in this section the dependence on the frequency ω from the notation. We assume the following model for the data

$$A^{\mathrm{meas}} \approx \sum_{l=1}^{L} \tau_l \, g(x_l) g(x_l)^T + \sigma_{\mathrm{noise}} W \; ,$$

where $\{x_l\}$ is a sampling of the boundary of the extended target. Here L is an estimated signal space dimension, $W \in \mathbb{C}^{N \times N}$ models additive noise and is a matrix of independent and identically distributed random coefficients, that have circularly symmetric Gaussian with variance one

$$\mathbb{E}[(\Re e W_{nm})^2] = \mathbb{E}[(\Im m W_{nm})^2] = 1/2 \; ,$$

$g(x)$ is defined by (14.3), and

$$g(x_j) \perp g(x_l), j \neq l, \quad \text{i.e., } \langle g(x_j), g(x_l) \rangle = 0 \; .$$

Since the unperturbed response matrix is symmetric, we symmetrize the measured response, which reduces the variance of the noisy coefficients: Recall that the measured response matrix is symmetrized, so the additive noise also undergoes the same transformation:

$$A^{\mathrm{meas,s}} = \frac{1}{2}\big(A^{\mathrm{meas}} + (A^{\mathrm{meas}})^T\big) \approx \sum_{l=1}^{L} \tau_l \, g(x_l) g(x_l)^T + \sigma_{\mathrm{noise}} W^{\mathrm{s}} \; ,$$

where $W^{\mathrm{s}} \in \mathbb{C}^{N \times N}$ is a symmetric matrix of random coefficients that have circularly symmetric Gaussian with variance one on the diagonal and $1/2$ off the diagonal:

$$\mathbb{E}[(\Re e W_{nn})^2] = \mathbb{E}[(\Im m W_{nn})^2] = 1/2, \qquad n = 1, \ldots, N \; ,$$
$$\mathbb{E}[(\Re e W_{nm})^2] = \mathbb{E}[(\Im m W_{nm})^2] = 1/4, \qquad 1 \leq n < m \leq N \; .$$

It is also worth emphasizing that the orthogonality assumption

$$g(x_j) \perp g(x_l), \quad j \neq l$$

is ideal. In fact, if the sampling points x_j are well-separated, the distance between the array and the target is large and the illumination is uniform in the angle space, then this orthogonality assumption holds approximately and can be used to provide a good initial guess.

Given the observations $A^{\text{meas,s}}$, we find by using Bayes' theorem with the Jeffreys prior for the parameters (a non-informative prior distribution) that the likelihood function of the parameters $X = (x_j)_{j=1,\ldots,L}$, $\tau = (\tau_1,\ldots,\tau_L)$ and σ_{noise}^2 is proportional to

$$l_0\left(X,\tau,\sigma_{\text{noise}}^2 \mid A^{\text{meas,s}}\right) = \frac{1}{\sigma_{\text{noise}}^{N^2+N+1}}$$
$$\times \exp\left(-\frac{\left\|A^{\text{meas,s}} - \sum_{l=1}^{L} \tau_l g(x_l)g(x_l)^T\right\|_F^2}{\sigma_{\text{noise}}^2}\right).$$
$$(14.7)$$

The maximum likelihood estimate of X and the nuisance parameters σ_{noise} and τ are found by maximizing the likelihood function (14.7) with respect to these:

$$\left(\hat{X},\hat{\tau},\hat{\sigma}_{\text{noise}}^2\right) = \underset{X,\tau,\sigma_{\text{noise}}^2 \mid g(x_j)\perp g(x_l), j\neq l}{\text{argmax}} \; l_0\left(X,\tau,\sigma_{\text{noise}}^2 \mid A^{\text{meas,s}}\right).$$

We first eliminate σ_{noise}^2 by requiring

$$\frac{\partial l_0\left(X,\tau,\sigma_{\text{noise}}^2 \mid A^{\text{meas,s}}\right)}{\partial \sigma_{\text{noise}}} = 0.$$

This gives

$$\hat{\sigma}_{\text{noise}}^2 = 2\frac{\left\|A^{\text{meas,s}} - \sum_{l=1}^{L} \tau_l g(x_l)g(x_l)^T\right\|_F^2}{N^2 + N + 1},$$

and the likelihood ratio is proportional to

$$l_0\left(X,\tau,\hat{\sigma}_{\text{noise}}^2 \mid A^{\text{meas,s}}\right) \approx \left\|A^{\text{meas,s}} - \sum_{l=1}^{L} \tau_l g(x_l)g(x_l)^T\right\|_F^{-(N^2+N+1)/2}.$$

Denote by $A^{\text{meas,s}} = \sum_{j=1}^{N} \sigma_{\text{meas}}^{(j)} v_{\text{meas}}^{(j)} {v_{\text{meas}}^{(j)}}^T$ the symmetric singular value decomposition of the symmetric matrix $A^{\text{meas,s}}$. We have

$$\left\|A^{\text{meas,s}} - \sum_{l=1}^{L} \tau_l g(x_l)g(x_l)^T\right\|_F^2 = \left\|\tilde{v} - \sum_{l=1}^{L} \tau_l \tilde{g}^{(l)}\right\|_2^2$$

for $\tilde{v}$ and $\tilde{g}^{(l)}$ the N^2-dimensional vector defined by

$$\tilde{v}_{(m-1)N+n} = \sum_{j=1}^{N} \sigma_{\text{meas}}^{(j)} v_{\text{meas},m}^{(j)} \otimes v_{\text{meas},n}^{(j)}, \qquad n, m = 1, \ldots, N$$

$$\tilde{g}_{(m-1)N+n}^{(l)} = g_m(x_l) \otimes g_n(x_l), \qquad n, m = 1, \ldots, N ,$$

and $\|\tilde{v}\|_2^2 = \sum_{n=1}^{N^2} |\tilde{v}_n|^2$. Using that $\tilde{g}(x_j) \perp \tilde{g}(x_l)$ for $j \neq l$ we obtain

$$\left\| A^{\text{meas,s}} - \sum_{l=1}^{L} \tau_l g(x_l) g(x_l)^T \right\|_F^2 = \sum_{l=1}^{L} \| \tilde{v} - \tau_l \tilde{g}(x_l) \|_2^2 - (L-1)\|\tilde{v}\|_2^2 .$$

This allows us to identify the minimizer in τ:

$$\hat{\tau} = \operatorname*{argmin}_{\tau} \sum_{l=1}^{L} \| \tilde{v} - \tau_l \tilde{g}(x_l) \|_2^2 = \big(\Re \langle \tilde{v}, \tilde{g}(x_l) \rangle \big)_{l=1,\ldots,L} ,$$

where we have taken into account the fact that $\|\tilde{g}(x_l)\|_2 = 1$. We therefore conclude that the estimate $\hat{X}$ derives from

$$\hat{X} = \operatorname*{argmin}_{X | g(x_j) \perp g(x_l), j \neq l} \sum_{l=1}^{L} \| \tilde{v} - \Re \langle \tilde{v}, \tilde{g}(x_l) \rangle \tilde{g}(x_l) \|_2^2 .$$

Note that

$$\sum_{l=1}^{L} \| \tilde{v} - \Re \langle \tilde{v}, \tilde{g}(x_l) \rangle \tilde{g}(x_l) \|_2^2 = L\|\tilde{v}\|_2^2 - \sum_{l=1}^{L} |\Re \langle \tilde{g}(x_l), \tilde{v} \rangle|^2$$

$$= L\|\tilde{v}\|_2^2 - \sum_{l=1}^{L} \left| \Re \sum_{l'=1}^{N} \sigma_{\text{meas}}^{(l')}[\omega] \langle g(x_l, \omega), v_{\text{meas}}^{(l')}[\omega] \rangle^2 \right|^2 .$$

From this representation we find that the estimates of the locations $X = (x_l)_{l=1,\ldots,L}$ can be expressed in terms of the weighted subspace migration $\mathcal{I}_{\text{SM}}$ with the weights $w^{(1)} = (\sigma_{\text{meas}}^{(l)})_{l=1,\ldots,L}$, which is the KM functional $\mathcal{I}_{\text{KM}}$ by (14.5):

$$\hat{X} = \operatorname*{argmax}_{X | g(x_j) \perp g(x_l), j \neq l} \sum_{l=1}^{L} |\Re \mathcal{I}_{\text{KM}}(x_l, \omega)|^2 . \qquad (14.8)$$

This gives then an algorithm for the prior guess:

(i) Compute the KM map $\mathcal{I}_{\text{KM}}(x, \omega)$;
(ii) By parameterizing the curve corresponding to the illuminated part of the boundary of the inclusion with L points separated by approximately $\lambda/2$, and by maximizing $\sum_{l=1}^{L} |\Re e \, \mathcal{I}_{\text{KM}}(x_l, \omega)|^2$ over the positions of the L points, we obtain the initial guess. Here, λ is the wavelength.

Note that the weighted subspace migration with the weights $w^{(1)}$, corresponding to KM, is more appropriate for the initial guess than the weighted subspace migration $\mathcal{I}_{\text{SM}}$ with the weights $w^{(2)}$, corresponding to MUSIC.

We remark that the implementation regarding the identification of the points approximating the boundary may be carried out recursively. It is then relevant to project the signal space and illumination vectors on the complement of the illumination space associated with the points already identified. That is, one may implement the prior guess identification as:

(i') Identify $\hat{x}_1$ as the spatial location maximizing $|\mathcal{I}_{\text{KM}}(x, \omega)|^2$.
(ii') Given $\hat{x}_1, \cdots, \hat{x}_{k-1}$ identify $\hat{x}_k$ as the location separated approximately by $\lambda_0/2$ from previously identified points and maximizing the imaging function associated with the projected signal space and illumination vectors:

$$\breve{A}^{\text{meas},\text{s}} = \Pi_{1,k-1} A^{\text{meas},\text{s}} \Pi_{1,k-1}, \qquad \breve{g}(x_k) = \Pi_{1,k-1} g(x_k) , \quad (14.9)$$

for

$$\Pi_{1,k-1} = I - \sum_{j=1}^{k-1} g(\hat{x}_j) g(\hat{x}_j)^T .$$

14.3.3 Measurements at Multiple Frequencies

In the case of measurements at multiple frequencies one can use the following imaging functional

$$\mathcal{I}_{\text{SM}}(x, w) = \frac{1}{P} \sum_{p=1}^{P} \mathcal{I}_{\text{SM}}(x, \omega_p, w) , \qquad (14.10)$$

where $\mathcal{I}_{\text{SM}}(x, \omega_p, w)$ is given by (14.4) and P is the number of used frequencies, in order to get an initial guess of the illuminated part of the inclusion D. This can be justified by the following approximate calculations. For any smooth function $\tilde{a}(y)$ and boundary ∂D,

$$\sum_{m,n=1}^{N} \frac{1}{P} \sum_{p=1}^{P} e^{-i\omega_p(|x-y_m|+|x-y_n|)} \int_{\partial D} \tilde{a}(y) e^{i\omega_p(|y-y_m|+|y-y_n|)} \, d\sigma(y)$$

$$= \sum_{m,n=1}^{N} \int_{\partial D} \tilde{a}(y) \left[\frac{1}{P} \sum_{p=1}^{P} e^{-i\omega_p(|x-y_m|-|y-y_m|+|x-y_n|-|y-y_n|)} \right] d\sigma(y)$$

$$\approx \sum_{m,n=1}^{N} \int_{\partial D} \tilde{a}(y) \delta(|x-y_m| + |x-y_n| - |y-y_m| - |y-y_n|) \, d\sigma(y)$$

$$\approx N^2 \int_{\partial D} \tilde{a}(y) \delta_x(y) \, d\sigma(y) \approx \begin{cases} N^2 \tilde{a}(x) & \text{if } x \in \partial D, \\ 0 & \text{elsewhere.} \end{cases}$$

It is possible to do a detailed analysis of the previous sum along the same lines as in [82]. It would exhibit that the final Dirac distribution is in fact a sharp peak whose width depends on the bandwidth and on the geometry of the array. Here this approximate calculation is sufficient to justify that (14.10) gives a reasonable initial guess. Therefore, it follows from (2.103) that in order to construct an initial guess one can use (14.10). This is good in absence of additive noise. In the presence of additive Gaussian white noise (which gives independent noises for each frequency), we can repeat the Bayesian arguments of the previous subsection, and we find the following algorithm for the prior guess:

(i) Compute the KM map $\mathcal{I}_{\mathrm{KM}}(x, \omega)$;
(ii) By parameterizing the curve corresponding to the illuminated part of the boundary of the inclusion with L points separated by approximately $\lambda_0/2$, and by maximizing $\frac{1}{P} \sum_{p=1}^{P} \sum_{l=1}^{N} |\Re e\, \mathcal{I}_{\mathrm{KM}}(x_l, \omega_p)|^2$ over the positions of the L points, we obtain the initial guess. Here, λ_0 is the central wavelength.

Note that we should look for the maximum of the sum of the square moduli of the KM functionals in order to exploit the multi-frequency information optimally. The fact that the relevant operation is the sum of the squares comes from the fact that the additive noise matrices are assumed to be independent for different frequencies.

14.4 A General Remark on Least-Squares Imaging and Bayesian Approach

In this section we would like to connect the least-squares imaging strategy and the Bayesian approach for imaging in the presence of additive noise. Let us consider a general situation where the unperturbed $N \times N$ MSR matrix $A^0[f]$ (i.e., in the absence of additive noise) depends on the quantity of interest f in a general way, for instance $A^0[D]$ that depends on the shape of the inclusion D (see (13.3)), or $A^0[V]$ that depends on the reflectivity function $V(z)$ (see (9.1)), or $A^0[(z_j)_{j=1,\ldots,r}, (\rho_j)_{j=1,\ldots,r}]$ that depends on the locations $(z_j)_{j=1,\ldots,r}$ and reflectivity coefficients $(\rho_j)_{j=1,\ldots,r}$ of a collection of r point reflectors (see (6.15)). Let us denote by f the quantity of interest. In the presence of additive noise, the measured MSR matrix is

$$A^{\mathrm{meas}} = A^0[f] + W ,$$

where W is a matrix of independent and identically distributed complex random coefficients with Gaussian distribution, mean zero, and variance $\sigma_{\mathrm{noise}}^2$. This means that, given f, the PDF of the matrix A^{meas} is

$$p(A^{\mathrm{meas}}|f) = c_{N,\sigma} \exp\left(-\frac{\|A^{\mathrm{meas}} - A^0[f]\|_F^2}{\sigma_{\mathrm{noise}}^2} \right) ,$$

where $\| \cdot \|_F$ is the Frobenius norm and $c_{N,\sigma} = \pi^{-N^2} \sigma_{\mathrm{noise}}^{-2N^2}$ is a normalizing constant. The Bayesian theory tells us that, given the observation A^{meas}, the best estimate $\hat{f}$ (called maximum likelihood estimator) of the quantity of interest f is the maximizer of the likelihood function

$$p(f|A^{\mathrm{meas}}) = p(A^{\mathrm{meas}}|f)\pi(f) ,$$

which is obtained from Bayes's formula (1.52) and which depends on the prior distribution $\pi(f)$. This prior distribution models the typical behavior that we expect from the unknown quantity f, for instance that it belongs to a Banach space H with norm $\| \cdot \|_H$. Then the prior distribution takes the form

$$\pi(f) = \exp\left(-\alpha \|f\|_H^2 \right) ,$$

for some positive constant α whose role will be clarified below. As a consequence the best estimate $\hat{f}$ is

$$\hat{f} = \operatorname*{argmax}_{f \in H} \exp\left(-\frac{\|A^{\mathrm{meas}} - A^0[f]\|_F^2}{\sigma_{\mathrm{noise}}^2} - \alpha \|f\|_H^2 \right) ,$$

which is equivalent to

$$\hat{f} = \operatorname*{argmin}_{f \in H} \left\{ \|A^{\mathrm{meas}} - A^0[f]\|_F^2 + \alpha \sigma_{\mathrm{noise}}^2 \|f\|_H^2 \right\} .$$

Therefore, it appears that the Bayesian approach gives an estimation of the quantity of interest that is the solution of a regularized least-squares problem. In this minimization problem the parameter α plays the role of a regularization parameter. The resolution of the minimization problem can be done iteratively as we discuss in this chapter or in a one-shot approximate method (by computing an approximate solution without any iteration).

Bibliography and Discussion

The results of this chapter are from [21]. Discrepancy functionals similar to $\mathcal{J}_2$ and $\mathcal{J}_3$ introduced in the previous chapter can be used [23] in order to enhance the resolution of the reconstructed image. The optimal control approach can be interpreted as a Landweber iteration scheme and its convergence can be proved using the recent results in [96].

Part VII
Invisibility

Chapter 15
GPT- and S-Vanishing Structures for Near-Cloaking

The concept of GPTs is used in Chaps. 4 and 11 for resolved imaging, identification, and tracking of small targets. In this chapter, we use the concept of GPTs for constructing cloaking devices.

Cloaking is to make a target invisible with respect to probing by electromagnetic waves. An extensive work has been produced on cloaking in the context of conductivity and electromagnetism. Many schemes are under active current investigations. These include exterior cloaking [12, 132, 133], active cloaking [89, 90], transmission line cloaking [155], and interior cloaking [34, 35, 84, 85, 118, 119, 122], which is the focus of this chapter.

In interior cloaking, the difficulty is to construct material parameter distributions of a cloaking structure such that any target placed inside the structure is undetectable to waves. One approach is to use transformation optics [58, 85, 94, 144, 146, 156]. It takes advantage of the fact that the equations governing electrostatics, electromagnetism, and acoustics have transformation laws under change of variables. This allows one to design structures that bend waves around a hidden region, returning them to their original path on the far side. The change of variables based cloaking method uses a singular transformation to boost the material properties so that it makes a cloaking region look like a point to outside measurements. However, this transformation induces the singularity of material constants in the transversal direction (also in the tangential direction in two dimensions), which causes difficulty both in the theory and applications. To overcome this weakness, so called 'near-cloaking' is naturally considered, which is a regularization or an approximation of singular cloaking. In [119], instead of the singular transformation, the authors use a regular one to push forward the material constant in the conductivity equation describing the static limit of electromagnetism, in which a small ball is blown up to the cloaking region. In [118], this regularization point of view is adopted for the Helmholtz equation. See also [125, 136].

H. Ammari et al., *Mathematical and Statistical Methods for Multistatic Imaging*, Lecture Notes in Mathematics 2098, DOI 10.1007/978-3-319-02585-8_15, © Springer International Publishing Switzerland 2013

In [34, 35], a new cancellation technique designed to achieve enhanced cloaking from measurements of the Dirichlet-to-Neumann map in electrostatics and the scattering cross section in electromagnetism is proposed. The approach is to first design a multi-coated structure around a small perfect insulator to significantly reduce its effect on boundary or scattering cross section measurements. The multi-coating cancels the generalized polarization tensors or the scattering coefficients of the cloaking device. One then obtains a near-cloaking structure by pushing forward the multi-coated structure around a small object via the standard blow-up transformation technique.

The purpose of this chapter is to review this cancellation technique. We first design a structure coated around an inclusion to have vanishing GPTs of lower orders and show that the order of perturbation due to a small inclusion can be reduced significantly. We then obtain near-cloaking structure by pushing forward the multi-coated structure around a small object via the usual blow-up transformation. For the conductivity equation, we show that, in two dimensions, the order of near-cloaking is δ^{2N} using N coatings, which is a significant improvement over δ^2 approximation obtained in [119]. It is worth noticing that similar type of reduction of scattering was considered in [2, 3]. However, in [2, 3] materials of negative refractive index are needed.

When considering near-cloaking for the Helmholtz equation, we construct structures such that their first scattering coefficients vanish. Analogously to the conductivity case, we prove that, after a transformation optics, structures with vanishing scattering coefficients enhance near-cloaking. We also show that near-cloaking for the Helmholtz equation becomes increasingly difficult as the cloaked object becomes bigger or the operating frequency becomes higher. The difficulty scales inversely proportionally to the object diameter or the frequency.

This chapter is organized as follows. In the next section we characterize structures with vanishing GPTs. In Sect. 15.1 we show that the near-cloaking is enhanced (to δ^{2N}) if such structures are used. In Sect. 15.2 we extend to scattering problems the cancellation technique. We design a structure coated around a perfect insulator to have vanishing scattering coefficients of lower orders and show that the order of the scattering cross section of a small perfect insulator can be reduced significantly. We then analogously obtain near-cloaking structure by pushing forward the multi-coated structure around a small object via the standard blow-up transformation. We emphasize that such a structure achieves near-cloaking for a band of frequencies.

Even though we consider only two dimensional conductivity equation in this chapter, the same argument can be applied to the conductivity and Helmholtz equations in three dimensions.

15.1 Near-Cloaking for the Conductivity Equation

To explain the principle of construction of cloaking structures, we review the results on the conductivity equation obtained in [34].

Let Ω be a domain in $\mathbb{R}^2$ containing 0 possibly with multiple components with smooth boundary. For a given harmonic function H in $\mathbb{R}^2$, consider

$$\begin{cases} \nabla \cdot \left(\sigma_0 \chi(\mathbb{R}^2 \setminus \overline{\Omega}) + \sigma \chi(\Omega) \right) \nabla u = 0 & \text{in } \mathbb{R}^2 , \\ u(x) - H(x) = O(|x|^{-1}) & \text{as } |x| \to \infty , \end{cases} \qquad (15.1)$$

where σ_0 and σ are conductivities (positive constants) of $\mathbb{R}^2 \setminus \overline{\Omega}$ and Ω, respectively.

If the harmonic function H admits the expansion

$$H(x) = H(0) + \sum_{n=1}^{\infty} r^n \left(a_n^c(H) \cos n\theta + a_n^s(H) \sin n\theta \right)$$

with $x = (r \cos \theta, r \sin \theta)$, then, analogously to (7.12), we have the following formula

$$(u - H)(x) = - \sum_{m=1}^{\infty} \frac{\cos m\theta}{2\pi m r^m} \sum_{n=1}^{\infty} \left(M_{mn}^{cc} a_n^c(H) + M_{mn}^{cs} a_n^s(H) \right)$$
$$- \sum_{m=1}^{\infty} \frac{\sin m\theta}{2\pi m r^m} \sum_{n=1}^{\infty} \left(M_{mn}^{sc} a_n^c(H) + M_{mn}^{ss} a_n^s(H) \right) \quad \text{as } r = |x| \to \infty ,$$
$$(15.2)$$

where $M_{mn}^{cc}, M_{mn}^{cs}, M_{mn}^{sc}$, and M_{mn}^{sc} are the contracted generalized polarization tensors defined by (7.6)–(7.9).

In order to make u look like H for large $|x|$, we construct structures with vanishing contracted generalized polarization tensors for all $|n|, |m| \leq N$. We call such structures GPT-vanishing structures of order N. For doing so, we use a disc with multiple coatings. Let Ω be a disc of radius r_1. For a positive integer N, let $0 < r_{N+1} < r_N < \ldots < r_1$ and define

$$A_j := \{ r_{j+1} < r = |x| \leq r_j \}, \quad j = 1, 2, \ldots, N . \qquad (15.3)$$

Let $A_0 = \mathbb{R}^2 \setminus \overline{\Omega}$ and $A_{N+1} = \{ r \leq r_{N+1} \}$. Set σ_j to be the conductivity of A_j for $j = 1, 2, \ldots, N+1$, and $\sigma_0 = 1$. Let

$$\sigma = \sum_{j=0}^{N+1} \sigma_j \chi(A_j) . \qquad (15.4)$$

Because of the symmetry of the disc, one can easily see that

$$M_{mn}^{cs}[\sigma] = M_{mn}^{sc}[\sigma] = 0 \quad \text{for all } m, n , \tag{15.5}$$

$$M_{mn}^{cc}[\sigma] = M_{mn}^{ss}[\sigma] = 0 \quad \text{if } m \neq n , \tag{15.6}$$

and

$$M_{nn}^{cc}[\sigma] = M_{nn}^{ss}[\sigma] \quad \text{for all } n . \tag{15.7}$$

Let $M_n = M_{nn}^{cc}$, $n = 1, 2, \ldots$, for the simplicity of notation. Let

$$\zeta_j := \frac{\sigma_j - \sigma_{j-1}}{\sigma_j + \sigma_{j-1}}, \quad j = 1, \ldots, N+1 . \tag{15.8}$$

The following is a characterization of GPT-vanishing structures. See [34].

Proposition 15.1. *If there are nonzero constants $\zeta_1, \ldots, \zeta_{N+1}$ ($|\zeta_j| < 1$) and $r_1 > \ldots > r_{N+1} > 0$ such that*

$$\prod_{j=1}^{N+1} \begin{bmatrix} 1 & \zeta_j r_j^{-2l} \\ \zeta_j r_j^{2l} & 1 \end{bmatrix} \text{ is an upper triangular matrix for } l = 1, 2, \ldots, N, \tag{15.9}$$

then (Ω, σ), given by (15.3), (15.4), and (15.8), is a GPT-vanishing structure of order N, i.e., $M_l = 0$ for $l \leq N$. More generally, if there are nonzero constants $\zeta_1, \zeta_2, \zeta_3, \ldots$ ($|\zeta_j| < 1$) and $r_1 > r_2 > r_3 > \ldots$ such that r_n converges to a positive number, say $r_\infty > 0$, and

$$\prod_{j=1}^{\infty} \begin{bmatrix} 1 & \zeta_j r_j^{-2l} \\ \zeta_j r_j^{2l} & 1 \end{bmatrix} \text{ is an upper triangular matrix for every } l, \tag{15.10}$$

then (Ω, σ), given by (15.3), (15.4), and (15.8), is a GPT-vanishing structure with $M_l = 0$ for all l.

Let (Ω, σ) be a GPT-vanishing structure of order N of the form (15.4). We take $r_1 = 2$ so that Ω is the disk of radius 2, and $r_{N+1} = 1$. We assume that $\sigma_{N+1} = 0$ which amounts to that the structure is insulated along ∂B_1. For small $\delta > 0$, let

$$\Psi_{\frac{1}{\delta}}(x) = \frac{1}{\delta} x, \quad x \in \mathbb{R}^2 . \tag{15.11}$$

Then, $(B_{2\delta}, \sigma \circ \Psi_{\frac{1}{\delta}})$ is a GPT-vanishing structure of order N and it is insulated on ∂B_δ.

For a given domain Ω and a subdomain $B \subset \Omega$, we introduce the DtN map $\Lambda_{\Omega,B}[\sigma]$ as

$$\Lambda_{\Omega,B}[\sigma](f) = \sigma \frac{\partial u}{\partial \nu}\Big|_{\partial\Omega} , \tag{15.12}$$

where u is the solution to

$$\begin{cases} \nabla \cdot \sigma \nabla u = 0 & \text{in } \Omega \setminus \overline{B} , \\ \dfrac{\partial u}{\partial \nu} = 0 & \text{on } \partial B , \\ u = f & \text{on } \partial\Omega , \end{cases} \tag{15.13}$$

where ν is the outward normal to ∂B. Note that with $\Omega = B_2$, $\Lambda_{\Omega,B_\delta}[\sigma \circ \Psi_{\frac{1}{\delta}}]$ may be regarded as small perturbation of $\Lambda_{\Omega,\emptyset}[1]$ if $M_l = 0$ for all $l \leq N$. In fact, a complete asymptotic expansion of $\Lambda_{\Omega,B_\delta}[\sigma \circ \Psi_{\frac{1}{\delta}}]$ as $\delta \to 0$ is obtained and it is proved that

$$\left\| \Lambda_{B_2,B_\delta}\left[\sigma \circ \Psi_{\frac{1}{\delta}}\right] - \Lambda_{B_2,\emptyset}[1] \right\| \leq C\delta^{2N+2}$$

for some constant C independent of δ, where the norm is the operator norm from $H^{1/2}(\partial\Omega)$ into $H^{-1/2}(\partial\Omega)$. We then push forward $\sigma \circ \Psi_{\frac{1}{\delta}}$ by the change of variables F_δ,

$$F_\delta(x) := \begin{cases} \left(\dfrac{3-4\delta}{2(1-\delta)} + \dfrac{1}{4(1-\delta)}|x|\right)\dfrac{x}{|x|} & \text{for } 2\delta \leq |x| \leq 2 , \\ \left(\dfrac{1}{2} + \dfrac{1}{2\delta}|x|\right)\dfrac{x}{|x|} & \text{for } \delta \leq |x| \leq 2\delta , \\ \dfrac{x}{\delta} & \text{for } |x| \leq \delta , \end{cases} \tag{15.14}$$

in other words,

$$(F_\delta)_\star(\sigma \circ \Psi_{\frac{1}{\delta}}) = \frac{(DF_\delta)(\sigma \circ \Psi_{\frac{1}{\delta}})(DF_\delta)^T}{|\det(DF_\delta)|} \circ F_\delta^{-1} . \tag{15.15}$$

Note that F_δ maps $|x| = \delta$ onto $|x| = 1$, and is the identity on $|x| = 2$. So by invariance of the DtN map, we have

$$\Lambda_{B_2,B_1}\left[(F_\delta)_\star(\sigma \circ \Psi_{\frac{1}{\delta}})\right] = \Lambda_{B_2,B_\delta}\left[\sigma \circ \Psi_{\frac{1}{\delta}}\right] . \tag{15.16}$$

Thus we obtain the following theorem from [34], which shows that, using GPT-vanishing structures we achieve enhanced near-cloaking.

Theorem 15.2. *Let the conductivity profile σ be a GPT-vanishing structure of order N such that $\sigma_{N+1} = 0$. There exists a constant C independent of δ such that*

$$\left\| \Lambda_{B_2,B_1}\left[(F_\delta)_*(\sigma \circ \Psi_{\frac{1}{\delta}})\right] - \Lambda_{B_2,\emptyset}[1] \right\| \leq C\delta^{2N+2} . \tag{15.17}$$

15.2 Near-Cloaking for the Helmholtz Equation

15.2.1 Scattering Coefficients

Let D be a bounded domain in $\mathbb{R}^2$ with smooth boundary ∂D, and let (ε_0, μ_0) be the pair of electromagnetic parameters (permittivity and permeability) of $\mathbb{R}^2 \setminus \bar{D}$ and (ε_*, μ_*) be that of D. Then the permittivity and permeability distributions are given by

$$\varepsilon = \varepsilon_0 \chi(\mathbb{R}^2 \setminus \bar{D}) + \varepsilon_* \chi(D) \quad \text{and} \quad \mu = \mu_0 \chi(\mathbb{R}^2 \setminus \bar{D}) + \mu_* \chi(D) . \tag{15.18}$$

Given a frequency ω, set $k_* = \omega\sqrt{\varepsilon_*\mu_*}$ and $k_0 = \omega\sqrt{\varepsilon_0\mu_0}$. For a function U satisfying $(\Delta + k_0^2)U = 0$ in $\mathbb{R}^2$, we consider the scattered wave u, i.e., the solution to (5.9).

Suppose that U is given by a plane wave $e^{ik_0\xi\cdot x}$ with ξ being on the unit circle, then (5.23) yields

$$u(x) - e^{ik_0\xi\cdot x} = -\frac{i}{4}\sum_{n\in\mathbb{Z}} H_n^{(1)}(k_0|x|)e^{in\theta_x} \sum_{m\in\mathbb{Z}} W_{nm} e^{im(\frac{\pi}{2}-\theta_\xi)} \quad \text{as } |x| \to \infty , \tag{15.19}$$

where W_{nm}, given by (5.15), are the scattering coefficients, $\xi = (\cos\theta_\xi, \sin\theta_\xi)$, and $x = (|x|, \theta_x)$.

We now show that the scattering coefficients are basically the Fourier coefficients of the far-field pattern (the scattering amplitude) which is 2π-periodic function in two dimensions. The far-field pattern $A_\infty[\varepsilon, \mu, \omega]$, when the incident field is given by $e^{ik_0\xi\cdot x}$, is defined to be

$$u(x) - e^{ik_0\xi\cdot x} = -ie^{-\frac{\pi i}{4}}\frac{e^{ik_0|x|}}{\sqrt{8\pi k_0|x|}} A_\infty[\varepsilon, \mu, \omega](\theta_\xi, \theta_x) + o(|x|^{-\frac{1}{2}}) \quad \text{as } |x| \to \infty . \tag{15.20}$$

Recall that

$$H_0^{(1)}(t) \sim \sqrt{\frac{2}{\pi t}} e^{i(t-\frac{\pi}{4})} \quad \text{as } t \to \infty , \tag{15.21}$$

where $\sim$ indicates that the difference between the right-hand and left-hand side is $O(t^{-1})$. If $|x|$ is large while $|y|$ is bounded, then we have

$$|x - y| = |x| - |y| \cos(\theta_x - \theta_y) + O(\frac{1}{|x|}),$$

and hence

$$H_0^{(1)}(k_0|x - y|) \sim e^{-\frac{\pi i}{4}} \sqrt{\frac{2}{\pi k_0 |x|}} e^{ik_0(|x| - |y| \cos(\theta_x - \theta_y))} \quad \text{as } |x| \to \infty .$$

Thus, from (5.10), we get

$$u(x) - e^{ik_0 \xi \cdot x} \sim -ie^{-\frac{\pi i}{4}} \frac{e^{ik_0|x|}}{\sqrt{8\pi k_0 |x|}} \int_{\partial D} e^{-ik_0|y| \cos(\theta_x - \theta_y)} \psi(y) \, d\sigma(y) \quad (15.22)$$

as $|x| \to \infty$ and infer that the far-field pattern is given by

$$A_\infty(\theta_\xi, \theta_x) = \int_{\partial D} e^{-ik_0|y| \cos(\theta_x - \theta_y)} \psi(y) \, d\sigma(y) , \qquad (15.23)$$

where ψ is given by (5.25).

Let

$$A_\infty(\theta_\xi, \theta_x) = \sum_{n \in \mathbb{Z}} b_n(\theta_\xi) e^{in\theta_x}$$

be the Fourier series of $A_\infty(\theta_\xi, \cdot)$. From (15.23) it follows that

$$b_n(\theta_\xi) = \frac{1}{2\pi} \int_0^{2\pi} \int_{\partial D} e^{-ik_0|y| \cos(\theta_x - \theta_y)} \psi(y) \, d\sigma(y) \, e^{-in\theta_x} \, d\theta_x$$

$$= \frac{1}{2\pi} \int_{\partial D} \int_0^{2\pi} e^{-ik_0|y| \cos(\theta_x - \theta_y)} e^{-in\theta_x} \, d\theta_x \, \psi(y) \, d\sigma(\theta_y) .$$

Since

$$\frac{1}{2\pi} \int_0^{2\pi} e^{-ik_0|y| \cos(\theta_x - \theta_y)} e^{-in\theta_x} \, d\theta_x = J_n(k_0|y|) e^{-in(\theta_y + \frac{\pi}{2})} ,$$

we deduce that

$$b_n(\theta_\xi) = \int_{\partial D} J_n(k_0|y|) e^{-in(\theta_y + \frac{\pi}{2})} \psi(y) \, d\sigma(\theta_y) .$$

Using (5.25) we now arrive at the following theorem.

Theorem 15.1. *Let θ and θ' be respectively the incident and scattered direction. Then we have*

$$A_\infty[\varepsilon, \mu, \omega](\theta, \theta') = \sum_{n,m \in \mathbb{Z}} i^{(m-n)} e^{in\theta'} W_{nm}[\varepsilon, \mu, \omega] e^{-im\theta} , \qquad (15.24)$$

where the scattering coefficients W_{nm} are defined by (5.15).

We emphasize that the series in (15.24) is well-defined provided that k_0^2 is not a Dirichlet eigenvalue for $-\Delta$ on D. Moreover, it converges uniformly in θ and θ' thanks to (5.16). Furthermore, there exists $\delta_0 > 0$ such that for any $\delta \leq \delta_0$ the series expansion of $A_\infty[\varepsilon, \mu, \delta\omega](\theta, \theta')$ is well-defined and its convergence of is uniform in δ. This is the key point of our construction of near-cloaking structures. We also note that if U is given by (5.21) then the scattering amplitude, which we denote by $A_\infty[\varepsilon, \mu, \omega](U, \theta')$, is given by

$$A_\infty[\varepsilon, \mu, \omega](U, \theta') = \sum_{n \in \mathbb{Z}} i^{-n} e^{in\theta'} \sum_{m \in \mathbb{Z}} W_{nm} a_m(U) . \qquad (15.25)$$

The conversion of the far-field to the near field is achieved via formula (15.19).

The scattering cross section $S[\varepsilon, \mu, \omega]$ is defined by

$$S[\varepsilon, \mu, \omega](\theta') := \int_0^{2\pi} \left| A_\infty[\varepsilon, \mu, \omega](\theta, \theta') \right|^2 d\theta . \qquad (15.26)$$

See [54, 153]. As an immediate consequence of Theorem 15.1 we obtain the following corollary.

Corollary 15.3. *We have*

$$S[\varepsilon, \mu, \omega](\theta') = 2\pi \sum_{m \in \mathbb{Z}} \left| \sum_{n \in \mathbb{Z}} i^{-n} W_{nm}[\varepsilon, \mu, \omega] e^{in\theta'} \right|^2 . \qquad (15.27)$$

It is worth mentioning that the optical theorem [54, 153] leads to a natural constraint on W_{nm}. In fact, we have

$$\Im m \, A_\infty[\varepsilon, \mu, \omega](\theta', \theta') = -\sqrt{\frac{\omega}{8\pi}} S[\varepsilon, \mu, \omega](\theta'), \quad \forall \, \theta' \in [0, 2\pi] , \qquad (15.28)$$

or equivalently,

$$\Im m \sum_{n,m \in \mathbb{Z}} i^{m-n} e^{i(n-m)\theta'} W_{nm}[\varepsilon, \mu, \omega] = -\sqrt{\frac{\pi\omega}{2}} \sum_{m \in \mathbb{Z}} \left| \sum_{n \in \mathbb{Z}} i^{-n} W_{nm}[\varepsilon, \mu, \omega] e^{in\theta'} \right|^2 ,$$

$$(15.29)$$

$\forall \, \theta' \in [0, 2\pi]$.

In the next subsection, we compute the scattering coefficients of multiply coated inclusions and provide structures whose scattering coefficients vanish. Such structures will be used to enhance near-cloaking. Any target placed inside such structures will have nearly vanishing scattering cross section S, uniformly in the direction θ'.

15.2.2 S-Vanishing Structures

The purpose of this subsection is to construct multiply layered structures whose scattering coefficients vanish. We call such structures *S-vanishing structures*. We design a multi-coating around an insulated inclusion D, for which the scattering coefficients vanish. The computations of the scattering coefficients of multi-layered structures (with multiple phase electromagnetic materials) follow in exactly the same way as in Sect. 5.3. The system of two equations (5.11) should be replaced by a system of $2\times$ the number of phase interfaces (-1 if the core is perfectly insulating).

For positive numbers $r_1, \ldots, r_{L+1}$ with $2 = r_1 > r_2 > \cdots > r_{L+1} = 1$, let

$$A_j := \{x : r_{j+1} \le |x| < r_j\}, \quad j = 1, \ldots, L, \quad A_0 := \mathbb{R}^2 \setminus \overline{A_1},$$

and

$$A_{L+1}(= D) := \{x : |x| < 1\}.$$

Let (μ_j, ε_j) be the pair of permeability and permittivity of A_j for $j = 0, 1, \ldots, L+1$. Set $\mu_0 = 1$ and $\varepsilon_0 = 1$. Let

$$\mu = \sum_{j=0}^{L+1} \mu_j \chi(A_j) \quad \text{and} \quad \varepsilon = \sum_{j=0}^{L+1} \varepsilon_j \chi(A_j). \tag{15.30}$$

In this case the scattering coefficient $W_{nm} = W_{nm}[\mu, \varepsilon, \omega]$ can be defined using (5.22). In fact, if u is the solution to

$$\nabla \cdot \frac{1}{\mu} \nabla u + \omega^2 \varepsilon u = 0 \quad \text{in } \mathbb{R}^2 \tag{15.31}$$

with the outgoing radiation condition on $u - U$ where U is given by (5.21), then $u - U$ admits the asymptotic expansion (5.22) with $k_0 = \omega \sqrt{\varepsilon_0 \mu_0}$.

Exactly like the conductivity case, one can show using symmetry that

$$W_{nm} = 0 \quad \text{if } m \ne n. \tag{15.32}$$

Let us define W_n by

$$W_n := W_{nn} \; . \tag{15.33}$$

Our purpose is to design, given N and ω, μ and ε so that $W_n[\mu, \varepsilon, \omega] = 0$ for $|n| \le N$. We call such a structure (μ, ε) an *S-vanishing structure of order N at frequency ω*. Since $H_{-n}^{(1)} = (-1)^n H_n^{(1)}$ and $J_{-n} = (-1)^n J_n$, we have

$$W_{-n} = W_n \; , \tag{15.34}$$

and hence it suffices to consider W_n only for $n \ge 0$.

Note that (15.29) leads to

$$\Im m \sum_{n \in \mathbb{Z}} W_n[\varepsilon, \mu, \omega] = -\sqrt{\frac{\pi \omega}{2}} \sum_{n \in \mathbb{Z}} \left| W_n[\varepsilon, \mu, \omega] \right|^2 . \tag{15.35}$$

Let $k_j := \omega \sqrt{\mu_j \varepsilon_j}$ for $j = 0, 1, \ldots, L$. We assume that $\mu_{L+1} = +\infty$, which amounts to that the solution satisfies the zero Neumann condition on $|x| = r_{L+1} (= 1)$. To compute W_n for $n \ge 0$, we look for solutions u_n to (15.31) of the form

$$u_n(x) = a_j^{(n)} J_n(k_j r) e^{in\theta} + b_j^{(n)} H_n^{(1)}(k_j r) e^{in\theta}, \quad x \in A_j, \quad j = 0, \ldots, L , \tag{15.36}$$

with $a_0^{(n)} = 1$. Note that

$$W_n = 4 i b_0^{(n)} \; . \tag{15.37}$$

The solution u_n satisfies the transmission conditions

$$u_n|_+ = u_n|_- \quad \text{and} \quad \frac{1}{\mu_{j-1}} \frac{\partial u_n}{\partial \nu}\Big|_+ = \frac{1}{\mu_j} \frac{\partial u_n}{\partial \nu}\Big|_- \quad \text{on } |x| = r_j$$

for $j = 1, \ldots, L$, which reads

$$\begin{bmatrix} J_n(k_j r_j) & H_n^{(1)}(k_j r_j) \\ \sqrt{\dfrac{\varepsilon_j}{\mu_j}} J_n'(k_j r_j) & \sqrt{\dfrac{\varepsilon_j}{\mu_j}} \left(H_n^{(1)} \right)' (k_j r_j) \end{bmatrix} \begin{bmatrix} a_j^{(n)} \\ b_j^{(n)} \end{bmatrix}$$

$$= \begin{bmatrix} J_n(k_{j-1} r_j) & H_n^{(1)}(k_{j-1} r_j) \\ \sqrt{\dfrac{\varepsilon_{j-1}}{\mu_{j-1}}} J_n'(k_{j-1} r_j) & \sqrt{\dfrac{\varepsilon_{j-1}}{\mu_{j-1}}} \left(H_n^{(1)} \right)' (k_{j-1} r_j) \end{bmatrix} \begin{bmatrix} a_{j-1}^{(n)} \\ b_{j-1}^{(n)} \end{bmatrix} . \tag{15.38}$$

The Neumann condition $\frac{\partial u_n}{\partial \nu}|_+ = 0$ on $|x| = r_{L+1}$ amounts to

$$\begin{bmatrix} 0 & 0 \\ J_n'(k_L) & \left(H_n^{(1)}\right)'(k_L) \end{bmatrix} \begin{bmatrix} a_L^{(n)} \\ b_L^{(n)} \end{bmatrix} = \begin{bmatrix} 0 \\ 0 \end{bmatrix} . \tag{15.39}$$

Combining (15.38) and (15.39), we obtain

$$\begin{bmatrix} 0 \\ 0 \end{bmatrix} = P^{(n)}[\varepsilon, \mu, \omega] \begin{bmatrix} a_0^{(n)} \\ b_0^{(n)} \end{bmatrix} , \tag{15.40}$$

where

$$P^{(n)}[\varepsilon, \mu, \omega] := \begin{bmatrix} 0 & 0 \\ p_{21}^{(n)} & p_{22}^{(n)} \end{bmatrix} = \left(-\frac{\pi}{2}i\omega\right)^L \left(\prod_{j=1}^{L} \mu_j r_j\right) \begin{bmatrix} 0 & 0 \\ J_n'(k_L) & \left(H_n^{(1)}\right)'(k_L) \end{bmatrix}$$

$$\times \prod_{j=1}^{L} \begin{bmatrix} \sqrt{\frac{\varepsilon_j}{\mu_j}} \left(H_n^{(1)}\right)'(k_j r_j) & -H_n^{(1)}(k_j r_j) \\ -\sqrt{\frac{\varepsilon_j}{\mu_j}} J_n'(k_j r_j) & J_n(k_j r_j) \end{bmatrix}$$

$$\times \begin{bmatrix} J_n(k_{j-1} r_j) & H_n^{(1)}(k_{j-1} r_j) \\ \sqrt{\frac{\varepsilon_{j-1}}{\mu_{j-1}}} J_n'(k_{j-1} r_j) & \sqrt{\frac{\varepsilon_{j-1}}{\mu_{j-1}}} \left(H_n^{(1)}\right)'(k_{j-1} r_j) \end{bmatrix} .$$

In order to have a structure whose scattering coefficients W_n vanishes up to the N-order, we need to have $b_0^{(n)} = 0$ (when $a_0^{(n)} = 1$) for $n = 0, \ldots, N$, which amounts to

$$p_{21}^{(n)} = 0 \quad \text{for } n = 0, \ldots, N , \tag{15.41}$$

because of (15.40). We emphasize that $p_{22}^{(n)} \neq 0$. In fact, if $p_{22}^{(n)} = 0$, then (15.40) can be fulfilled with $a_0^{(n)} = 0$ and $b_0^{(n)} = 1$. It means that there exists (μ, ε) on $\mathbb{R}^2 \setminus D$ such that the following problem has a solution:

$$\begin{cases} \nabla \cdot \frac{1}{\mu} \nabla u + \omega^2 \varepsilon u = 0 & \text{in } \mathbb{R}^2 \setminus \overline{D} , \\ \frac{\partial u}{\partial \nu}\Big|_+ = 0 & \text{on } \partial D , \\ u(x) = H_n^{(1)}(k_0 r) e^{in\theta} & \text{for } |x| = r > 2 , \end{cases} \tag{15.42}$$

which is not possible.

We note that (15.41) is a set of conditions on (μ_j, ε_j) and r_j for $j = 1, \ldots, L$. In fact, $p_{21}^{(n)}$ is a nonlinear algebraic function of μ_j, ε_j and r_j, $j = 1, \ldots, L$. We are not able to show existence of (μ_j, ε_j) and r_j, $j = 1, \ldots, L$, satisfying (15.41) even if it is quite important to do so. But the solutions (at fixed frequency) can be computed numerically in the same way as in the conductivity case.

We now consider the S-vanishing structure for all (low) frequencies. Let ω be fixed and we look for a structure (μ, ε) such that

$$W_n[\mu, \varepsilon, \delta\omega] = 0 \quad \text{for all } |n| \leq N \text{ and } \delta \leq \delta_0 \tag{15.43}$$

for some δ_0. Such a structure may not exist. Even numerically, it does not seem to exist. So instead we look for a structure such that

$$W_n[\mu, \varepsilon, \delta\omega] = o(\delta^{2N}) \quad \text{for all } |n| \leq N \text{ and } \delta \to 0 . \tag{15.44}$$

We call such a structure an *S-vanishing structure of order N at low frequencies*.

To investigate the behavior of $W_n[\mu, \varepsilon, \delta\omega]$ as $\delta \to 0$, we need the asymptotic expansions (1.27) and (1.28) of Bessel functions for small arguments. Plugging formulas (1.27) and (1.28) into (15.40), we obtain

$$P^{(0)}[\varepsilon, \mu, \delta\omega]$$

$$= \left(-\frac{\pi}{2}i\delta\omega\right)^L \left(\prod_{j=1}^{L} \mu_j r_j\right) \begin{bmatrix} 0 & 0 \\ -\dfrac{k_L}{2}\delta + O(\delta^3) & \dfrac{2i}{\pi k_L}\delta^{-1} + O(\delta\log\delta) \end{bmatrix}$$

$$\times \prod_{j=1}^{L} \begin{bmatrix} \dfrac{2i}{\pi\omega\mu_j r_j}\delta^{-1} + O(\delta\log\delta) & \dfrac{4}{\pi^2}\left(\dfrac{1}{\omega\mu_{j-1}r_j} - \dfrac{1}{\omega\mu_j r_j}\right)\dfrac{\log\delta}{\delta} + O(\delta^{-1}) \\ \dfrac{r_j}{2}\omega\varepsilon_j\left(1 - \dfrac{\varepsilon_{j-1}}{\varepsilon_j}\right)\delta + O(\delta^3) & \dfrac{2i}{\pi\omega\mu_{j-1}r_j}\delta^{-1} + O(\delta\log\delta) \end{bmatrix}$$

$$= \delta^{-1}\begin{bmatrix} 0 & 0 \\ O(\delta^2) & \dfrac{2i}{\pi k_L}\displaystyle\prod_{j=1}^{L}\dfrac{\mu_j}{\mu_{j-1}} + O(\delta) \end{bmatrix}, \tag{15.45}$$

and, for $n \geq 1$,

$$P^{(n)}[\varepsilon, \mu, \delta\omega]$$

$$= \left(-i\frac{\pi}{2}\delta\omega\right)^L \left(\prod_{j=1}^{L} \mu_j r_j\right)$$

$$\times \begin{bmatrix} 0 & 0 \\ \dfrac{nk_L^{n-1}}{2^n\Gamma(n+1)}\delta^{n-1} + O(\delta^n) & \dfrac{i2^n\Gamma(n+1)}{\pi k_L^{n+1}}\delta^{-n-1} + O(\delta^{-n}) \end{bmatrix}$$

$$\times \prod_{j=1}^{L} \left[\begin{array}{cc} \sqrt{\dfrac{\varepsilon_j}{\mu_j}} \dfrac{i2^n \Gamma(n+1)}{\pi(k_j r_j)^{n+1}} \delta^{-n-1} + O(\delta^{-n}) & \dfrac{i2^n \Gamma(n)}{\pi(k_j r_j)^n} \delta^{-n} + O(\delta^{-n+1}) \\[2ex] -\sqrt{\dfrac{\varepsilon_j}{\mu_j}} \dfrac{n(k_j r_j)^{n-1}}{2^n \Gamma(n+1)} \delta^{n-1} + O(\delta^n) & \dfrac{(k_j r_j)^n}{2^n \Gamma(n+1)} \delta^n + O(\delta^{n+1}) \end{array} \right]$$

$$\times \left[\begin{array}{cc} \dfrac{(k_{j-1} r_j)^n}{2^n \Gamma(n+1)} \delta^n + O(\delta^{n+1}) & -\dfrac{i2^n \Gamma(n)}{\pi(k_{j-1} r_j)^n} \delta^{-n} + O(\delta^{-n+1}) \\[2ex] \sqrt{\dfrac{\varepsilon_{j-1}}{\mu_{j-1}}} \dfrac{n(k_{j-1} r_j)^{n-1}}{2^n \Gamma(n+1)} \delta^{n-1} + O(\delta^n) & \sqrt{\dfrac{\varepsilon_{j-1}}{\mu_{j-1}}} \dfrac{i2^n \Gamma(n+1)}{\pi(k_{j-1} r_j)^{n+1}} \delta^{-n-1} + O(\delta^{-n}) \end{array} \right] ,$$

and hence

$$P^{(n)}[\varepsilon, \mu, \delta\omega] = \frac{1}{2^L} \left[\begin{array}{cc} 0 & 0 \\[1ex] \dfrac{nk_L^{n-1}}{2^n \Gamma(n+1)} \delta^{n-1} + O(\delta^n) & \dfrac{i2^n \Gamma(n+1)}{\pi k_L^{n+1}} \delta^{-n-1} + O(\delta^{-n}) \end{array} \right]$$

$$\times \prod_{j=1}^{L} \left[\begin{array}{cc} a_j(b_j+1) + o(1) & c_j(b_j-1)\delta^{-2n} + o(\delta^{-2n}) \\[1ex] \dfrac{b_j-1}{c_j} \delta^{2n} + o(\delta^{2n}) & \dfrac{b_j+1}{a_j} + o(1) \end{array} \right] , \tag{15.46}$$

where

$$a_j := \left(\frac{k_{j-1}}{k_j} \right)^n , \quad b_j := \frac{\mu_j}{\mu_{j-1}}, \quad c_j := \frac{i2^{2n} \Gamma(n)\Gamma(n+1)}{\pi(k_{j-1} k_j r_j^2)^n} ,$$

with $k_j = \omega\sqrt{\varepsilon_j \mu_j}$.

From the above calculations of the leading order terms of $P^{(n)}[\varepsilon, \mu, \delta\omega]$ and the expansion formula of $J_n(t)$ and $Y_n(t)$, we see that $p_{21}^{(n)}$ and $p_{22}^{(n)}$ admit the following expansions:

$$p_{21}^{(n)}(\mu, \varepsilon, t) = t^{n-1} \left(f_0^{(n)}(\mu, \varepsilon) + \sum_{l=1}^{(N-n)} \sum_{j=0}^{L+1} f_{l,j}^{(n)}(\mu, \varepsilon) t^{2l} (\log t)^j + o(t^{2N-2n}) \right) \tag{15.47}$$

and

$$p_{22}^{(n)}(\mu, \varepsilon, t) = t^{-n-1} \left(g_0^{(n)}(\mu, \varepsilon) + \sum_{l=1}^{(N-n)} \sum_{j=0}^{L+1} g_{l,j}^{(n)}(\mu, \varepsilon) t^{2l} (\log t)^j + o(t^{2N-2n}) \right) \tag{15.48}$$

for $t = \delta\omega$ and some functions $f_0^{(n)}, g_0^{(n)}, f_{l,j}^{(n)}$, and $g_{l,j}^{(n)}$ independent of t.

Lemma 15.2. *For any pair of (μ, ε), we have*

$$g_0^{(n)}(\mu, \varepsilon) \neq 0 . \tag{15.49}$$

Proof. For $n = 0$, it follows from (15.45) that

$$g_0^{(0)}(\mu, \varepsilon) = \frac{2i}{\pi \sqrt{\varepsilon_L \mu_L}} \prod_{j=1}^{L} \frac{\mu_j}{\mu_{j-1}} \neq 0 .$$

Suppose $n > 0$. Assume that there exists a pair of (μ, ε) such that $g_0^{(n)}(\mu, \varepsilon) = 0$. Then the solution given by (15.36) with $a_0^{(n)} = 0$ and $b_0^{(n)} = 1$ satisfies

$$\begin{cases} \nabla \cdot \dfrac{1}{\mu} \nabla u + \delta^2 \omega^2 \varepsilon u = 0 & \text{in } \mathbb{R}^2 \setminus \overline{D} , \\[2mm] \dfrac{\partial u}{\partial \nu}\bigg|_+ = o(\delta^{-n}) & \text{on } \partial D , \\[2mm] u(x) = H_n^{(1)}(\delta k_0 r) e^{in\theta} & \text{for } |x| = r > 2 . \end{cases} \tag{15.50}$$

Let $v(x) := \lim_{\delta \to 0} \delta^n u(x)$. Then using (1.26) it follows that v satisfies

$$\begin{cases} \nabla \cdot \dfrac{1}{\mu} \nabla v = 0 & \text{in } \mathbb{R}^2 \setminus \overline{D} , \\[2mm] \dfrac{\partial v}{\partial \nu}\bigg|_+ = 0 & \text{on } \partial D , \\[2mm] v(x) = -\dfrac{i 2^n \Gamma(n)}{\pi k_0^n} r^{-n} e^{in\theta} & \text{for } |x| = r > 2 , \end{cases} \tag{15.51}$$

which is impossible. Thus $g_0^{(n)}(\mu, \varepsilon) \neq 0$, as desired and the proof is complete.
□

Equations (15.47) and (15.48) together with the above lemma give us the following proposition.

Proposition 15.4. *For $n \geq 1$, let W_n be defined by (15.33). We have*

$$W_n[\mu, \varepsilon, t] = t^{2n} \left(W_n^0[\mu, \varepsilon] + \sum_{l=1}^{(N-n)} \sum_{j=0}^{M_l} W_n^{l,j}[\mu, \varepsilon] t^{2l} (\log t)^j \right) + o(t^{2N}) , \tag{15.52}$$

where $t = \delta \omega$, $M_l := (L + 1)l$ (L being the number of layers), and the coefficients $W_n^0[\mu, \varepsilon]$ and $W_n^{l,j}[\mu, \varepsilon]$ are independent of t.

To construct an S-vanishing structure of order N at low frequencies, we need to have a pair (μ, ε) of the form (15.30) satisfying

$$W_n^0[\mu, \varepsilon] = 0, \text{ and } W_n^{l,j}[\mu, \varepsilon] = 0 \quad \text{for } 0 \le n \le N, \ 1 \le l \le (N-n), 1 \le j \le M_l \ . \tag{15.53}$$

As in the conductivity case, it should be emphasized that one does not know if a solution exists for any order N. Nevertheless, numerical constructions of such structures for small N are given in the last section.

15.2.3 Enhancement of Near-Cloaking

In this section we show that the S-vanishing structures (after a transformation optics) enhance the near-cloaking.

Let (μ, ε) be a S-vanishing structure of order N at low frequencies, i.e., (15.53) holds, and it is of the form (15.30). It follows from (5.17), Theorem 15.1, and Proposition 15.4 that

$$A_\infty[\mu, \varepsilon, \delta\omega](\theta, \theta') = o(\delta^{2N}) \tag{15.54}$$

uniformly in (θ, θ') if $\delta \le \delta_0$ for some δ_0.

Let

$$\Psi_\delta(x) = \frac{1}{\delta}x, \quad x \in \mathbb{R}^2 \ . \tag{15.55}$$

Then we have

$$A_\infty\left[\mu \circ \Psi_\delta, \varepsilon \circ \Psi_\delta, \omega\right] = A_\infty[\mu, \varepsilon, \delta\omega] \ . \tag{15.56}$$

To see this, let u be the solution to

$$\begin{cases} \nabla \cdot \dfrac{1}{(\mu \circ \Psi_\delta)(x)} \nabla u(x) + \omega^2(\varepsilon \circ \Psi_\delta)(x)u(x) = 0 \quad \text{in } \mathbb{R}^2 \setminus \overline{B_\delta} \ , \\ \dfrac{\partial u}{\partial \nu} = 0 \quad \text{on } \partial B_\delta \ , \\ (u - U) \text{ satisfies the outgoing radiation condition,} \end{cases} \tag{15.57}$$

where $U(x) = e^{ik_0(\cos\theta, \sin\theta)\cdot x}$. Here B_δ is the disk of radius δ centered at 0. Define for $y = x/\delta$

$$\tilde{u}(y) := \left(u \circ \Psi_\delta^{-1}\right)(y) = \left(u \circ \Psi_{\frac{1}{\delta}}\right)(y) \quad \text{and} \quad \tilde{U}(y) = \left(U \circ \Psi_{\frac{1}{\delta}}\right)(y) \ .$$

Then, we have

$$
\begin{cases}
\nabla \cdot \dfrac{1}{\mu(y)} \nabla_y \tilde{u}(y) + \delta^2 \omega^2 \varepsilon(y) \tilde{u}(y) = 0 \quad \text{in } \mathbb{R}^2 , \\[2mm]
\dfrac{\partial \tilde{u}}{\partial \nu} = 0 \quad \text{on } \partial B_1 , \\[2mm]
(\tilde{u} - \tilde{U}) \text{ satisfies the outgoing radiation condition.}
\end{cases}
\tag{15.58}
$$

From the definition of the far-field pattern A_∞, we get

$$
(u - U)(x) \sim -ie^{-\frac{\pi i}{4}} \frac{e^{ik_0|x|}}{\sqrt{8\pi k_0 |x|}} A_\infty \Big[\mu \circ \Psi_\delta, \varepsilon \circ \Psi_\delta, \omega \Big](\theta, \theta') \quad \text{as } |x| \to \infty ,
$$

and

$$
(\tilde{u} - \tilde{U})(y) \sim -ie^{-\frac{\pi i}{4}} \frac{e^{i\delta k_0|y|}}{\sqrt{8\pi \delta k_0 |y|}} A_\infty [\mu, \varepsilon, \delta\omega](\theta, \theta') \quad \text{as } |y| \to \infty ,
$$

where $x = |x|(\cos\theta', \sin\theta')$. So, we have (15.56). It then follows from (15.54) that

$$
A_\infty \Big[\mu \circ \Psi_\delta, \varepsilon \circ \Psi_\delta, \omega \Big](\theta, \theta') = o(\delta^{2N}) .
\tag{15.59}
$$

We also obtain from (15.27)

$$
S \Big[\mu \circ \Psi_\delta, \varepsilon \circ \Psi_\delta, \omega \Big](\theta') = o(\delta^{4N}) .
\tag{15.60}
$$

It is worth emphasizing that $(\mu \circ \Psi_\delta, \varepsilon \circ \Psi_\delta)$ is a multi-coated structure of radius 2δ.

We now apply a transformation to the structure $(\mu \circ \Psi_\delta, \varepsilon \circ \Psi_\delta)$ to blow up the small disk of radius δ. Let us recall the following well-known lemma (see, for instance, [85]).

Lemma 15.3. *Let F be a diffeomorphism of $\mathbb{R}^2$ onto $\mathbb{R}^2$ such that $F(x)$ is identity for $|x|$ large enough. If v is a solution to*

$$
\nabla \cdot \frac{1}{\mu} \nabla v + \omega^2 \varepsilon v = 0 \quad \text{in } \mathbb{R}^2 ,
$$

subject to the outgoing radiation condition, then w defined by $w(y) = v(F^{-1}(y))$ satisfies

$$
\nabla \cdot (F_\star \mu) \nabla w + \omega^2 (F_\star \varepsilon) w = 0 \quad \text{in } \mathbb{R}^2 ,
\tag{15.61}
$$

together with the outgoing radiation condition, where

$$(F_\star\mu)(y) = \frac{DF(x))DF^T(x)}{detDF(x)}\frac{1}{\mu(x)} \quad and \quad (F_\star\varepsilon)(y) = \frac{\varepsilon(x)}{detDF(x)} \quad (15.62)$$

with $x = F^{-1}(y)$ and T being the transpose.

For a small number δ, let F_δ be the diffeomorphism defined by

$$F_\delta(x) := \begin{cases} x & \text{for } |x| \geq 2, \\ \left(\dfrac{3-4\delta}{2(1-\delta)} + \dfrac{1}{4(1-\delta)}|x|\right)\dfrac{x}{|x|} & \text{for } 2\delta \leq |x| \leq 2, \\ \left(\dfrac{1}{2} + \dfrac{1}{2\delta}|x|\right)\dfrac{x}{|x|} & \text{for } \delta \leq |x| \leq 2\delta, \\ \dfrac{x}{\delta} & \text{for } |x| \leq \delta . \end{cases} \quad (15.63)$$

We then get from (15.59), (15.60), and Lemma 15.3 the main result of this section.

Theorem 15.4. *If (μ, ε) is a S-vanishing structure of order N at low frequencies, then there exists δ_0 such that*

$$A_\infty\Big[(F_\delta)_\star(\mu \circ \Psi_\delta), (F_\delta)_\star(\varepsilon \circ \Psi_\delta), \omega\Big](\theta, \theta') = o(\delta^{2N}) \quad (15.64)$$

and

$$S\Big[(F_\delta)_\star(\mu \circ \Psi_\delta), (F_\delta)_\star(\varepsilon \circ \Psi_\delta), \omega\Big](\theta') = o(\delta^{4N}) \quad (15.65)$$

for all $\delta \leq \delta_0$, uniformly in θ and θ'. Moreover, the cloaking enhancement, given by (15.64) and (15.65), is achieved for all frequencies smaller than ω.

Since (15.44) holds if we replace ω by $\omega' \leq \omega$, the cloaking enhancement is achieved for all the frequencies smaller than ω. Then it is worth comparing (15.64) with (15.54). In (15.54), (μ, ε) is a multiply layered structure between radius 1 and 2 in which each layer is filled with an isotropic material, and enhanced near-cloaking is achieved for low frequencies $\delta\omega$ with $\delta \leq \delta_0$. On the other hand, in (15.64) the frequency ω does not have to be small. In fact, (15.64) says that for any frequency ω there is a radius δ which yields the enhanced near-cloaking up to $o(\delta^{2N})$.

Bibliography and Discussion

The results of this chapter are from [34] and [35]. Our results on near-cloaking for the Helmholtz equation differ from those in [83, 86, 87, 118, 122]. One of the main reasons is that we are here interested in cloaking with respect to cross-section measurements. Frequencies related to "interior eigenvalues" do not play any role in our construction. Moreover, it seems that asking the question whether the cloaking can be achieved for frequencies from zero to a frequency below some band seems more natural and may be physically more realizable than looking for cloaking devices at all frequencies [84]. We should also emphasize that $(F_\delta)_\star(\mu \circ \Psi_\delta)$ and $(F_\delta)_\star(\varepsilon \circ \Psi_\delta)$ are anisotropic permittivity and permeability distributions. It is not clear whether we can achieve enhanced near-cloaking at high frequencies by using isotropic layers as done in (15.54). Furthermore, real materials are dispersive and thus the broadband cloaking would be challenging to implement. The approach of this chapter can be generalized for the full Maxwell equations [36, 43].

Chapter 16
Anomalous Resonance Cloaking

We consider the dielectric problem with a source term αf, proportional to f, which models the quasi-static (zero-frequency) transverse magnetic regime. The cloaking of the source is achieved in a region external to a plasmonic structure. The plasmonic structure consists of a shell having relative permittivity $-1 + i\delta$ with δ modeling losses.

The cloaking issue is directly linked to the existence of anomalous localized resonance (ALR), which is tied to the fact that an elliptic system of equations can exhibit localization effects near the boundary of ellipticity. The plasmonic structure exhibits ALR if, as the loss parameter δ goes to zero, the magnitude of the quasi-static in-plane electric field diverges throughout a specific region (with sharp boundary not defined by any discontinuities in the relative permittivity), called the anomalous resonance region, but converges to a smooth field outside that region.

To state the problem, let Ω be a bounded domain in $\mathbb{R}^2$ and let D be a domain whose closure is contained in Ω. Throughout this chapter, we assume that Ω and D are smooth. For a given loss parameter $\delta > 0$, the permittivity distribution in $\mathbb{R}^2$ is given by

$$\varepsilon_\delta = \begin{cases} 1 & \text{in } \mathbb{R}^2 \setminus \overline{\Omega}\,, \\ -1 + i\delta & \text{in } \Omega \setminus \overline{D}\,, \\ 1 & \text{in } D\,. \end{cases} \tag{16.1}$$

We may consider the configuration as a core with permittivity 1 coated by the shell $\Omega \setminus \overline{D}$ with permittivity $-1 + i\delta$. For a given function f compactly supported in $\mathbb{R}^2$ satisfying

$$\int_{\mathbb{R}^2} f\, dx = 0 \tag{16.2}$$

H. Ammari et al., *Mathematical and Statistical Methods for Multistatic Imaging*, Lecture Notes in Mathematics 2098, DOI 10.1007/978-3-319-02585-8_16, © Springer International Publishing Switzerland 2013

(which physically is required by conservation of charge), we consider the following dielectric problem:

$$\nabla \cdot \varepsilon_\delta \nabla V_\delta = \alpha f \quad \text{in } \mathbb{R}^2 , \tag{16.3}$$

with the decay condition $V_\delta(x) \to 0$ as $|x| \to \infty$.

A fundamental problem is to identify those sources f such that when $\alpha = 1$ then first

$$E_\delta := \int_{\Omega \backslash \overline{D}} \delta |\nabla V_\delta|^2 \, dx \to \infty \quad \text{as } \delta \to 0 . \tag{16.4}$$

and second V_δ remains bounded outside some radius a:

$$|V_\delta(x)| < C, \quad \text{when } |x| > a \tag{16.5}$$

for some constants C and a independent of δ (which requires that the ball B_a contains the entire region of anomalous localized resonance). The quantity E_δ is proportional to the electromagnetic power dissipated into heat by the time harmonic electrical field averaged over time. Hence (16.4) implies an infinite amount of energy dissipated per unit time in the limit $\delta \to 0$ which is unphysical. If instead we choose $\alpha = 1/\sqrt{E_\delta}$ then the source αf will produce the same power independent of δ and the new associated solution V_δ (which is the previous solution V_δ multiplied by α) will approach zero outside the radius a: cloaking due to anomalous localized resonance (CALR) occurs. The conditions (16.4) and (16.5) are sufficient to ensure CALR: a necessary and sufficient condition is that (with $\alpha = 1$) $V_\delta/\sqrt{E_\delta}$ goes to zero outside some radius as $\delta \to 0$. We also consider a weaker blow-up of the energy dissipation, namely,

$$\limsup_{\delta \to 0} E_\delta = \infty . \tag{16.6}$$

We say that weak CALR takes place if (16.6) holds (in addition to (16.5)). Then the (renormalized) source $f/\sqrt{E_\delta}$ will be essentially invisible for an infinite sequence of small values of δ tending to zero (but would be visible for values of δ interspersed between this sequence if CALR does not additionally hold).

The aim of this chapter is to review a general method based on the potential theory to study cloaking due to anomalous resonance. Using layer potential techniques, we reduce the problem to a singularly perturbed system of integral equations. The system is non-self-adjoint. A symmetrization technique can be applied in the general case [12]. In the case of an annulus (D is the disk of radius r_i and Ω is the concentric disk of radius r_e), it is known [132] that there exists a critical radius (the cloaking radius)

$$r_\star = \sqrt{r_e^3 r_i^{-1}} . \tag{16.7}$$

such that any finite collection of dipole sources located at fixed positions within the annulus $B_{r_*} \setminus \overline{B}_e$ is cloaked. We show that if f is an integrable function supported in $E \subset B_{r_*} \setminus \overline{B}_e$ satisfying (16.2) and the Newtonian potential of f does not extend as a harmonic function in B_{r_*}, then weak CALR takes place. Moreover, we show that if the Fourier coefficients of the Newtonian potential of f satisfy a mild gap condition, then CALR takes place. Conversely we show that if the source function f is supported outside B_{r_*} then (16.4) does not happen and no cloaking occurs.

This chapter is organized as follows. In Sect. 16.1 we transform the problem into a system of integral equations using layer potentials. In Sect. 16.2, we treat the special case of an annulus.

16.1 Layer Potential Formulation

As in Chap. 2, for ∂D or $\partial \Omega$, we denote, respectively, the single and double layer potentials of a function $\phi \in L^2$ as $\mathcal{S}_D[\phi]$ and $\mathcal{D}_\Omega[\phi]$. We also introduce the associated Neumann-Poincaré operators $\mathcal{K}_D$ and $\mathcal{K}_\Omega$.

Let F be the Newtonian potential of f, i.e.,

$$F(x) = \int_{\mathbb{R}^2} \Gamma(x, y) f(y) \, dy, \quad x \in \mathbb{R}^2 . \tag{16.8}$$

Then F satisfies $\Delta F = f$ in $\mathbb{R}^2$, and the solution V_δ to (16.3) may be represented as

$$V_\delta(x) = F(x) + \mathcal{S}_D[\phi_i](x) + \mathcal{S}_\Omega[\phi_e](x) \tag{16.9}$$

for some functions $\phi_i \in L_0^2(\partial D)$ and $\phi_e \in L_0^2(\partial \Omega)$ (L_0^2 is the collection of all square integrable functions with the integral zero). The transmission conditions along the interfaces $\partial \Omega$ and ∂D satisfied by V_δ read

$$(-1 + i\delta) \frac{\partial V_\delta}{\partial \nu} \bigg|_+ = \frac{\partial V_\delta}{\partial \nu} \bigg|_- \quad \text{on } \partial D ,$$

$$\frac{\partial V_\delta}{\partial \nu} \bigg|_+ = (-1 + i\delta) \frac{\partial V_\delta}{\partial \nu} \bigg|_- \quad \text{on } \partial \Omega .$$

Hence the pair of potentials (ϕ_i, ϕ_e) is the solution to the following system of integral equations:

$$\begin{cases} (-1 + i\delta) \dfrac{\partial \mathcal{S}_D[\phi_i]}{\partial \nu_i} \bigg|_+ - \dfrac{\partial \mathcal{S}_D[\phi_i]}{\partial \nu_i} \bigg|_- + (-2 + i\delta) \dfrac{\partial \mathcal{S}_\Omega[\phi_e]}{\partial \nu_i} = (2 - i\delta) \dfrac{\partial F}{\partial \nu_i} & \text{on } \partial D , \\[2mm] (2 - i\delta) \dfrac{\partial \mathcal{S}_D[\phi_i]}{\partial \nu_e} + \dfrac{\partial \mathcal{S}_\Omega[\phi_e]}{\partial \nu_e} \bigg|_+ - (-1 + i\delta) \dfrac{\partial \mathcal{S}_\Omega[\phi_e]}{\partial \nu_e} \bigg|_- = (-2 + i\delta) \dfrac{\partial F}{\partial \nu_e} & \text{on } \partial \Omega . \end{cases}$$

Note that we have used the notation ν_i and ν_e to indicate the outward normal on ∂D and $\partial \Omega$, respectively. Using the jump formula (2.23) for the normal derivative of the single layer potentials, the above equations can be rewritten as

$$
\begin{bmatrix} -z_\delta I + \mathcal{K}_D^* & \dfrac{\partial}{\partial \nu_i}\mathcal{S}_\Omega \\[2mm] \dfrac{\partial}{\partial \nu_e}\mathcal{S}_D & z_\delta I + \mathcal{K}_\Omega^* \end{bmatrix} \begin{bmatrix} \phi_i \\ \phi_e \end{bmatrix} = - \begin{bmatrix} \dfrac{\partial F}{\partial \nu_i} \\[2mm] \dfrac{\partial F}{\partial \nu_e} \end{bmatrix} \tag{16.10}
$$

on $L_0^2(\partial D) \times L_0^2(\partial \Omega)$, where we set

$$
z_\delta = \frac{i\delta}{2(2 - i\delta)}. \tag{16.11}
$$

Note that the operator in (16.10) can be viewed as a compact perturbation of the operator

$$
R_\delta := \begin{bmatrix} -z_\delta I + \mathcal{K}_D^* & 0 \\ 0 & z_\delta I + \mathcal{K}_\Omega^* \end{bmatrix}. \tag{16.12}
$$

From Lemma 2.9, it follows that the eigenvalues of $\mathcal{K}_D^*$ and $\mathcal{K}_\Omega^*$ lie in the interval $]-\frac{1}{2}, \frac{1}{2}]$. Observe that $z_\delta \to 0$ as $\delta \to 0$ and that there are sequences of eigenvalues of $\mathcal{K}_D^*$ and $\mathcal{K}_\Omega^*$ approaching 0 since $\mathcal{K}_D^*$ and $\mathcal{K}_\Omega^*$ are compact. So 0 is the essential singularity of the operator valued meromorphic function

$$
\lambda \in \mathbb{C} \mapsto (\lambda I + \mathcal{K}_\Omega^*)^{-1}.
$$

This causes a serious difficulty in dealing with (16.10). We emphasize that $\mathcal{K}_\Omega^*$ is not self-adjoint in general. In fact, $\mathcal{K}_\Omega^*$ is self-adjoint only when $\partial \Omega$ is a circle or a sphere.

Let $\mathcal{H} = L^2(\partial D) \times L^2(\partial \Omega)$. We write (16.10) in a slightly different form. We first apply the operator

$$
\begin{bmatrix} -I & 0 \\ 0 & I \end{bmatrix} : \mathcal{H} \to \mathcal{H}
$$

to (16.10). Then the equation becomes

$$
\begin{bmatrix} z_\delta I - \mathcal{K}_D^* & -\dfrac{\partial}{\partial \nu_i}\mathcal{S}_\Omega \\[2mm] \dfrac{\partial}{\partial \nu_e}\mathcal{S}_D & z_\delta I + \mathcal{K}_\Omega^* \end{bmatrix} \begin{bmatrix} \phi_i \\ \phi_e \end{bmatrix} = \begin{bmatrix} \dfrac{\partial F}{\partial \nu_i} \\[2mm] -\dfrac{\partial F}{\partial \nu_e} \end{bmatrix}. \tag{16.13}
$$

Let the Neumann-Poincaré-type operator $\mathbb{K}^* : \mathcal{H} \to \mathcal{H}$ be defined by

$$
\mathbb{K}^* := \begin{bmatrix} -\mathcal{K}_D^* & -\dfrac{\partial}{\partial \nu_i} \mathcal{S}_\Omega \\[2ex] \dfrac{\partial}{\partial \nu_e} \mathcal{S}_D & \mathcal{K}_\Omega^* \end{bmatrix} ,
\tag{16.14}
$$

and let

$$
\Phi := \begin{bmatrix} \phi_i \\ \phi_e \end{bmatrix} , \quad g := \begin{bmatrix} \dfrac{\partial F}{\partial \nu_i} \\[2ex] -\dfrac{\partial F}{\partial \nu_e} \end{bmatrix} .
\tag{16.15}
$$

Then, (16.13) can be rewritten in the form

$$
(z_\delta \mathbb{I} + \mathbb{K}^*)\Phi = g ,
\tag{16.16}
$$

where $\mathbb{I}$ is given by

$$
\mathbb{I} = \begin{bmatrix} I & 0 \\ 0 & I \end{bmatrix} .
\tag{16.17}
$$

16.2 Anomalous Resonance in an Annulus

In this section we consider the anomalous resonance when the domains Ω and D are concentric disks. We calculate the explicit form of the limiting solution. Throughout this section, we set $\Omega = B_e = \{|x| < r_e\}$ and $D = B_i = \{|x| < r_i\}$, where $r_e > r_i$.

Because of (2.19) it follows that

$$
\mathbb{K}^* = \begin{bmatrix} 0 & -\dfrac{\partial}{\partial \nu_i} \mathcal{S}_\Omega \\[2ex] \dfrac{\partial}{\partial \nu_e} \mathcal{S}_D & 0 \end{bmatrix} ,
$$

and hence we have from (2.18) that

$$
\mathbb{K}^* \begin{bmatrix} e^{in\theta} \\ 0 \end{bmatrix} = \frac{1}{2} \rho^{|n|+1} \begin{bmatrix} 0 \\ e^{in\theta} \end{bmatrix}
\tag{16.18}
$$

and

$$
\mathbb{K}^* \begin{bmatrix} 0 \\ e^{in\theta} \end{bmatrix} = \frac{1}{2} \rho^{|n|-1} \begin{bmatrix} e^{in\theta} \\ 0 \end{bmatrix}
\tag{16.19}
$$

for all $n \neq 0$, where

$$\rho = \frac{r_i}{r_e} \; .$$

Thus $\mathbb{K}^*$ as an operator on $\mathcal{H}$ has the trivial kernel, i.e.,

$$\operatorname{Ker} \mathbb{K}^* = \{0\} \; . \tag{16.20}$$

According to (16.18) and (16.19), if Φ is given by

$$\Phi = \sum_{n \neq 0} \begin{bmatrix} \phi_i^n \\ \phi_e^n \end{bmatrix} e^{in\theta} \; ,$$

then

$$\mathbb{K}^* \Phi = \sum_{n \neq 0} \begin{bmatrix} \dfrac{\rho^{|n|-1}}{2} \phi_e^n \\ \dfrac{\rho^{|n|+1}}{2} \phi_i^n \end{bmatrix} e^{in\theta} \; .$$

Thus, if g is given by

$$g = \sum_{n \neq 0} \begin{bmatrix} g_i^n \\ g_e^n \end{bmatrix} e^{in\theta} \; ,$$

the integral equations (16.16) are equivalent to

$$\begin{cases} z_\delta \phi_i^n + \dfrac{\rho^{|n|-1}}{2} \phi_e^n = g_i^n \; , \\ z_\delta \phi_e^n + \dfrac{\rho^{|n|+1}}{2} \phi_i^n = g_e^n \; , \end{cases} \tag{16.21}$$

for every $|n| \geq 1$. It is readily seen that the solution $\Phi = (\phi_i, \phi_e)$ to (16.21) is given by

$$\phi_i = 2 \sum_{n \neq 0} \frac{2 z_\delta g_i^n - \rho^{|n|-1} g_e^n}{4 z_\delta^2 - \rho^{2|n|}} e^{in\theta} \; ,$$

$$\phi_e = 2 \sum_{n \neq 0} \frac{2 z_\delta g_e^n - \rho^{|n|+1} g_i^n}{4 z_\delta^2 - \rho^{2|n|}} e^{in\theta} \; .$$

If the source is located outside the structure, i.e., f is supported in $\mathbb{R}^2 \setminus \overline{B}_e$, then the Newtonian potential of f, F, is harmonic in B_e and

$$F(x) = c - \sum_{n \neq 0} \frac{g_e^n}{|n| r_e^{|n|-1}} r^{|n|} e^{in\theta} , \tag{16.22}$$

for $|x| \leq r_e$, where g is defined by (16.15). Thus we have

$$g_i^n = -g_e^n \rho^{|n|-1} . \tag{16.23}$$

Here, g_e^n is the Fourier coefficient of $-\frac{\partial F}{\partial \nu_e}$ on Γ_e, or in other words,

$$-\frac{\partial F}{\partial \nu_e} = \sum_{n \neq 0} g_e^n e^{in\theta} . \tag{16.24}$$

We then get

$$\begin{cases} \phi_i = -2 \sum_{n \neq 0} \dfrac{(2z_\delta + 1)\rho^{|n|-1} g_e^n}{4z_\delta^2 - \rho^{2|n|}} e^{in\theta} , \\[4mm] \phi_e = 2 \sum_{n \neq 0} \dfrac{(2z_\delta + \rho^{2|n|}) g_e^n}{4z_\delta^2 - \rho^{2|n|}} e^{in\theta} . \end{cases} \tag{16.25}$$

Therefore, from (2.17) we find that

$$\mathcal{S}_D[\phi_i](x) + \mathcal{S}_\Omega[\phi_e](x) = \sum_{n \neq 0} \frac{2(r_i^{2|n|} - r_e^{2|n|}) z_\delta}{|n| r_e^{|n|-1} (4z_\delta^2 - \rho^{2|n|})} \frac{g_e^n}{r^{|n|}} e^{in\theta}, \quad r_e < r = |x| , \tag{16.26}$$

and

$$\mathcal{S}_D[\phi_i](x) = -\sum_{n \neq 0} \frac{r_i^{2|n|}(2z_\delta + 1)}{|n| r_e^{|n|-1} (\rho^{2|n|} - 4z_\delta^2)} \frac{g_e^n}{r^{|n|}} e^{in\theta}, \quad r_i < r = |x| < r_e , \tag{16.27}$$

$$\mathcal{S}_\Omega[\phi_e](x) = \sum_{n \neq 0} \frac{(2z_\delta + \rho^{2|n|})}{|n| r_e^{|n|-1} (\rho^{2|n|} - 4z_\delta^2)} g_e^n r^{|n|} e^{in\theta}, \quad r_i < r = |x| < r_e . \tag{16.28}$$

We next obtain the following lemma which provides essential estimates for the investigation of this section.

Lemma 16.1. *There exists δ_0 such that*

$$E_\delta := \int_{B_e \setminus \overline{B_i}} \delta |\nabla V_\delta|^2 \approx \sum_{n \neq 0} \frac{\delta |g_e^n|^2}{|n|(\frac{\delta^2}{4} + \rho^{2|n|})} \tag{16.29}$$

uniformly in $\delta \leq \delta_0$.

Proof. Using (16.22), (16.27), and (16.28), one can see that

$$V_\delta(x) = c + r_e \sum_{n \neq 0} \left[\frac{r_i^{2|n|}}{r^{|n|}} (2z_\delta + 1) - (4z_\delta^2 + 2z_\delta) r^{|n|} \right] \frac{g_e^n e^{in\theta}}{|n| r_e^{|n|} (4z_\delta^2 - \rho^{2|n|})}.$$

Then straightforward computations yield that

$$E_\delta \approx r_e^2 \sum_{n \neq 0} \delta(1 - \rho^{2|n|}) \left| \frac{2z_\delta + 1}{4z_\delta^2 + \rho^{2|n|}} \right|^2 (4|z_\delta|^2 - \rho^{2|n|}) \frac{|g_e^n|^2}{|n|}.$$

If δ is sufficiently small, then one can also easily show that

$$|4z_\delta^2 - \rho^{2|n|}| \approx \frac{\delta^2}{4} + \rho^{2|n|}.$$

Therefore we get (16.29) and the proof is complete.   □

We next investigate the behavior of the series in the right hand side of (16.29). Let

$$N_\delta = \frac{\log(\delta/2)}{\log \rho}. \tag{16.30}$$

If $|n| \leq N_\delta$, then $(\delta/2) \leq \rho^{|n|}$, and hence

$$\sum_{n \neq 0} \frac{\delta |g_e^n|^2}{|n|(\frac{\delta^2}{4} + \rho^{2|n|})} \geq \sum_{0 \neq |n| \leq N_\delta} \frac{\delta |g_e^n|^2}{|n|(\frac{\delta^2}{4} + \rho^{2|n|})} \geq \frac{1}{2} \sum_{0 \neq |n| \leq N_\delta} \frac{\delta |g_e^n|^2}{|n| \rho^{2|n|}}. \tag{16.31}$$

Suppose that

$$\limsup_{|n| \to \infty} \frac{|g_e^n|^2}{|n| \rho^{|n|}} = \infty. \tag{16.32}$$

Then there is a subsequence $\{n_k\}$ with $|n_1| < |n_2| < \cdots$ such that

$$\lim_{k \to \infty} \frac{|g_e^{n_k}|^2}{|n_k| \rho^{|n_k|}} = \infty. \tag{16.33}$$

If we take $\delta = 2\rho^{|n_k|}$, then $N_\delta = |n_k|$ and

$$\sum_{0 \neq |n| \leq N_\delta} \frac{\delta |g_e^n|^2}{|n| \rho^{2|n|}} = \rho^{|n_k|} \sum_{0 \neq |n| \leq |n_k|} \frac{|g_e^n|^2}{|n| \rho^{2|n|}} \geq \frac{|g_e^{|n_k|}|^2}{|n_k| \rho^{|n_k|}}. \tag{16.34}$$

Thus we obtain from (16.29) that

$$\lim_{k \to \infty} E_{\rho^{|n_k|}} = \infty . \tag{16.35}$$

We emphasize that (16.32) is not enough to guarantee (16.4). We now impose an additional condition for CALR to occur. We assume that $\{g_e^n\}$ satisfies the following gap property:

GP: There exists a sequence $\{n_k\}$ with $|n_1| < |n_2| < \cdots$ such that

$$\lim_{k \to \infty} \rho^{|n_{k+1}| - |n_k|} \frac{|g_e^{n_k}|^2}{|n_k| \rho^{|n_k|}} = \infty .$$

If GP holds, then we immediately see that (16.32) holds, but the converse is not true. If (16.32) holds, i.e., there is a subsequence $\{n_k\}$ with $|n_1| < |n_2| < \cdots$ satisfying (16.33) and the gap $|n_{k+1}| - |n_k|$ is bounded, then GP holds. In particular, if

$$\lim_{n \to \infty} \frac{|g_e^n|^2}{|n| \rho^{|n|}} = \infty , \tag{16.36}$$

then GP holds.

Assume that $\{g_e^n\}$ satisfies GP and $\{n_k\}$ is such a sequence. Let $\delta = 2\rho^\alpha$ for some α and let $k(\alpha)$ be the number such that

$$|n_{k(\alpha)}| \le \alpha < |n_{k(\alpha)+1}|.$$

Then, we have

$$\sum_{0 \ne |n| \le N_\delta} \frac{\delta |g_e^n|^2}{|n| \rho^{2|n|}} = \rho^\alpha \sum_{0 \ne |n| \le \alpha} \frac{|g_e^n|^2}{|n| \rho^{2|n|}} \ge \rho^{|n_{k(\alpha)+1}| - |n_{k(\alpha)}|} \frac{|g_e^{n_{k(\alpha)}}|^2}{|n_{k(\alpha)}| \rho^{|n_{k(\alpha)}|}} \to \infty , \tag{16.37}$$

as $\alpha \to \infty$.

We obtain the following lemma.

Lemma 16.2. *If (16.32) holds, then*

$$\limsup_{\delta \to 0} E_\delta = \infty . \tag{16.38}$$

If $\{g_e^n\}$ satisfies the condition GP, then

$$\lim_{\delta \to 0} E_\delta = \infty . \tag{16.39}$$

Suppose that the source function is supported inside the radius $r_\star = \sqrt{r_e^3 r_i^{-1}}$. Then its Newtonian potential cannot be extended harmonically in $|x| < r_\star$ in general. So, if F is given by

$$F = c - \sum_{n \neq 0} a_n r^{|n|} e^{in\theta}, \quad r < r_e, \tag{16.40}$$

then the radius of convergence is less than $r_\star$. Thus we have

$$\limsup_{|n| \to \infty} |n||a_n|^2 r_\star^{2|n|} = \infty, \tag{16.41}$$

i.e., (16.32) holds. The GP condition is equivalent to that there exists $\{n_k\}$ with $|n_1| < |n_2| < \cdots$ such that

$$\lim_{k \to \infty} \rho^{|n_{k+1}| - |n_k|} |n_k||a_{n_k}|^2 r_\star^{2|n_k|} = +\infty. \tag{16.42}$$

The following is the main theorem of this section.

Theorem 16.3. *Let f be a source function supported in $\mathbb{R}^2 \setminus \overline{B}_e$ and F be the Newtonian potential of f.*

(i) *If F does not extend as a harmonic function in $B_{r_\star}$, then weak CALR occurs, i.e.,*

$$\limsup_{\delta \to 0} E_\delta = \infty \tag{16.43}$$

and (16.5) holds with $a = r_e^2/r_i$.

(ii) *If the Fourier coefficients of F satisfy (16.42), then CALR occurs, i.e.,*

$$\lim_{\delta \to 0} E_\delta = \infty \tag{16.44}$$

and (16.5) holds with $a = r_e^2/r_i$.

(iii) *If F extends as a harmonic function in a neighborhood of $\overline{B_{r_\star}}$, then CALR does not occur, i.e.,*

$$E_\delta < C \tag{16.45}$$

for some C independent of δ.

Proof. If F does not extend as a harmonic function in $B_{r_\star}$, then (16.32) holds. Thus we have (16.43). If (16.42) holds, then (16.44) holds by Lemma 16.2. Moreover, by (16.26), we see that

$$|V_\delta| \leq |F| + \sum_{n \neq 0} \left| \frac{2(r_i^{2|n|} - r_e^{2|n|})z_\delta}{|n|r_e^{|n|-1}(4z_\delta^2 - \rho^{2|n|})} \frac{g_e^n}{r^{|n|}} \right| \leq |F| + C \sum_{n \neq 0} \frac{\delta r_e^{|n|}}{(\frac{\delta^2}{4} + \rho^{2|n|})|n|r^{|n|}}$$

$$\leq |F| + C \sum_{n \neq 0} \frac{r_e^{2|n|}}{|n|r_i^{|n|}r^{|n|}} < C, \quad \text{if} \quad r = |x| > \frac{r_e^2}{r_i}$$

for some constants C which may differ at each occurrence.

If F extends as a harmonic function in a neighborhood of $\overline{B_{r_\star}}$, then the power series of F, which is given by (16.22), converges for $r < r_\star + 2\epsilon$ for some $\epsilon > 0$. Therefore there exists a constant C such that

$$\frac{|g_e^n|}{|n|r_e^{|n|-1}} \leq C \frac{1}{(r_\star + \epsilon)^{|n|}}$$

for all n. It then follows that

$$|g_e^n| \leq C(r_e^2 \rho^{-1} + r_e\epsilon)^{-|n|/2} r_e^{|n|} \leq (\rho^{-1} + \epsilon)^{-|n|/2} \tag{16.46}$$

for all n. This tells us that

$$\sum_{n \neq 0} \frac{\delta|g_e^n|^2}{|n|(\delta^2 + \rho^{2|n|})} \leq \sum_{n \neq 0} \frac{|g_e^n|^2}{2|n|\rho^{|n|}} \leq \sum_{n \neq 0} \frac{1}{2|n|(1 + \epsilon\rho)^{|n|}}.$$

This completes the proof.   □

If f is a dipole in $B_{r_\star} \setminus \overline{B}_e$, i.e., $f(x) = a \cdot \nabla\delta_y(x)$ for a vector a and $y \in B_{r_\star} \setminus \overline{B}_e$ where δ_y is the Dirac delta function at y. Then $F(x) = a \cdot \nabla\Gamma(x, y)$. From the expansion (7.11) of the fundamental solution of the Laplacian, we have

$$\Gamma(x, y) = \sum_{n=1}^{\infty} \frac{-1}{2\pi n} \left[\frac{\cos n\theta_y}{r_y^n} r^n \cos n\theta + \frac{\sin n\theta_y}{r_y^n} r^n \sin n\theta \right] + C . \tag{16.47}$$

Then we see that the Fourier coefficients of F has the growth rate r_y^{-n} and satisfies (16.42), and hence CALR takes place. Similarly CALR takes place for a sum of dipole sources at different fixed positions in $B_{r_\star} \setminus \overline{B}_e$. We mention that this fact was found in [132].

If f is a quadrapole, i.e.,

$$f(x) = A : \nabla\nabla\delta_y(x) = \sum_{i,j=1}^{2} a_{ij} \frac{\partial^2}{\partial x_i \partial x_j} \delta_y(x)$$

for a 2×2 matrix $A = (a_{ij})$ and $y \in B_{r_\star} \setminus \overline{B}_e$. Then

$$F(x) = \sum_{i,j=1}^{2} a_{ij} \frac{\partial^2 \Gamma(x,y)}{\partial x_i \partial x_j} \ .$$

Thus CALR takes place. This is in agreement with the numerical result in [139].

If f is supported in $\mathbb{R}^2 \setminus \overline{B}_{r_\star}$, then F is harmonic in a neighborhood of $\overline{B}_{r_\star}$, and hence CALR does not occur by Theorem 16.3. In fact, we can say more about the behavior of the solution V_δ as $\delta \to 0$ which is related to the observation in [133, 138] that in the limit $\delta \to 0$ the annulus itself becomes invisible to sources that are sufficiently far away.

Theorem 16.4. *If f is supported in $\mathbb{R}^2 \setminus \overline{B}_{r_\star}$, then (16.45) holds (with $\alpha = 1$ in (16.3)). Moreover, we have*

$$\sup_{|x| \geq r_\star} |V_\delta(x) - F(x)| \to 0 \quad as \quad \delta \to 0 \ . \tag{16.48}$$

Proof. Since $\operatorname{supp} f \subset \mathbb{R}^2 \setminus \overline{B}_{r_\star}$, the power series of F, which is given by (16.22), converges for $r < r_\star + 2\epsilon$ for some $\epsilon > 0$.

According to (16.26), if $r_e < r = |x|$, then we have

$$V_\delta(x) - F(x) = \sum_{n \neq 0} \frac{2(r_e^{2|n|} - r_i^{2|n|})z_\delta}{|n|r_e^{|n|-1}(\rho^{2|n|} - 4z_\delta^2)} \frac{g_e^n}{r^{|n|}} e^{in\theta} \ .$$

If $|x| = r_\star$, then the identity

$$\frac{(r_e^{2|n|} - r_i^{2|n|})z_\delta}{|n|r_e^{|n|-1}(\rho^{2|n|} - 4z_\delta^2)} \frac{g_e^n}{r_\star^{|n|}} = \frac{(1 - \rho^{2|n|})z_\delta}{(\rho^{|n|} - 4z_\delta^2 \rho^{-|n|})} \frac{g_e^n r_\star^{|n|}}{|n|r_e^{|n|-1}}$$

holds and

$$\left| \frac{(1 - \rho^{2|n|})z_\delta}{(\rho^{|n|} - 4z_\delta^2 \rho^{-|n|})} \right| \leq \left| \frac{1}{(z_\delta^{-1}\rho^{|n|} - z_\delta \rho^{-|n|})} \right|$$

$$\leq \left| \frac{1}{\Im m(z_\delta^{-1}\rho^{|n|} - z_\delta \rho^{-|n|})} \right| = \left(\frac{\delta}{4 + \delta^2} \rho^{-|n|} + \frac{1}{\delta} \rho^{|n|} \right)^{-1} \ .$$

It then follows from (16.46) that

$$|V_\delta(x) - F(x)| \leq 2 \sum_{n \neq 0} \left(\frac{\delta}{4 + \delta^2} \rho^{-|n|} + \frac{1}{\delta} \rho^{|n|} \right)^{-1} \frac{r_e}{|n|} \left(\frac{\rho^{-1}}{\rho^{-1} + \epsilon} \right)^{|n|/2} ,$$

and hence

$$|V_\delta(x) - F(x)| \to 0 \quad \text{as } \delta \to 0 .$$

Since $V_\delta - F$ is harmonic in $|x| > r_e$ and tends to 0 as $|x| \to \infty$, we obtain (16.48) by the maximum principle. This completes the proof. $\qquad\square$

Theorem 16.4 shows that any source supported outside B_{r_*} cannot make the blow-up of the power dissipation happen and is not cloaked. In fact, it is known that we can recover the source f from its Newtonian potential F outside B_{r_*} since f is supported outside $\overline{B}_{r_*}$ (see [100]). Therefore we infer from (16.48) that f may be recovered approximately by observing V_δ outside B_{r_*}.

Bibliography and Discussion

The convergence to a smooth field outside the region was shown in [138], where the first numerical evidence for ALR was also presented. A proof of ALR for a dipolar source outside a plasmonic annulus was given in [133]. The condition for CALR in the annulus case was also derived in [12]. A necessary and sufficient condition on the source term to be cloaked in the general case was derived in [12]. It is based on a symmetrization principle for the associated boundary integral formulation. It is worth mentioning that if the real part of the permittivity of the shell is different from -1, then CALR does not occur [13]. The results of this chapter were extended in [117] to the case when the core D is not radial by a different method based on a variational approach. On the other hand, it was shown in [13] that in three dimensions CALR does not occur. The occurrence of CALR is in fact determined by the eigenvalue distribution of the Neumann-Poincaré-type operator associated with the structure [13]. However, using a shell with a specially designed anisotropic dielectric constant, it is possible to make CALR occur in three dimensions [14].

Part VIII
Numerical Implementations and Results

Chapter 17
Numerical Implementations

This chapter provides MATLAB codes for the main algorithms described in this book.

17.1 Implementation of $\mathcal{K}_D^*$

Recall the definition (2.14) of the Neumann-Poincaré operator $\mathcal{K}_D^*$ which acts on $\phi \in L^2(\partial D)$. Assume that D is represented by N boundary points $\{x_i\}_{i=1\dots N}$, then (2.14) can be approximated by a quadrature formula. This can be implemented using the following MATLAB function.

```matlab
function
Ks = Kstar(D,tvec,normal,avec,Sigma)
% The operator $\Kcal_D^*$.
% Inputs:
% D,tvec,normal,avec,Sigma: the boundary and its parameterizations
% Output:
% Ks: a matrix obtained by quadrature that approximates the integral
% definition of K^*

M = size(D,2); norm_tvec_square = tvec(1,:).^2 + tvec(2,:).^2 ; Ks
= zeros(M, M);

for j = 1:M
    xdoty = (D(1,j)-D(1,:))*normal(1,j)...
                 +(D(2,j)-D(2,:))*normal(2,j);
    norm_xy_square = (D(1,j)-D(1,:)).^2+(D(2,j)-D(2,:)).^2;

    Ks(j, 1:j-1) = 1/(2*pi)*xdoty(1:j-1).*...
                       (Sigma(1:j-1)./norm_xy_square(1:j-1));
    Ks(j, j+1:M) = 1/(2*pi)*xdoty(j+1:M).*...
                       (Sigma(j+1:M)./norm_xy_square(j+1:M));
end

for j = 1:M
    Ks(j,j) = 1/(2*pi)*(-1)/2*avec(:,j)'*normal(:,j)/...
                  norm_tvec_square(j)*Sigma(j);
end
```

H. Ammari et al., *Mathematical and Statistical Methods for Multistatic Imaging*,
Lecture Notes in Mathematics 2098, DOI 10.1007/978-3-319-02585-8_17,
© Springer International Publishing Switzerland 2013

17.2   MSR Matrices

17.2.1   Calculation of the MSR Matrix for Conductivity Problem

Recall that the unique solution u_s of the conductivity problem with a point source at x_s satisfies:

$$u_s(x) - \Gamma_s(x) = \int_{\partial D} \Gamma(x,y)(\lambda I - \mathcal{K}_D^*)^{-1}\left[\frac{\partial \Gamma_s}{\partial \nu}\Big|_{\partial D}\right](y)d\sigma(y) , \qquad (17.1)$$

with $\Gamma_s(x) := \Gamma(x,x_s) = (1/2\pi)\log|x - x_s|$. For each source point x_s, the MSR matrix entry V_{sr} is (17.1) evaluated at the receiver position x_r:

$$V_{sr} = u_s(x_r) - \Gamma_s(x_r) . \qquad (17.2)$$

The MSR matrix V can be simulated by a discretization of (17.1).

```
     function
     MSR = MSR_simulation(kappa, D, tvec, normal, avec, Sigma,
     Xs, Xr)
     % Generate the multi-static response (MSR) matrix of the
5    % conductivity problem (with an inclusion D) using its integral
     % definition.
     % Inputs:
     % kappa: conductivity of the inclusion
     % D, tvec, normal, avec, Sigma: boundary of the inclusion and related
10   % vectors
     % Xs, Xr: the positions of sources and receivers
     % Output:
     % MSR matrix ~ Ns X Nr, MSR(s,r) is the r-th measurement for the s-th
     % source
15
     Ns = size(Xs,2); Nr = size(Xr,2); MSR = zeros(Ns,Nr) ;

     if kappa~=1                              % MSR is zero when k==1
         M = size(D, 2);
20       lambda = (kappa+1)/2/(kappa-1) ;

         Ks = Kstar(D,tvec,normal,avec,Sigma) ;
         A = lambda*eye(M,M)-Ks ;

25       [L,U] = lu(A); % LU decomposition of A

         xy1_D = repmat(D(1,:),size(Xr,2),1)-repmat(Xr(1,:)',1,size(D,2)) ;
         xy2_D = repmat(D(2,:),size(Xr,2),1)-repmat(Xr(2,:)',1,size(D,2)) ;

30       for s=1:Ns
         % Resolution of integral equation
             xy = D-repmat(Xs(:,s),1,M) ;
             source = 1/2/pi*(xy(1,:).*normal(1,:) +...
                     xy(2,:).*normal(2,:))./(xy(1,:).^2+xy(2,:).^2) ;
35           phi = U\(L\source');
             S_phi = 1/4/pi * log(xy1_D.^2 + xy2_D.^2)*(phi.*Sigma(:)) ;

             MSR(s,:) = S_phi';
```

```
         end
40   end
```

17.2.2   SVD of the MSR for the Helmholtz Equation

Let u_s be the wave produced in the presence of a small electromagnetic inclusion D of permittivity ε and permeability μ by a point source at x_s operating at the frequency ω. The wave u_s satisfies

$$\nabla \cdot \left(1 + \left(\frac{1}{\mu} - 1 \right) \chi(D) \right) \nabla u_s(x) + \omega^2 (1 + (\varepsilon - 1)\chi(D)) u_s(x) = -\delta_{x_s}(x) ,$$
(17.3)

together with the outgoing radiation condition. Define the MSR data matrix V as

$$V_{sr} = u_s(x_r) - G_s(x_r) ,$$
(17.4)

with $G_s(x) = \Gamma_{k_0}(x, x_s)$ being the Green function of the Helmholtz equation. Here, $k_0 = \omega\sqrt{\varepsilon_0\mu_0}$ and x_r is the r-th receiver. When there are well-separated multiple inclusions, the MSR data matrix is the sum of the individual MSR matrices. The following function constructs the singular values and the corresponding singular vectors of the MSR matrix, which is generated using the asymptotic formalism of Chap. 7.

```
   function
   [sigmath,Uth]=SVD_Helmholtz(Xs,omega,epsilon,mu,z,delta,theta,a,b)
   % Compute the theoretical singular values and singular vectors of the MSR
   % matrix.
5  % Inputs:
   % Xs: the coordinate of the coincident sources/receivers
   % omega: operating frequency
   % epsilon, mu: permittivity and permeability of the inclusion
   % z: center of the ellipse
10 % delta: characteristic size of the inclusion
   % theta: orientation of the ellipse
   % a,b: semi-major and semi-minor axes of the ellipse
   % Outputs:
   % sigmath: singular values
15 % Uth: matrix of left singular vectors, each column is a singular vector

   N = size(z,2); sigmath = zeros(3*N,1); Uth = zeros(length(Xs),
   3*N);

20 for i=1:N
       G = Green2D_Helmholtz(omega, Xs, z(:,i));
       sigmath((i-1)*3+1) = abs(omega^2*delta(i)^2*pi*a(i)*b(i)*...
                            (epsilon(i)-1)*norm(G)^2);

25     dG = GradGreen2D_Helmholtz(omega, Xs, z(:,i));
       dG = dG.';
       v1=cos(theta(i))*dG(:,1)+sin(theta(i))*dG(:,2);
```

```
       v2=-sin(theta(i))*dG(:,1)+cos(theta(i))*dG(:,2);

30     sigmath((i-1)*3+2)  = abs((1-mu(i))*(a(i)+b(i))/(a(i)*mu(i)+b(i))*...
                             pi*a(i)*b(i)*delta(i)^2*norm(v1)^2);
       sigmath((i-1)*3+3)  = abs((1-mu(i))*(a(i)+b(i))/(a(i)+b(i)*mu(i))*...
                             pi*a(i)*b(i)*delta(i)^2*norm(v2)^2);

35     Uth(:, (i-1)*3+1)   = G / norm(G);
       Uth(:, (i-1)*3+2)   = v1 / norm(v1);
       Uth(:, (i-1)*3+3)   = v2 / norm(v2);
   end
```

17.3 GPTs

17.3.1 Computation of the GPTs

The real contracted GPTs are defined by (7.6)–(7.9). In (7.6)–(7.9) the integrals can be approximated by quadrature formulas.

```
function CGPT =
theoretical_CGPT(D,tvec,normal,avec,Sigma,kappa,ord)
% Compute the CGPTs of the inclusion D up to a given order.
% Inputs:
5   % D, tvec...Sigma: inclusion and its parametrization
    % kappa: contrast
    % ord: maximum order of CGPTs to be computed
    % Outputs:
    % CGPT: a matrix of dimension (2*ord) X (2*ord) with the
10  % o-o(odd-odd), o-e, e-o, e-e(even-even) entries corresponding to
    % respectively CC, CS, SC, and SS matrices.

len_D = size(D, 2); lambda = (kappa+1)/2/(kappa-1) ;

15  A = lambda*eye(len_D, len_D) - Kstar(D,tvec,normal,avec,Sigma) ;
[L,U] = lu(A); % LU decomposition of A

CC = zeros(ord); CS = zeros(ord); SC = zeros(ord); SS =
zeros(ord);
20
for m=1:ord
    dm = (D(1,:) + 1i*D(2,:)).^(m-1);
    gzm = m * [dm; 1i*dm];              % grad(z^m)
    nu_grad = normal(1,:) .* gzm(1,:) + normal(2,:) .* gzm(2,:) ;
25  phi = U\(L\nu_grad(:));

    for n=1:ord
        zn = (D(1,:) + 1i*D(2,:)).^n;

30      CC(m, n) = (real(zn).*real(phi).')*Sigma(:);
        CS(m, n) = (imag(zn).*real(phi).')*Sigma(:);
        SC(m, n) = (real(zn).*imag(phi).')*Sigma(:);
        SS(m, n) = (imag(zn).*imag(phi).')*Sigma(:);
    end
35  end

CGPT = zeros(2*ord);

CGPT(1:2:end, 1:2:end) = CC; CGPT(1:2:end, 2:2:end) = CS;
```

```
40  CGPT(2:2:end, 1:2:end) = SC; CGPT(2:2:end, 2:2:end) = SS;

    % Symmetrization
    CGPT = (CGPT+CGPT')/2;
```

17.3.2 Reconstruction from CGPTs

In this subsection we give the main ingredients of the GPTs matching
algorithm described in Chap. 4. Computation of CGPTs and the corresponding shape derivative function are as follows:

```
    function [CGPT CGPT_vec  SHF_vec] =
    theoretical_CGPT(D,tvec,normal,
                                           avec,Sigma, kappa,ord)
    % Compute the CGPTs of the inclusion D up to a given order
 5  % and the corresponding shape derivative
    % Inputs:
    % D, tvec...Sigma: inclusion and its parametrization
    % kappa: contrast
    % ord: maximum order of CGPTs to be computed
10  % Outputs:
    % CGPT: a matrix of dimension (2*ord) X (2*ord) with the
    % o-o(odd-odd), o-e, e-o, e-e(even-even) entries corresponding to
    % respectively CC, CS, SC, and SS matrices.
    % CGPT_vec: vectorwise stored CGPT
15  % fvec: shape derivative
    SHF_vec = []; CGPT_vec = []; CC = zeros(ord); CS = zeros(ord); SC
    = zeros(ord); SS =zeros(ord); for Hord = 1:(ord-1)
        for Ford = 1:(ord-Hord)
            [a f] = CGPT_HF(D,tvec,normal,avec,Sigma,kappa,Hord, 1, Ford, 1);
20          CC(Hord, Ford) =a; SHF_vec = [SHF_vec f]; CGPT_vec = [CGPT_vec a];
            [a f] =  CGPT_HF(D,tvec,normal,avec,Sigma,kappa,Hord, 1, Ford, 2);
            CS(Hord, Ford) =a; SHF_vec = [SHF_vec f]; CGPT_vec = [CGPT_vec a];
            [a f] =  CGPT_HF(D,tvec,normal,avec,Sigma,kappa,Hord, 2, Ford, 1);
            SC(Hord, Ford) =a; SHF_vec = [SHF_vec f]; CGPT_vec = [CGPT_vec a];
25          a=  CGPT_HF(D,tvec,normal,avec,Sigma,kappa,Hord, 2, Ford, 2);
            SS(Hord, Ford)=a; SHF_vec = [SHF_vec f];CGPT_vec = [CGPT_vec a];
        end
    end CGPT = zeros(2*ord); CGPT(1:2:end, 1:2:end) = CC;
    CGPT(1:2:end, 2:2:end) = CS; CGPT(2:2:end, 1:2:end) = SC;
30  CGPT(2:2:end, 2:2:end) = SS;
    % Symmetrization
    CGPT_mat = (CGPT+CGPT')/2; function [CGPT_HF SHF] =
    CGPT_HF(D,tvec,normal,avec,Sigma,kappa,
                                           Hord, Hcs, Ford, Fcs)
35  % Hord, Ford: order of a harmonic function
    % Hcs, Fcs: 1 if cosine, and 2 if sine
    len_D = size(D, 2); lambda = (kappa+1)/2/(kappa-1) ; A =
    lambda*eye(len_D, len_D) - Kstar(D,tvec,normal,avec,Sigma) ;
    [L,U] = lu(A); % LU decomposition of A
40  %%
    m = Hord; n=Ford;
    zm = (D(1,:) + 1i*D(2,:)).^m;%H
    dm = (D(1,:) + 1i*D(2,:)).^(m-1);
    gzm = m * [dm; 1i*dm];                      % grad(z^m)
45  nu_gradm = normal(1,:) .* gzm(1,:) + normal(2,:) .* gzm(2,:) ;
    phi = U\(L\nu_gradm(:)); %H
    zn = (D(1,:) + 1i*D(2,:)).^n;%F
    dn = (D(1,:) + 1i*D(2,:)).^(n-1);
    gzn = n * [dn; 1i*dn];                      % grad(z^m)
```

```
50   nu_gradn = normal(1,:) .* gzn(1,:) + normal(2,:) .* gzn(2,:) ; psi
     = U\(L\nu_gradn(:)); %%
     if (Hcs == 1)
         phi = real(phi); H=real(zm);
     elseif (Hcs == 2)
55       phi = imag(phi); H=imag(zm);
     else end if (Fcs == 1)
         psi = real(psi); F = real(zn);
     elseif (Fcs == 2)
         psi = imag(psi); F= imag(zn);
60   else end CGPT_HF = (F.*phi.')*Sigma(:);
     SLphi = SingleL(D, D)*phi(1:len_D); % single layer potential
     SLpsi = SingleL(D, D)*psi(1:len_D); SHF0 =
     kappa/(kappa-1)*psi'.*phi'  +  (kappa-1)*(TDeri(F, Sigma)
            + TDeri(SLpsi, Sigma)).*(TDeri(H, Sigma) + TDeri(SLphi, Sigma));
65   aa = fft(SHF0);
     aa(11:length(aa)-9)=0; % low—pass filter
     SHF = ifft(aa); SHF = SHF';
```

Initial Guess from the CGPTs of orders 2 and 3

```
     function D_ellipse = Initial_Guess(CGPT, kappa, nbPoints)
     % make initial guess grom CGPT of order 2 and 3
     [v d] = eig(CGPT(1:2, 1:2)); e =
     (d(2,2)-kappa*d(1,1))/(d(1,1)-kappa*d(2,2));
5    AreaE = d(1,1)*(e+kappa)/(e+1)/(kappa-1); %AreaE = pi*a*b=pi*e*x^2, a=e,
     % b-1
     b = sqrt(AreaE/pi/e); a = e*b;
     %
     angle1 = atan(-v(1,2)/(v(1,1)+0.0001)); if (angle1 < 0)
10       angle1 = angle1+pi;
     end
     % assuming the initial guess is a disk
     center(1) =   .5*CGPT(1,3)*(kappa+1)/(4*(kappa-1)*pi*a*b);
     center(2) = -.5*CGPT(2,3)*(kappa+1)/(4*(kappa-1)*pi*a*b);
15   D_ellipse = make_ellipse(center, a, b, angle1, nbPoints);
```

Update the initial guess

```
     function Dre = Modify_boundary(CGPT_vec, CGPT1_vec, SHF_vec,
                                    D, normal,Sigma, delta)
     % Modify the inclusion's boundary
     % Inputs:
5    % D, tvec...Sigma: initial guess and its parametrization
     % CGPT_vec: CGPT of target
     % CGPT1_vec: CGPT of initial guess
     % SHF_vec: shape derivative
     % Output: modified domain
10   CGPT_diff = CGPT1_vec - CGPT_vec; fvec = []; data_CGPT = []; for
     i= 1:length(CGPT_vec)
         [v c] = GramSch([fvec SHF_vec(:, i)],Sigma);
         vv = v(:, size(v,2))'.^2*Sigma';
         if ( vv > 0.1)
15           fvec = [fvec SHF_vec(:,i)/max(abs(CGPT_vec(i)),1) ];
             data_CGPT = [data_CGPT CGPT_diff(i)/max(abs(CGPT_vec(i)), 1)  ];
         else
         end
     end
20   [SBasis c] = GramSch(fvec, Sigma); %Orthogonalization
     ShapeBasis = []; for l = 1:size(SBasis,2)
         if (sqrt((abs(SBasis(:,l)).^2)'*Sigma') > 1)
             ShapeBasis = [ShapeBasis SBasis(:,l)];
         else
25       end
     end %%
     for i = 1:size(fvec,2)
```

```
          for j=1:size(ShapeBasis,2)
              Fmat(i,j) = (fvec(:,i).*ShapeBasis(:,j))'*Sigma';
30    end
      end coeff=(Fmat'*Fmat+delta*eye(size(Fmat,2)))\(Fmat'*data_CGPT'));
      perD =  ShapeBasis*coeff;
      if (max(abs(perD))>0.2) %
          perD = (0.2/max(abs(perD)))*perD;
35    else end Dre(1,:)=D(1,:) - 1*perD'.*normal(1,:); Dre(2,:)=D(2,:) -
      1*perD'.*normal(2,:); for i = 1:2
          aa = fft(Dre(i,:)); aa(11:length(aa)-9)=0; % low-pass filter
          aa2 = ifft(aa); Dre(i,:) = aa2';
      end
40    function [v c] = GramSch(v, S) %v(right) = v(left)*c, c: coefficient
      c = eye(size(v,2)); c(1,1) = 1; v(:,1) = v(:,1); for ii =
      2:size(v,2)
          for j = 1:ii-1;
              if (abs(v(:,j)).^2 < 0.0001)
45                c(j,ii) = 0;
              else
                  c(j, ii) = v(:, j)'*(S'.*v(:,ii))/(v(:,j)'*(S'.*v(:,j)));
                  v(:, ii) = v(:,ii) - c(j,ii)*v(:,j);
              end
50        end
      end
```

17.3.3 Reconstruction of GPTs from MSR Data

The MSR matrix of the conductivity problem can be related to the CGPT matrix of the shape through the linear operator L defined by (10.1). In the case of a circular configuration with uniformly distributed and coincident sources/receivers, we recall that $A = CDMDC^T$, where the matrices C and D are given by (10.3). Then one can use an analytical formula for reconstructing the CGPTs from MSR data.

```
      function [C,D] = make_matrix_CD(Xs, Z, order)
      % Construct the matrix C and D involved in the linear operator L
      % Inputs:
      % Xs: coordinates of sources/receivers
5     % Z: reference center
      % order: highest order of CGPTs

      N = size(Xs,2); C = zeros(N, 2*order);

10    for n=1:N
          toto = Xs(:,n)-Z(:);
          [T,R] = cart2pol(toto(1), toto(2));

          for m=1:order
15            C(n, 2*m-1:2*m)=[cos(m*T),sin(m*T)];
          end
      end

      tt = (1:order); dd = diag(1./(2*pi*tt.*(R.^tt))); D = kron(dd,
20    eye(2));
```

```
      function CGPT=reconstruct_CGPT_full_aov(MSR, Rs, Ns, ord)
      % Reconstruct the CGPTs up to a certain order. This function works only
      % when the sources and receivers are equally distributed on some
```

```
     % circles around the small inclusion.
5    % Inputs:
     % MSR: data
     % Rs: the distances between the sources and receivers to the reference
     % point
     % Ns: the number of sources/receivers
10   % ord: the maximum (possible) order of CGPTs to be reconstructed
     % Outputs:
     % CGPT: reconstructed CGPTs

     Ts = 2*pi/Ns*(0:Ns-1);
15   Xs = Rs * [reshape(cos(Ts),1,[]); reshape(sin(Ts),1,[])];

     [C,D] = make_matrix_CD(Xs, [0,0]', ord); % construct the matrix C and D
     dd = diag(D);
     iD = diag(1./dd);
20   CGPT = 4*iD*C'*MSR*C*iD/Ns/Ns;
```

In a more general situation, e.g., a limited-view configuration or a
nonuniformly distributed sources/receivers situation, one can solve a least-
squares problem to reconstruct the CGPTs from MSR data using an iterative
method.

```
     function Y = CGPT_op(X,S,R,transp_flag)
     % Linear operator acting on the CGPTs:
     % L(M) := S M R^T
     % S and R are matrices corresponding to the source and receiver arrays
5    % The adjoint of L is:
     % L^T(D) := S^T D R

     [Ms,Ns]=size(S);
     [Mr,Nr]=size(R);
10
     if strcmp(transp_flag,'notransp')
         Y = S * reshape(X, Ns, Nr) * R';
         Y=Y(:);
     elseif strcmp(transp_flag,'transp')
15       Y = S' * reshape(X, Ms, Mr) * R;
         Y=Y(:);
     end
```

```
     function CGPT=reconstruct_CGPT_lsqr(S, R, MSR, maxiter, tol)
     % Reconstruct the contracted GPT (CGPT) by solving the linear system
     % S X R^t = MSR, using the LSQR method.
     % Inputs:
5    % S, R: measurement matrices related to the arrays of sources(S) and
     % receivers(R)
     % MSR: data returned by the function MSR_simulation()
     % maxiter, tol: parameters for LSQR
     % Outputs:
10   % CGPT: reconstructed CGPTs in matrix form

     [Ms,Ns]=size(S);
     [Mr,Nr]=size(R);

15   A = @(x,tflag)CGPT_op(x,S,R,tflag); % The operator L(X)=S*X*R^t

     X = lsqr(A, MSR(:), tol, maxiter);

     CGPT = reshape(X, Ns, Nr);
```

17.3.4 Target Identification Using GPTs

The relation between the complex CGPTs of a shape D and those associated to the rotated, scaled, and translated domain $T_z s R_\theta D$ is implemented in the following function:

```
function [Z1, Z2] = CCGPT_transform(Y1, Y2, T0, S0, Phi0)
% On the CCGPT Y1, Y2, apply first the rotation Phi0,
% then the scaling S0, and finally the translation T0.
% The outputs Z1, Z2 are the transformed Complex CGPTs.

ord = size(Y1, 1); R = repmat((1:ord)', 1, ord); Comb =
zeros(ord);

for n=1:ord
    for m=1:n
        Comb(n,m) = nchoosek(n,m);
    end
end

t0 = (T0(1) + T0(2) * j); Ct = Comb .* (t0.^(R - R'));
Ct(isnan(Ct)) = 0; Ct(isinf(Ct)) = 0;

Gy = diag((S0*exp(j*Phi0)).^(1:ord));

Z1 = Ct * Gy * Y1 * Gy * (Ct.'); Z2 = conj(Ct) * conj(Gy) * Y2 *
Gy * (Ct.');
```

And the inverse transform is as follows:

```
function [Z1, Z2] = CCGPT_inverse_transform(Y1, Y2, T0, S0, Phi0)
% On the CCGPT Y1, Y2, apply first the translation -T0,
% then the scaling 1/S0, and the rotation -Phi0.
% The outputs Z1, Z2 are the transformed Complex CGPT.

ord = size(Y1, 1); R = repmat((1:ord)', 1, ord); Comb =
zeros(ord);

for n=1:ord
    for m=1:n
        Comb(n,m) = nchoosek(n,m);
    end
end

if length(T0)>1 % if a vector is passed
    T0 = (T0(1) + T0(2) * 1i);
end

Ct = Comb .* ((-T0).^(R - R'));

Gy = diag((1/S0*exp(-1i*Phi0)).^(1:ord));

Z1 = Gy * Ct * Y1 * Ct.' * Gy; Z2 = conj(Gy) * conj(Ct) * Y2 *
Ct.' * Gy;
```

Using these relations, we can construct the transform invariant shape descriptors.

```
function [I1,I2] = ShapeDescriptorCGPT(N1, N2)
% [I1,I2] = ShapeDescriptorCGPT(N1, N2);
% Invariant shape descriptor based on CGPTs
% Inputs:
% N1, N2: the complex CGPTs associated to the shape
```

```
   % Outputs:
   % I1, I2: the invariant shape descriptor based on CGPTs

   [M,N] = size(N1); [T1, T2] =
10 CCGPT_inverse_transform(N1,N2,N2(1,2)/N2(1,1)/2,1,0); D =
   diag(1./sqrt(abs(diag(T2))));

   S1 = D * T1 * D; S2 = D * T2 * D; I1 = abs(S1); I2 = abs(S2);
```

17.3.5 Tracking Using GPTs

In order to apply the Extended Kalman Filter for location and orientation tracking using GPTs, we need to implement and linearize the function h defined in (1.68).

```
   function Y = func_h(X, Sm, Rm, M0)
   % Evaluate the h function of the observation equation in EKF
   % Inputs:
   % X: system state vector
 5 % Sm, Rm: measurement matrices corresponding to sources and receivers
   % M0: CGPT matrix
   % Output:
   % Y: observation vector (MSR)
   T0 = X(3:4); Phi0 = X(5);

10
   % Apply first the translation and rotation on original CGPTs
   Mc = CGPT_transform(M0, T0, 1, Phi0);

   Ns=size(Sm,1); Nr=size(Rm,1);
15
   Y=zeros(Ns, Nr);

   % Apply then the acquisition operator on new CGPTs to get the MSR
   Y = Sm * Mc * Rm';
20 Y=Y(:); % Put the result into vector form
```

```
   function Y = func_dh(X, Sm, Rm, M0, nrmlcst)
   % Evaluate the partial derivative of h as a function of the observation
   % equation in Kalman Filter
   % Inputs:
 5 % X: system state vector
   % Sm, Rm: source and receiver matrix returned by matrix_SR_CGPT
   % M0: CGPT matrix
   % Output:
   % Y: partial derivatives

10
   ord = size(M0,1)/2;

   % T0: translation, 2D vector
   % Phi0: rotation, scalar
15 T0 = (X(3) + X(4) * j); Phi0 = X(5);

   % construct the embedding matrix U
   u = [1;j]; U = kron(eye(ord), u); RU = real(U); IU = imag(U);

20 F = zeros(ord); dt1_F = F; dt2_F = F; dp_F = F; for m=1:ord
       for n=m:ord
           Cmn = nchoosek(n,m); % Lower triangle matrix
           F(m,n) = Cmn * T0^(n-m) * exp(j*m*Phi0);
```

```
           if n>m
25             % Parital derivative wrt x axis
               dt1_F(m,n) = Cmn * (n-m) * T0^(n-m-1) * exp(j*m*Phi0);
               % Parital derivative wrt y axis
               dt2_F(m,n) = Cmn * (n-m) * T0^(n-m-1) * exp(j*m*Phi0) * j;
           end
30         % Parital derivative wrt angle
           dp_F(m,n) = Cmn * T0^(n-m) * exp(j*m*Phi0) * j * m;
       end
   end

35 % Building blocks
   RJ = real(U*F); IJ = imag(U*F); % J = U*F;
   dt1_RJ = real(U * dt1_F); dt1_IJ = imag(U * dt1_F); dt2_RJ =
   real(U * dt2_F); dt2_IJ = imag(U * dt2_F); dp_RJ  = real(U *
   dp_F); dp_IJ   = imag(U * dp_F );
40
   dt1_RMR = dt1_RJ' * MO * RJ + RJ' * MO * dt1_RJ; dt1_RMI = dt1_RJ'
   * MO * IJ + RJ' * MO * dt1_IJ; dt1_IMR = dt1_IJ' * MO * RJ + IJ' *
   MO * dt1_RJ; dt1_IMI = dt1_IJ' * MO * IJ + IJ' * MO * dt1_IJ;

45 dt2_RMR = dt2_RJ' * MO * RJ + RJ' * MO * dt2_RJ; dt2_RMI = dt2_RJ'
   * MO * IJ + RJ' * MO * dt2_IJ; dt2_IMR = dt2_IJ' * MO * RJ + IJ' *
   MO * dt2_RJ; dt2_IMI = dt2_IJ' * MO * IJ + IJ' * MO * dt2_IJ;

   dp_RMR = dp_RJ' * MO * RJ + RJ' * MO * dp_RJ; dp_RMI = dp_RJ' * MO
50 * IJ + RJ' * MO * dp_IJ; dp_IMR = dp_IJ' * MO * RJ + IJ' * MO *
   dp_RJ; dp_IMI = dp_IJ' * MO * IJ + IJ' * MO * dp_IJ;

   dt1h = RU * dt1_RMR * RU' + RU * dt1_RMI * IU' +...
          IU * dt1_IMR * RU' + IU * dt1_IMI * IU';
55 dt2h = RU * dt2_RMR * RU' + RU * dt2_RMI * IU' +...
          IU * dt2_IMR * RU' + IU * dt2_IMI * IU';

   dph  = RU * dp_RMR * RU' + RU * dp_RMI * IU' +...
          IU * dp_IMR * RU' + IU * dp_IMI * IU';
60
   % Apply then the acquisition operator
   Ns=size(Sm,1); Nr=size(Rm,1);

   Y=zeros(Ns*Nr, 5); Y(:, 3) = reshape(Sm * dt1h * Rm', [], 1); Y(:,
65 4) = reshape(Sm * dt2h * Rm', [], 1); Y(:, 5) = reshape(Sm * dph *
   Rm', [], 1);
```

The EKF algorithm is implemented below:

```
function  X_tt = Extended_Kalman_Filter(Y, Ntime, F, Q, h, dh, R,
X0, P0, verbose)
% Extended Kalman Filter
% Inputs:
5 % Y: data stream of dimension ? X Nt, with Nt >= Ntime
% Ntime: processing time
% F: system state matrix
% Q: system noise covariance matrix
% h, dh: observation function and its derivatives, function handles
10 % R: observation noise covariance matrix
% X0: guess for initial state
% P0: guess for covariance matrix of the initial state
X_tt = zeros(length(X0), Ntime);

15 Q0 = zeros(size(Q,1), 1); R0 = zeros(size(R,1), 1);

   if nargin <= 8
       P0 = zeros(size(Q));
   end
20
   if nargin <= 9
```

```
        verbose = 0;
    end

25  P_tt = P0;

    for n=1:Ntime
        if verbose
            disp(['Tracking time ', num2str(n), ' of ',num2str(Ntime)]);
30      end

        % Prediction
        if n==1
            X_tn = F * X0;
35      else
            X_tn = F * X_tt(:, n-1);
        end

        yt = Y(:,n) - h(X_tn);
40      H = dh(X_tn);

        P_tn = F * P_tt * F' + Q;

        % Update
45      St = H * P_tn * H' + R;
        Kt = P_tn * H' * inv(St);
        X_tt(:, n) = X_tn + Kt * yt;
        P_tt = P_tn - Kt * H * P_tn;
    end
```

17.4 Implementation of the Imaging Functionals for Point Reflectors

We provide imaging functionals for the localization of point reflectors in free space using MSR data. The theoretical studies were detailed in Chap. 9. For simplicity, we consider the setting where the probing waves have a single frequency.

17.4.1 Reverse-Time Migration Imaging Functional

The following code implements the Reverse-Time migration imaging functional introduced in (9.5).

```
function Imag_Mat =
RTMImagFuncD(omega,XS,YS,U,SIGMA,V,xy_receiver)
% This is the Reverse Time Migration Imaging Functional for the point
% reflectors in the free space
5   % omega: frequency
% XS: x coordinate of a grid
% YS: y coordinate of a grid
% (U,SIGMA,V): SVD of the MSR
% xy_receiver: receiver locations
10
```

```
     Imag_Mat = zeros(size(XS)); K = length(SIGMA);

     for k = 1:K % summing over singular values
         factor_Renorm = zeros(size(Imag_Mat));
15       IM_left = zeros(size(Imag_Mat));
         IM_right = zeros(size(Imag_Mat));
         Sing_VecL = U(:,k);
         Sing_VecR = V(:,k);

20       for i=1:size(xy_receiver,1) % component(receiver)-wise
         G=Green2D(omega,XS,YS,xy_receiver(i,:));% matrix valued
         IM_left=IM_left+conj(G)*Sing_VecL(i);% Sing_vec(i) is scalar
         IM_right=IM_right+conj(G)*conj(Sing_VecR(i));
         factor_Renorm=factor_Renorm+conj(G).*G;
25       end

         Imag_Mat=Imag_Mat+SIGMA(k)*(IM_left.*IM_right)./factor_Renorm;

     end

30
     Imag_Mat = abs(Imag_Mat);
```

17.4.2 Kirchhoff Migration Imaging Functional

The Kirchhoff migration imaging functional (9.6) is realized by the code
below. It differs from the Reverse-Time migration functional in that it only
uses the phase, not the full information of the fundamental solution.

```
     function Imag_Mat = KMImagFuncD(omega,XS,YS,U,SIGMA,V,xy_receiver)
     % Kirchhoff migration imaging functional for dielectric point reflectors
     % omega: frequency (velocity = 1)
     % XS: x coordinates of a grid
5    % YS: y coordinates of a grid
     % (U,SIGMA,V): SVD of MSR. Note SIGMA is just the nonzero list of singular
     % values
     % xy_receiver: receiver positions
     %
10   % Imag_Mat: returned image
     % We note that KM does not account for the amplitude, it only accounts for
     % the phase.

     Imag_Mat = zeros(size(XS));
15
     N_receiver = length(xy_receiver);

     K = length(SIGMA);

20   for k = 1:K % summing over singular values
         IM_left = zeros(size(Imag_Mat));
         IM_right = zeros(size(Imag_Mat));
         Sing_VecL = U(:,k);
         Sing_VecR = V(:,k);
25
         for i=1:size(xy_receiver,1) % component(receiver)-wise
         dist = sqrt((XS-xy_receiver(i,1)).^2 + (YS-xy_receiver(i,2)).^2);
         d_Mat=exp(sqrt(-1)*omega*dist)./sqrt(N_receiver); % matrix valued
         IM_left=IM_left+conj(d_Mat)*Sing_VecL(i); % Sing_Vec(i) is scalar
30       IM_right=IM_right + conj(d_Mat)*conj(Sing_VecR(i));
         end
```

```
        Imag_Mat=Imag_Mat+SIGMA(k)*(IM_left.*IM_right);
     end
35
     Imag_Mat = abs(Imag_Mat);
```

17.4.3 MUSIC Imaging Functional

Finally, we provide the code for the MUSIC Imaging functional defined by (9.7). As discussed before, it relates to certain subspace migration algorithm.

```
function Imag_Mat = MUImagFuncD(omega,XS,YS,U,xy_receiver)
% The MUSIC scheme for imaging a dielectric point reflector
% omega: frequency (velocity = 1)
% XS: x coordinates of a grid
5  % YS: y coordiantes of a grid
% U: singular vectors chosen as subspace
% xy_receiver: the receiver location
%
% Imag_Mat: the image returned
10
Imag_Mat = zeros(size(XS));

n_sig = size(U,2);

15 nsq_proj = zeros(size(XS));

for j = 1:n_sig

IM_temp = zeros(size(Imag_Mat));
20
factor_Renorm = zeros(size(XS));

Sing_Vec = U(:,j);

25 for i=1:size(xy_receiver,1)% component(receiver)-wise
    d_Mat=Green2D(omega,XS,YS,xy_receiver(i,:));% matrix valued
    IM_temp=IM_temp+conj(d_Mat)*Sing_Vec(i);% Sing_Vec(i) is scalar
    factor_Renorm=factor_Renorm+conj(d_Mat).*d_Mat;
end nsq_proj = nsq_proj + abs(IM_temp).^2./factor_Renorm;
30
end Imag_Mat = 1./sqrt(1-nsq_proj);
```

17.5 Implementation of the Imaging Functionals for Inclusions

The imaging functionals for localizing small inclusions depend on mathematical modeling of the underlying physics. One can consider either the conductivity equation or the Helmholtz equation; these equations can be imposed on some bounded domain or in the free space.

Following the theoretical studies in Chap. 8, we consider the setting of Helmholtz equation on some bounded domain. We provide MATLAB codes for the MUSIC, the Backpropagation, the Kirchhoff migration, the Reverse Time migration and the Topological Derivative imaging functional. We also consider the Helmholtz equation in a random medium, i.e., with a clutter noise. For the Helmholtz equation with small inclusions in free space, which is studied in Chap. 6, those imaging functionals should be modified, and we refer the reader to [18, 24].

17.5.1 MUSIC Algorithm

As described above, we consider small inclusions on a bounded domain in the setting of Chap. 8. The response matrix is the weighted boundary measurements (8.9). The MUSIC imaging functional $\mathcal{I}_{\mathrm{MU}}$ is defined in (8.13); the following MATLAB code implements this scheme. Note that we suppose for simplicity that the inclusions are disks.

```
function IM = ImagMU(omega,theta,XS,YS,U)
% MUSIC imaging functional for disk inclusions on bounded domain
% omega: normalized frequency of wave
% theta: directions of plane waves
% XS,YS: searching grid
% U:      range space of the measurement matrix

n = size(theta,1);

d = size(theta,2);

IM = zeros(size(XS));

P = zeros(n,n); L = size(U,2);

for j=1:L
    P = P+U(:,j)*(U(:,j)');
end

for j=1:d%there are d+1 vectors g_j's
    IM_temp = zeros(size(XS));
    g_norm = zeros(size(XS));
    for k=1:n
    gjk = theta(k,j)*exp(1i*omega*(theta(k,1)*XS+theta(k,2)*YS))/sqrt(n);
    g_norm = g_norm+gjk.*conj(gjk);
        for l = 1:n
        gjl = theta(l,j)*exp(1i*omega*(theta(l,1)*XS+theta(l,2)*YS))/
        sqrt(n);
        IM_temp = IM_temp+P(k,l)*conj(gjk).*gjl;
        end
    end
end IM = IM+g_norm-IM_temp;

% now for j=d+1
IM_temp = zeros(size(XS));

g_norm = zeros(size(XS));

for k=1:n gnp1k =
```

```
40  exp(1i*omega*(theta(k,1)*XS+theta(k,2)*YS))/sqrt(n);

    g_norm = g_norm+gnp1k.*conj(gnp1k);
        for l = 1:n
        gnp1l = exp(1i*omega*(theta(l,1)*XS+theta(l,2)*YS))/sqrt(n);
45      IM_temp = IM_temp+P(k,l)*conj(gnp1k).*gnp1l;
        end
    end IM = abs(IM+g_norm-IM_temp);

    % finally take the inverse
50  IM = 1./sqrt(IM);
```

In the above code, the columns of U span the range of the MSR matrix. In practice, the first several left singular vectors are used.

17.5.2   Kirchhoff Migration

The Kirchhoff migration imaging functional $\mathcal{I}_{KM}$ is introduced in (8.16). The measurement matrix A for KM is the same as in MUSIC. This functional is implemented in the following code.

```
    function IM = ImagKM(omega,theta,XS,YS,A)
    % Kirchhoff migration imaging functional for disk inclusions bounded
    % domain
    % omega: normalized frequency of wave
 5  % theta: directions of plane waves
    % XS,YS: searching grid
    % A:     response matrix. One way to save computation is that we add up the
    % product of several vectors.

10  n = size(theta,1); IM = zeros(size(XS)); gjL = zeros(size(XS));
    gjR = zeros(size(XS));

    for j = 1:2
        IM_temp = zeros(size(XS));
15      for k = 1:n
            gjL = theta(k,j)*exp(1i*omega*(theta(k,1)*XS+theta(k,2)*YS))/
            sqrt(n);
            for l = 1:n
            gjR = theta(l,j)*exp(1i*omega*(theta(l,1)*XS+theta(l,2)*YS))/
20          sqrt(n);
            IM_temp = IM_temp+A(k,l)*conj(gjL).*gjR;
            end
        end
        IM = IM+IM_temp;
25  end

    IM_temp = zeros(size(XS));
        for k = 1:n
            gjL = exp(1i*omega*(theta(k,1)*XS+theta(k,2)*YS))/sqrt(n);
30          for l = 1:n
                gjR = exp(1i*omega*(theta(l,1)*XS+theta(l,2)*YS))/sqrt(n);
                IM_temp = IM_temp+A(k,l)*conj(gjL).*gjR;
            end
        end
35  IM = IM+IM_temp;

    IM = abs(IM);
```

17.5.3 Backpropagation

The Backpropagation imaging functional $\mathcal{I}_{BP}$ is defined in (8.14). It is important to notice that the measurement matrix for BP is different from that of MUSIC and KM. The following code implements this imaging functional.

```
function IM = ImagBP(omega,theta,XS,YS,A)
% Backpropagation imaging functional for disk inclusions bounded domain
% omega: normalized frequency of wave
% theta: directions of plane waves
% XS,YS: searching grid
% A:     diagonal response matrix
n = size(theta,1); IM = zeros(size(XS)); gj = zeros(size(XS));

for l=1:n
    gj = exp(-2i*omega*(theta(l,1)*XS+theta(l,2)*YS))/n;
    IM = IM+gj*A(l);
end

IM = abs(IM);
```

17.5.4 Topological Derivative

The Topological Derivative based imaging functional $\mathcal{I}_{TD}$ is introduced in (8.20). The following MATLAB file implements this imaging functional in the case when the contrast is in the electric permittivity only.

```
function Im=ImageTopologicalDerivative_K(geometry,umes,
                          omega,theta,K_D,K_Dstar)
% The Topological Derivative Imaging functional to localize inclusions
% with abnormal permittivity.
% geometry: parameterization of domain and structures of FEM
% umes:      boundary measurement;
% omega:     frequency;       theta:  directions of plane waves
% K_D, K_Dstar:   operators for preprocessing the data

g=geometry; Im=zeros(size(geometry.coordinates,1),1);
% boundary length element
hedge=norm(geometry.coordinates(geometry.edges(1,1),:)-...
    geometry.coordinates(geometry.edges(1,2),:));
s=zeros(size(geometry.coordinates,1),1);

Id=diag(ones(length(g.BoundNodes),1)); param.omega=omega;

for i=1:length(theta)
    param.theta=theta(i);
    g_neu=NeumannCondition(g,@PlaneWaveNormalDerivative,param);
    % calculate background solution
    UN=SolveHelmholtz(g.coordinates,g.K,g.M,omega,
                      g.FreeNodes,g.DirichletNodes,s,g_neu,[]);
    dwN=zeros(size(g.coordinates,1),1);
    % preprocessing the differential data
    dwN(g.BoundNodes)=(-Id/2+K_Dstar)*...
                conj((-Id/2+K_D)*(UN(g.BoundNodes).'-umes(:,i)));
    wN=SolveHelmholtz(g.coordinates,g.K,g.M,omega,
                      g.FreeNodes,g.DirichletNodes,s,hedge*dwN,[]);
```

```
30      Im=Im+omega^2*real((UN.*wN).'));
    end
```

The following MATLAB file implements the imaging functional $\mathcal{I}_{\mathrm{TD}}$ in the case when the contrast is in the magnetic permeability only.

```
function Im2=ImageTopologicalDerivative_rho(geometry,umes,
                               omega,theta,K_D,K_Dstar)
    % Topological Derivative Imaging functional to localize inclusions with
    % abnormal density.
5   % geometry: parameterization of domain and structures of FEM
    % umes:      boundary measurement;
    % omega:     frequency;        theta:  directions of plane waves
    % K_D, K_Dstar:  operators for preprocessing the data

10  g=geometry; Im=zeros(size(geometry.elements3,1),1);
    % boundary length element
    hedge=norm(geometry.coordinates(geometry.edges(1,1),:)-...
                           geometry.coordinates(geometry.edges(1,2),:)));
    s=zeros(size(geometry.coordinates,1),1);
15
    Id=diag(ones(length(g.BoundNodes),1));
    PT=diag([1 1]);
    param.omega=omega;

20  for i=1:length(theta)
        dwN=zeros(size(g.coordinates,1),1);
        param.theta=theta(i);
        g_neu=NeumannCondition(g,@PlaneWaveNormalDerivative,param);
        % calculate background solution
25      UN=SolveHelmholtz(g.coordinates,g.K,g.M,omega,
                          g.FreeNodes,g.DirichletNodes,s,g_neu,[]).';
        dwN(g.BoundNodes)=(-Id/2+K_Dstar)*...
                      conj((-Id/2+K_D)*(UN(g.BoundNodes)-umes(:,i)));
        wN=SolveHelmholtz(g.coordinates,g.K,g.M,omega,
30                        g.FreeNodes,g.DirichletNodes,s,hedge*dwN,[]);
        % TD formula for rho contrast
        [WNX WNY]=pdegrad(g.coordinates',g.elements3',transpose(wN));
        [UNX UNY]=pdegrad(g.coordinates',g.elements3',UN);

35      for j=1:size(g.elements3,1)
            Im(j)=Im(j)+real([UNX(j) UNY(j)]*PT*[WNX(j);WNY(j)]);
        end
    end Im2=pdeprtni(g.coordinates',g.elements3',transpose(Im));
```

In the above codes, the standard finite element method (FEM) is used to solve the Helmholtz equation on a bounded domain. The FEM scheme is implemented in the function SolveHelmholtz which is not given here. The variable geometry is a structure variable containing many parameters of the FEM scheme, in particular the triangulation of the domain, the stiffness matrix and the mass matrix. The parameters K_Dstar and K_D are the Neumann-Poincaré operator defined in (2.14) and its adjoint. They can be calculated from the parametrization of the boundary of the domain.

17.5.5 Simulation of Gaussian Random Field

To consider a clutter noise, we model it by a Gaussian random field as discussed in Sect. 1.8.6. Here is a MATLAB code that generates a stationary

Gaussian random field $\mu(x)$ on a two dimensional grid with mean zero and covariance function $C(x) = \exp(-|x|^2/\ell^2)$, where ℓ models the correlation length of the random medium.

```matlab
function M=RandomMediumGrid(xpoints,ypoints,corr_length)

m =max( ceil(log2(length(xpoints))),ceil(log(length(ypoints))));

nrc = 2^m; h = xpoints(2) - xpoints(1);
Z = (-nrc/2:nrc/2-1)*h;
[XY] = meshgrid(Z);
W = randn(nrc,nrc); C =exp(-(X/corr_length).^2
- (Y/corr_length).^2);
multiplier =fft2( fftshift(C) );
M = real(
ifft2(sqrt(multiplier).*fft2(W)) );
```

17.5.6 Numerical Experiments of Imaging Functionals

Finally, we present a MATLAB code that compares the performances of different imaging functionals introduced above. First, we consider pure data simulated using the FEM scheme.

```matlab
% load the FEM mesh and structures
clear load ./geometry/disk.mat g=geometry; omega=6;

% set the location and parameter of inclusion
za=[-0.5,0.5]; delta=0.05;
Inc=InclusionCharacteristicsMatrices(g,za,delta); rho=1; Bulk=1/2;
[K2 M2]=PerturbedMatrices(g,Inc,rho,Bulk);

% a set of probing plane waves
n_ang = 50; theta=linspace(0,2*pi,n_ang+1); theta=theta(1:n_ang);

% simulating boundary data
mes=GenerateMesHelmholtz(g,K2,M2,omega,theta);

% calculating the operators for preprocessing
K_Dstar=KDstar_Helmholtz(g,omega); K_D=KD_Helmholtz(g,omega);

%% TOPOLOGICAL DERIVATIVE IMAGING FUNCTIONAL
% choose probing plane waves for reconstruction
directions=[1 floor(length(theta)/4)];
ImRhoPrep=ImageTopologicalDerivative_K(g,mes(:,directions),...
                        omega,theta(directions),K_D,K_Dstar);
ImRhoNoPrep=ImageTopologicalDerivative_K(g,mes(:,directions),...
                        omega,theta(directions),0*K_D,0*K_Dstar);

figure(12) subplot 121
pdesurf(g.coordinates',g.elements3',ImRhoNoPrep); colormap jet
view([0 90]); title('Bulk contrast, no preprocessing, 2
directions') axis square

subplot 122 pdesurf(g.coordinates',g.elements3',ImRhoPrep);
colormap jet view([0 90]); title('Bulk contrast, preprocessing, 2
directions') axis square
```

```
35 | %% IMAGING USING MUSIC AND KM

   | % searching grid
   | XS=(g.coordinates(g.elements3(:,1),1)+g.coordinates(g.elements3(:,2),1)...
   |            + g.coordinates(g.elements3(:,3),1))/3;
40 | YS=(g.coordinates(g.elements3(:,1),2)+g.coordinates(g.elements3(:,2),2)...
   |            + g.coordinates(g.elements3(:,3),2))/3;

   | theta_coord = zeros(length(theta),2); theta_coord(:,1) =
   | cos(theta); theta_coord(:,2) = sin(theta);
45 |
   | param.omega = omega;
   | s=zeros(size(g.coordinates,1),1); % no source in our problem

   | % simulating the differential data
50 | dmes = zeros(size(mes)); for i=1:n_ang
   |      %In case of homogeneous background, one can cheat
   |      UN=exp(sqrt(-1)*omega*g.coordinates*[cos(theta(i));sin(theta(i))]);
   |      dmes(:,i) = UN(g.BoundNodes)-mes(:,i);
   | end
55 | % calculating the weighted boundary measurement
   | ARMnd = WeightBddMeas(omega,theta_coord,g,dmes,2); [U_ARMnd,
   | S_ARMnd, V_ARMnd] = svd(ARMnd); figure(100);
   | plot(diag(S_ARMnd),'*'); L_ARMnd=input('How many eigenvalues of
   | ARMnd should we use?')
60 |
   | %% MUSIC ALGORITHM
   | IM_MU = ImagMU(omega,theta_coord,XS,YS,U_ARMnd(:,1:L_ARMnd));
   | figure(13) IM_MU2 =
   | pdeprtni(g.coordinates',g.elements3',transpose(IM_MU));
65 | pdesurf(g.coordinates',g.elements3',IM_MU2);

   | colormap jet view([0 90]);

   | title('MUSIC using 30 directions');
70 |
   | axis square

   | %% KIRCHHOFF MIGRATION
   | ARMnd_modified = U_ARMnd(:,1:L_ARMnd)*...
75 |        S_ARMnd(1:L_ARMnd,1:L_ARMnd)*(V_ARMnd(:,1:L_ARMnd))';
   | IM_KM = ImagKM(omega,theta_coord,XS,YS,ARMnd_modified); figure(14)
   | IM_KM2 = pdeprtni(g.coordinates',g.elements3',transpose(IM_KM));
   | pdesurf(g.coordinates',g.elements3',IM_KM2); colormap jet view([0
   | 90]); title('Kirchhoff migration 30 directions') axis square
80 |
   | %% BACK—PROPAGATION
   | param.omega = omega; s=zeros(size(g.coordinates,1),1);

   | dmes = zeros(size(mes));
85 |
   | for i=1:n_ang
   |      % in case of homogeneous background, one can cheat
   |      UN=exp(sqrt(-1)*omega*g.coordinates*[cos(theta(i));sin(theta(i))]);
   |      dmes(:,i) = UN(g.BoundNodes)-mes(:,i);
90 | end ARMd = WeightBddMeas(omega,theta_coord,g,dmes,1);

   | IM_BP = ImagBP(omega,theta_coord,XS,YS,ARMd);

   | figure(17) IM_BP2 =
95 | pdeprtni(g.coordinates',g.elements3',transpose(IM_BP));
   | pdesurf(g.coordinates',g.elements3',IM_BP2);
   | colormap jet view([0
   | 90]); title('Backpropagation 30 directions') axis square
```

Both measurement noise and clutter noise can be considered. The first case is easier to simulate. Indeed, after getting the boundary measurement `mes` in Line 20 above, we simply add the following lines.

```
noise_level = 0.05;

[m_mes n_mes] = size(mes); mes = mes +
randn(m_mes,n_mes)*noise_level*max(max(abs(mes))));
```

These lines add independent Gaussian random noise to the measured boundary data. The noise has mean zero and its standard deviation is 5 % of the maximum amplitude of the pure data.

Simulating clutter noise is more involved. First, the MATLAB code in Sect. 17.5.5 is needed. Secondly, the Helmholtz equation is posed on a heterogeneous medium. Consequently, we need to modify the stiffness and mass matrices of `geometry`; moreover, the background solution `UN` in Lines 69 and 108 need to be calculated using `SolveHelmholtz` with modified `geometry`. We will not present the modified code here.

17.6 Implementation of Time-Reversal

In this section, we implement the modified time-reversal schemes for the inverse source problem introduced in Chap. 12.

17.6.1 Solving the Wave Equation

From the expression, we need to solve wave equations of the form (12.1). The following MATLAB scheme does this. It utilizes the Fourier method. After performing a Fourier transform in the spatial variable on the equation, one obtains an ordinary differential equation that can be explicitly solved; see Line 44. Since the spatial domain is in $\mathbb{R}^2$, the Fourier transform and its inverse are done using `fft2` and `ifft2`.

```
function I_radon=SolveWaveEq(f,L,T,N_R,x_k1,x_k2)
%This function calculates the solution of the Wave Equation with source
%f*d(Dirac Delta)/dt.
%L: size of domain; T: time duration; N_R: number of time steps
%(x_k1,x_k2): positions of (2D) wave recorders
%I_radon is a matrix of size N_k times N_R where N_k is the length
%of x_k1.
%Each column of I_radon records the wave field at the recorders at a time
%instance.

[N,N] = size(f); L1 = L; L2 = L;

N1 = N; N2 = N;
```

```
15  x = linspace(-L1/2,L1/2,N1+1); x = x(1:N1); X = (ones(N2,1)*x)';

    y = linspace(-L2/2,L2/2,N2+1); y = y(1:N2); Y = (ones(N1,1)*y);

    N_k = prod(size(x_k1)); I_radon = zeros(N_k,N_R); dt = T/N_R;
20
    %%%%%%%%%%%%%%% frequency variables %%%%%%%%%%%%%%
    k_x = [0:1:N1/2,N1/2-1:-1:1]; K_x =  (ones(N2,1)*k_x/L2)';

    k_y = [0:1:N2/2,N2/2-1:-1:1]; K_y =  ((ones(N1,1)*k_y/L1));
25
    %%%%%%%%%%%%%%%% indice x_k %%%%%%%%%%%%%%%%%%%%%
    indice_x_k1 = floor((x_k1+L1/2)*N1/L1+1); indice_x_k2 =
    floor((x_k2+L2/2)*N2/L2+1);

30  %%%%%%%%%% solve wave equation using Fourier method %%%%%%%%%%
    f_fourier = fft2(f); for i=1:N_R,

    U_fourier = cos(2*pi*sqrt(K_x.^2 +K_y.^2)*i*dt).*f_fourier; U =
    ifft2(U_fourier);
35
    for i_k = 1:N_k,
         I_radon(i_k,i) = U(indice_x_k1(i_k),indice_x_k2(i_k));
    end
    %%%%%%%%%%%%%%% showing the wavefield every 50 steps %%%%%%%%%%%%%%
40  if(mod(i,50)==1),
        imagesc(U);
        pause(0.01);
    end end
```

17.6.2 Time Reversal of Wave Equation

The next script provides the time-reversal scheme as discussed in Chap. 12.
It amounts to solving a series of wave equations with sources given by
the recorded wave fields at the boundary. Therefore, the MATLAB function
SolveWaveEq above is used repeatedly.

```
function I=TR_WaveSource(I_radon,L,T,x_k1,x_k2,N)
% This function reconstructs the source of the wave equation from a series
% of boundary measurement.
% I_radon: series of boundary measurement;      L:  size of box
5  % (x_k1,x_k2): coordinates of recorders;         T:  duration of time
% N: resolution of image;

L1 = L; L2 = L;

10  N1 = N; N2 = N;

x = linspace(-L1/2,L1/2,N1+1); x = x(1:N1); X = (ones(N2,1)*x)';

y = linspace(-L2/2,L2/2,N2+1); y = y(1:N2); Y = (ones(N1,1)*y);
15
[N_k,N_R] = size(I_radon); dt  = T/N_R;

%%%%%%%%%%%%%% frequency varialbes %%%%%%%%%%%%%%%%%%%
k_x = [0:1:N1/2,N1/2-1:-1:1]; K_x =  (ones(N2,1)*k_x/L2)'; k_y =
20  [0:1:N2/2,N2/2-1:-1:1]; K_y =  ((ones(N1,1)*k_y/L1));

%%%%%%%%%%%%%%%%% indice x_k %%%%%%%%%%%%%%%%%%%%%%%
indice_x_k1 = floor((x_k1+L1/2)*N1/L1+1); indice_x_k2 =
```

```
     floor((x_k2+L2/2)*N2/L2+1);
25
     %%%%%%%%%%%%%%%%%%%%%%%%%%%%%%%%%%%%%%%%%%%%%%%%
     U1_m = zeros(N,N); U = zeros(N,N);

     %%%%%%%%%%%%%%%%%%%%%%%%%%% calcul longueur Omega   %%%%%%%%%%%%%%%
30   Long_omega = 0; for i=1:N_k,
        i_moins = mod(i-2,N_k) + 1;
        i_plus = mod(i,N_k) +1;

        Long_omega = Long_omega + sqrt((x_k1(i_plus) - x_k1(i_moins))^2 + ...
35                          (x_k2(i_plus) - x_k2(i_moins))^2)/2;
     end

     %%%%%%%%%%%%%%%%%%%%%%%%%%%%%%%%%%%%%%%%%%%%%%%%%%%%%%%%%%%%%%%
     temp = zeros(N,N); for i_k = 1:N_k,
40         temp(indice_x_k1(i_k),indice_x_k2(i_k))= 1;
     end coef = Long_omega/(sum(sum(temp))/(N/L)^2);

     for i=1:N_R,%N_R-1,%reversed time

45   %%%% Solve wave equation with source on the boundary (recorded wave) %%%%
     for i_k = 1:N_k,
              U(indice_x_k1(i_k),indice_x_k2(i_k))= I_radon(i_k,N_R-(i-1));
     end

50   U_fourier = fft2(U); U_fourier_prim = cos(2*pi*sqrt(K_x.^2
     +K_y.^2)*(T-(i-1)*dt)).*U_fourier;

     U1_m = U1_m + coef*ifft2(U_fourier_prim)*T/N_R; end I =
     1.05*U1_m*pi/2;
```

17.6.3   Numerical Experiment

The following script is a numerical experiment for the above modified
time-reversal scheme for the inverse source problem. Letters "A", "I", and "P"
are used as the sources of wave equations, and boundary data are simulated
for a time duration $T = 2$.

```
     clear all;
     close all;
     N =2^9;    % resolution of image
     N_k =4*2^8; % number of recorders
5    N_R =4*2^8; % number of time steps
     L = 4; % image size
     T = 2; % time duration

     L1 = L; L2 = L; N1 = N; N2 = N;
10
     %%%%%% real source are letters A I P  %%%%%%%%%%
     source = zeros(N,N);
     temp = imread('A.png');%'I.png', 'P.png'
     temp = (temp-min(temp(:)))./(max(temp(:)) - min(temp(:)));
15   temp = abs(1-temp);
     [m_temp n_temp] = size(temp);
     source(N/2-m_temp/2+1:N/2+m_temp/2, N/2-n_temp/2+1:N/2+n_temp/2) = temp;
     %phantom is located in the middle
     x = linspace(-L/2,L/2,N+1);
20   x = x(1:N);
     X =  (ones(N,1)*x)';
```

```
    Y = X';

    %%%%%%%%%%%%% position of recorders %%%%%%%%%%%%%
25  theta = linspace(0,2*pi,N_k+1);

    theta = theta(1:N_k);

    x_k1 = (0.95 + 0.05*cos(8*theta)).*cos(theta);
30
    x_k2 = (0.95 + 0.05*cos(8*theta)).*sin(theta);

    indice_x_k1 = floor((x_k1+L/2)*N/L+1);

35  indice_x_k2 = floor((x_k2+L/2)*N/L+1);

    x_k1 = x(indice_x_k1);

    x_k2 = x(indice_x_k2);
40
    %%%%%%%%%%%%%%%%%%%  for display %%%%%%%%%%%%%%%
    rho = sqrt(X.^2 + Y.^2);

    domaine = (1-sign(rho - (0.95+0.05*cos(8*angle(X + 1i*Y)))))/2;
45
    %%%%%%%%%%% Simulating recorded wave field %%%%%%%%%%%%%
    I_radon=SolveWaveEq(source,L,T,N_R,x_k1,x_k2);

    %%%%%%%%%%%%%%%% Time reversed image of source %%%%%%%%%
50  I1 = TR_WaveSource(I_radon,L,T,x_k1,x_k2,N); temp = domaine.*I1;

    figure
    imagesc(x(N/4+1:3*N/4),x(N/4+1:3*N/4),temp(N/4+1:3*N/4,N/4+1:3*N/4));
     hold on;
55   for i=1:N_k,
        plot(x_k1(i),x_k2(i),'.w');
     end
     axis square;
    caxis([0,1]);
```

17.7   Implementation of Optimal Control Algorithms for Imaging Shape Deformations

The following is the numerical computation of the multistatic response matrix for the Helmholtz equation when the permeability of the inclusion is different from that of the background.

```
    function [MSR u Dn_u Dt_u] = MSR_Simulation(D,  normal, Sigma,
    tvec, avec, mu0, mu1, k0, k1, Xs, Xr)
    % Generate the multi-static response (MSR) matrix of the Helmholtz
    % problem (with an inclusion D) using its integral definition.
5   % Inputs:
    % mu: permeability of the inclusion
    % k : frequency, k^2 = omega^2*mu*epsilon
    % D,normal,avec,Sigma: the boundary and its parameterization
    % Xs, Xr: the positions of sources and receivers
10  % Output: MSR is symmetric matrix
    Ns = size(Xs, 2);

    Nr = size(Xr, 2);
```

```
15  M = size(D,2);

    t = 1:M; S0 = S_Helmholtz(k0, D, Sigma);

    S1 = S_Helmholtz(k1, D, Sigma);
20
    Ks0 = Kstar_Helmholtz(k0, D, normal, Sigma, tvec, avec);

    Ks1 = Kstar_Helmholtz(k1, D, normal, Sigma, tvec, avec);

25  A(t, t) = S0; A(t, M+t) = - S1;

    A(M+t, t)=1/mu0*(.5*eye(M) + Ks0);

    A(M+t, M+t)=-1/mu1*(-.5*eye(M) + Ks1);
30
    SR = zeros(size(Xr,2), M);

    for j = 1:M SR(:, j) = Sigma(j)*Green2D_Helmholtz(k0,Xr, D(:,j));
35  end u = zeros(M, Nr);

    Dn_u=zeros(M,Nr); Dt_u=zeros(M, Nr);

    source=zeros(2*M,1);
40
    MSR = zeros(Nr, Ns);

    for k = 1:Ns G =  Green2D_Helmholtz(k0,D,Xs(:,k));

45      [Gx Gy] = GradGreen2D_Helmholtz(k0,D,Xs(:,k));

        source(t,1) = -G;

        source(M+t,1) =- (Gx.*(normal(1,:).') + Gy.*(normal(2,:).'));
50
        phi = (A.'*A+0.001*eye(2*M))\(A.'*source);

        S_phi = SR*phi(t,1);       MSR(:, k) = S_phi.';
        %%% on the boundary of D
55      u(:, k)  = S0*phi(t,1) - source(t,1);
        Dn_u(:, k) = (.5*eye(M) + Ks0)*phi(t,1) - source(M+t,1);
        Dt_u(:, k) = TDeri(u(:,k), Sigma);
    end MSR = (MSR+MSR.')/2; function S = S_Helmholtz(omega, D, Sigma)
    % Single layer potential which corresponds
60  %to the operator laplacian + omega^2
    % Inputs:
    % D,Sigma: the boundary and its parameterizations
    % omega: frequency
    % Output:
65  % S: approximate integral representation of the single layer potential
    M = size(D,2); S= zeros(M, M);

    for j = 1:M
        S(:, j) = Sigma(j)*Green2D_Helmholtz(omega,D, D(:,j));
70
    end for j = 2:M-1
        S(j,j) = (S(j,j-1)+S(j,j+1))/2;

    end S(M,M)=S(M,M-1); S(1,1)= S(1,2);
75
    function Ks = Kstar_Helmholtz(omega, D,normal, Sigma, tvec, avec)
    %laplacian +omega^2
    % The adjoint operator of K.
    % Inputs:
80  % D, normal, Sigma: the boundary and its parameterizations
```

```
   % omega: frequency
   % Output:
   % Ks: a matrix obtained by quadrature that approximates the integral
   % definition of K^*
85 M = size(D,2); Ks = zeros(M, M); for j = 1:M
       [Gx Gy] = GradGreen2D_Helmholtz(omega, D, D(:, j));

       Ks(:, j) = Sigma(j)*(Gx.*(normal(1,:)') + Gy.*(normal(2,:)'));

90 end norm_tvec_square = tvec(1,:).^2 + tvec(2,:).^2 ;

   for j = 1:M Ks(j,j) =
   1/(2*pi)*(1)/2*avec(:,j)'*normal(:,j)/norm_tvec_square(j)*Sigma(j);
   end
```

The following is the code to modify the interface of an inclusion following the method described in Chap. 13. The resulting function is the shape deformation of the boundary of the inclusion in the normal direction.

```
   function hre =  Modify_boundary(MSR, D, normal, tvec, avec, Sigma, omega,
                              mu0, mu1, epsilon0, epsilon1, Xs, Xr)
   % Shape deformation algorithm
   % Input:
5  % MSR: Given multistatic matrix of target domain
   % D, normal, Sigma: Initial guess domain and its boundary information
   % mu1, epsilon1: permittivity and permeability of inclusion
   % mu0, epsilon0: permittivity and permeability of background

10 M = size(D,2);

   Ms = size(Xs,2);

   Mr = size(Xr, 2);

15
   k0 = sqrt(omega^2*mu0*epsilon0);

   k1 = sqrt(omega^2*mu1*epsilon1);

20 [MSR0  u Dt_u Dn_u] =  MSR_Simulation(D,  normal, Sigma, tvec,
   avec, mu0, mu1, k0, k1, Xs, Xr) ;
   MSRdiff = MSR - MSR0;
   B = zeros(M, Ms*Mr);
   for n = 1:Ms
25     for m = 1:Mr
           B(:, (n-1)*Ms+m) = (mu0/mu1-1)*(Dt_u(:,n).*(Dt_u(:,m)))...
               +(mu0/mu1-1)*(mu1/mu0)*(Dn_u(:,n)).*(Dn_u(:,m)))...
               +omega^2*(epsilon1-epsilon0)*mu0*u(:,n).*u(:,m);
       end
30 end
   %% shape basis and shape derivative
   SBasis = [];
   for k =1: size(B, 2)
       a = B(:,k);
35     a = a/sqrt(Sigma*(abs(a).^2));
       SBasis = [SBasis real(a) imag(a)];
   end
   [SBasis c] = GramSch(SBasis, Sigma); %Orthogonalization
   ShapeBasis = [];
40 for l = 1:size(SBasis,2)
       if (sqrt((abs(SBasis(:,l)).^2)'*Sigma') > 0.05)
           ShapeBasis = [ShapeBasis SBasis(:,l)];
       else
       end
45 end
   %% Reconstruction
   normsq = 0; hre = zeros(M,1);
```

```
     for l = 1:size(ShapeBasis,2)
          psi = ShapeBasis(:,l);
50
          psi = psi/sqrt((abs(psi).^2)'*Sigma');
          sderiv = shape_deriv(MSRdiff, Sigma, B, psi, Ms);
          hre = hre+ sderiv*psi;  normsq = normsq +sderiv^2;
     end
55   efunc = .5*sum(sum((abs(MSRdiff)).^2));

     hre = efunc/(normsq+10^(-8))*hre;
     if (max(abs(hre))>0.1)
          hre = (0.1/max(abs(hre)))*hre;
60   else
     end
     end
     function deriv = shape_deriv(MSRdiff, Sigma, B, psi, Ms) deriv =
     0; for n = 1:Ms
65        for m = 1:Ms
               Bnm = B(:, (n-1)*Ms+m);
               deriv = deriv +  real( MSRdiff(n,m)*(Sigma*(psi.*(conj(Bnm)))));
          end
     end end
```

17.8 Implementation of GPT-Vanishing Structures

The following is the numerical simulation to construct GPT-vanishing
structures. In this code we fix the radii of the concentric disks and find the
conductivity profile which is assumed to be constant in each layer.

```
    function   [sig GPTs] = cloaking_conductivity(N, sigcore, sig0)
    % Make the GPTs—vanishing structure
    % Structure is concentric disks with different conductivity
    % Radii of disks are fixed
5   % Input:
    % N: number of layer which is same as the numbers GPTs to be zero
    % sigcore: conductivity in the interior disk
    % sig0:  Exterior conductivity
    % Output: conductivity on each layer and GPTs upto order 15
10  r =2:-1/N:1; % r: radii of concentric disks % length(r)=N+1;
    sig = .5*ones(1,N); sig(2:2:N) = 2;% Initial guess
    IterN = 1000; %iteration number
    for nn = 1:IterN
        M21 =  GPTsMatrix(N, sig, sigcore, sig0, r);
15      dM21 = zeros(N,N);
        for i = 1:N
            h=0.001;
            sig_h=sig;
            sig_h(i) = sig(i) + h;
20          M21_2 =  GPTsMatrix(N, sig_h, sigcore, sig0, r);
            dM21(:, i) = (M21_2 - M21)/h;
        end
        phi = .5*pinv(dM21, 10^(-10))*(M21);
        sig = sig - phi';
25
        sig = max(sig, 0.01);

        sig = min(sig, 100);

30  end [M21 GPTs] = GPTsMatrix(15, sig, sigcore, sig0, r);
```

```matlab
function [M21 GPTs] = GPTsMatrix(GPTorder, sig, sigcore, sig0, r)
% compute GPTs upto order GPTorder of concentric disks
N = size(sig, 2);

lambda = ([sig sigcore] - [sig0 sig])./([sig sigcore] + [sig0
sig]);

M21 = zeros(N,1);

M22 = zeros(N,1);

for k = 1:GPTorder
    M = eye(2,2);
    for j = 1:length(r)
        M =[1 lambda(j)*(r(j)^(-2*k)); lambda(j)*(r(j)^(2*k)) 1]*M;
    end
    M21(k) = M(2,1);

    M22(k) = M(2,2);
end GPTs = M21./M22;
```

Chapter 18
Numerical Results

This chapter presents numerical illustrations using the codes described in Chap. 17 in order to highlight the performance and show the limitations of our numerical approaches for multistatic imaging.

18.1 Structure of the MSR Matrix

18.1.1 Conductivity Equation

SVD of the MSR Matrix for a Small Target

We consider a small conductivity target and compute the SVD of the MSR matrix. Figure 18.1 shows that there are 2 significant singular values.

SVD of the MSR Matrix for an Extended Target

Figure 18.2 shows the number of significant singular values for an extended target. The target is 5 times larger in this example than in Fig. 18.1. The significant singular values are larger than in Fig. 18.1. However, their number is the same.

18.1.2 Helmholtz Equation

SVD of the MSR Matrix for a Small Target

We compute the SVD of the MSR matrix for the Helmholtz equation. We first consider a small target. Figure 18.3 shows that the number of significant singular values is 3.

H. Ammari et al., *Mathematical and Statistical Methods for Multistatic Imaging*, 331
Lecture Notes in Mathematics 2098, DOI 10.1007/978-3-319-02585-8_18,
© Springer International Publishing Switzerland 2013

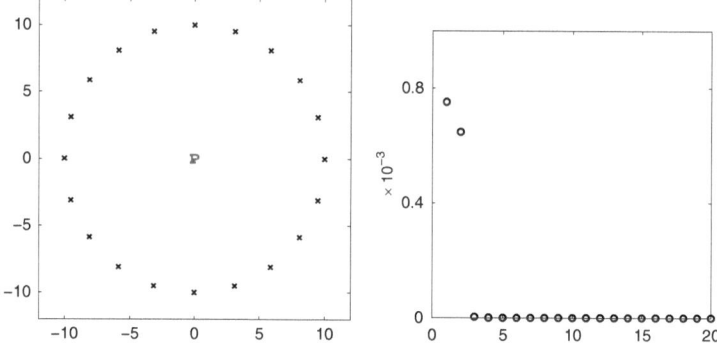

Fig. 18.1 Singular values of the MSR matrix with 20 transmitters and 20 receivers. The target is the letter "P" with conductivity $k_\star = 4$. The number of significant singular values is 2 out of 20

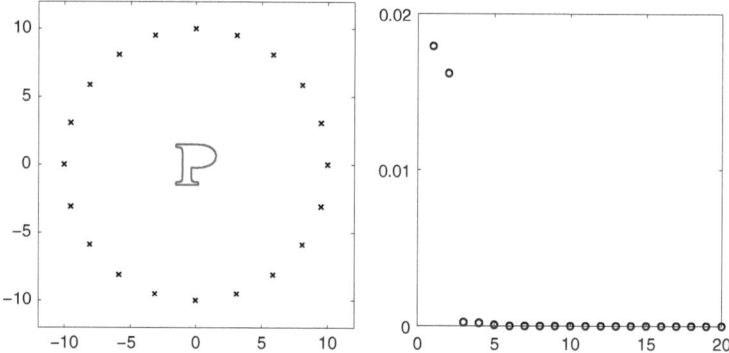

Fig. 18.2 Singular values of the MSR matrix with 20 transmitters and 20 receivers. The array configuration and the conductivity of the target are as in Fig. 18.1. The letter "P" is 5 times larger in this example than in Fig. 18.1. The number of significant singular values is 2 out of 20

SVD of the MSR Matrix for an Extended Target

Figure 18.4 shows the number of significant singular values for an extended target. The target is 5 times larger in this example than in Fig. 18.3. The number of significant singular values is 3 times more than in Fig. 18.3.

18.2   Matching GPTs Algorithm

In this section we perform some numerical experiments of recovering the shape of a domain from its GPTs. In all of the numerical examples presented in this section, we apply the level set approach presented in Chap. 4.

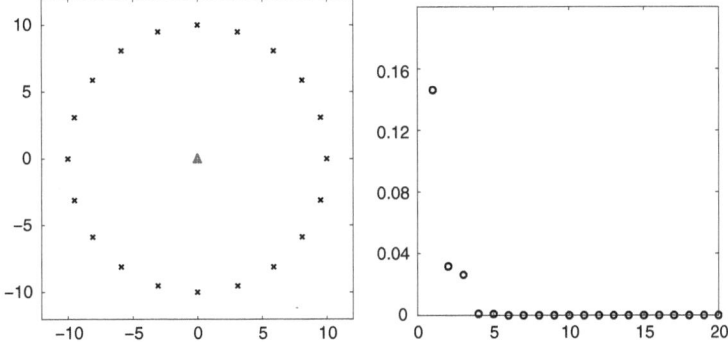

Fig. 18.3 Singular values of the MSR matrix with 20 transmitters and 20 receivers. The target is again the letter "A" with electromagnetic parameters $(\varepsilon_*, \mu_*) = (2, 5)$. The electromagnetic parameters of the background are $(\varepsilon_0, \mu_0) = (1, 1)$. The frequency is $\omega = 1$. The number of significant singular values is 3 out of 20

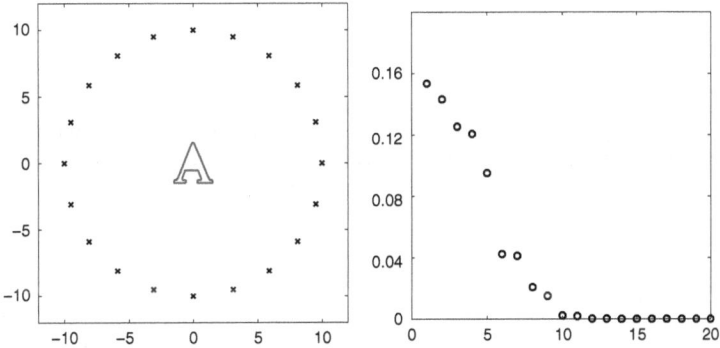

Fig. 18.4 Singular values of the MSR matrix with 20 transmitters and 20 receivers. The array configuration, the electromagnetic parameters of the target, and the operating frequency are as in Fig. 18.3. The letter "A" is 5 times larger in this example than in Fig. 18.3. The number of significant singular values is 9 out of 20

We emphasize that we do not make any a priori assumption on the number of connected components of the domain.

Figure 18.5 shows that the equivalent ellipse is separated into 2 pieces and gradually modified toward the target domain. The first image is the equivalent ellipse and the others are the reconstructed images after $20, 30, 40, 70, 90$ iterations. Figure 18.6 is the graph of the relative area difference $\frac{|D \triangle B|}{|D|}$, where B is the reconstructed domain.

The next example is to demonstrate that the more components the target has, the higher GPTs are required to separate them. Figure 18.7 shows that one can not separate the targets using $K = 3$, while $K = 4$ can. Figure 18.8 shows that using GPTs up to $K = 5$ may not be enough to separate 3 targets.

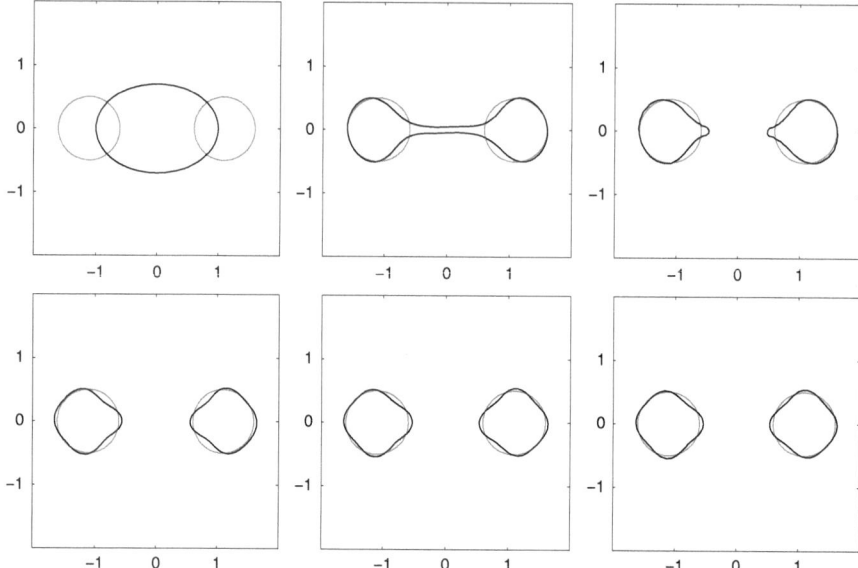

Fig. 18.5 The GPTs up to order $K = 4$ separate the inclusion of 2 pieces. The first image is the equivalent ellipse and the others are the reconstructed images after $20, 30, 40, 70, 90$ iterations. The *gray curve* is the target domain and the *black curve* is the reconstructed one

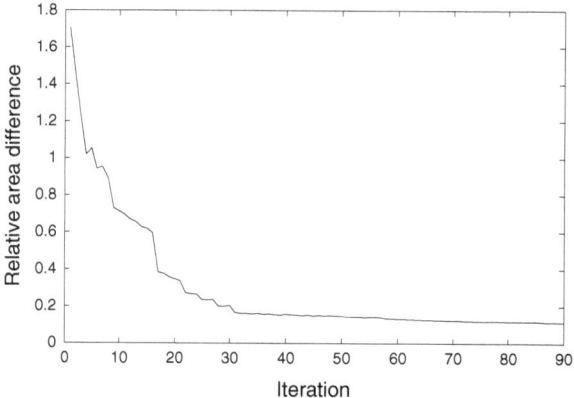

Fig. 18.6 The relative area difference of the example in Fig. 18.5

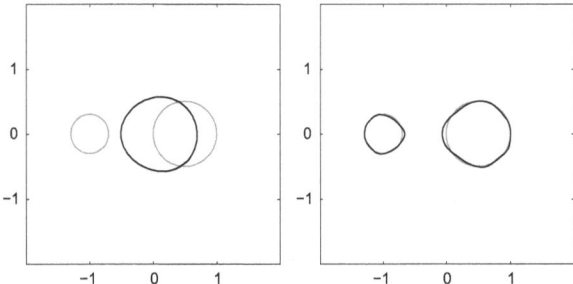

Fig. 18.7 Reconstructed images after 70 iterations. The figure on the left is obtained using GPTs of order up to $K = 3$, and the one on the right is of order up to $K = 4$. GPTs of order up to $K = 3$ cannot separate 2 targets

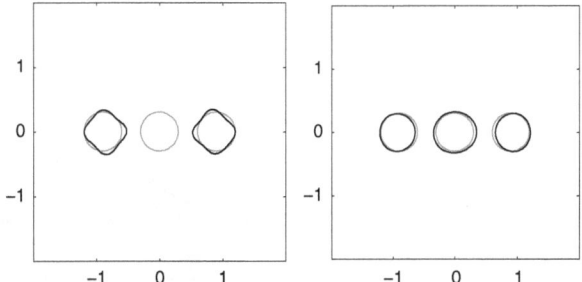

Fig. 18.8 Reconstructed images after 100 iterations. The figure on the left is obtained using GPTs of order up to $K = 5$, and the one on the right is of order up to $K = 6$

Figure 18.9 shows reconstructions of the letters "A", "I", and "P" from their GPTs of order up to $K = 6$. The conductivity $k_\star = 4$ in the letters.

18.3 Detection and Imaging

18.3.1 Realization of a Cluttered Noise

Figure 18.10 shows a typical realization of a medium noise and its projection on the finite-element mesh.

18.3.2 Comparison Between the Imaging Functions in the Continuum Approximation

Comparisons between the standard deviations of the localization error with respect to measurement and clutter noises for the imaging algorithms are

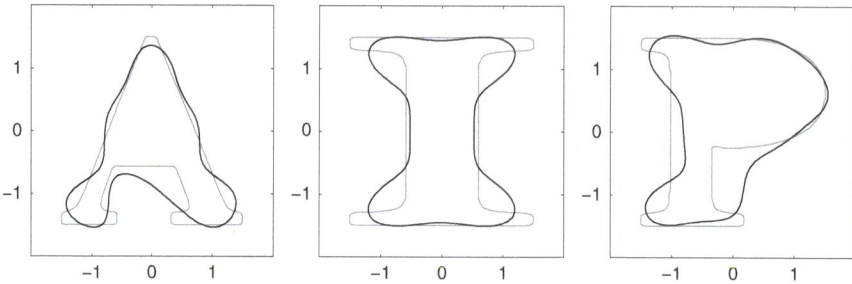

Fig. 18.9 Reconstructed images of the letters "A", "I", and "P" using GPTs of order up to $K = 6$. The conductivity $k_\star = 4$ in the letters

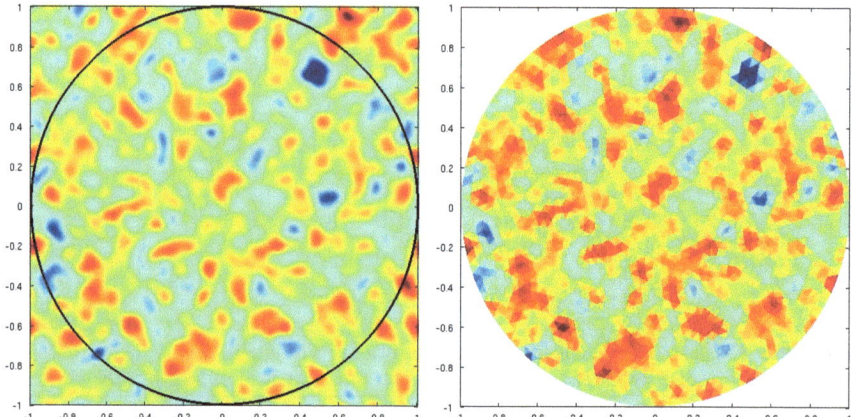

Fig. 18.10 Realization of a medium noise

given in Figs. 18.11 and 18.12. The topological derivative based functional performs as good as Kirchhoff migration and much better than MUSIC and backpropagation, specially at high levels of measurement noise and is the most robust functional with respect to medium noise.

18.4   Time-Reversal Imaging

Figure 18.13 shows time-reversal reconstructions of extended source terms. The sources are the letters "A" , "I", and "P".

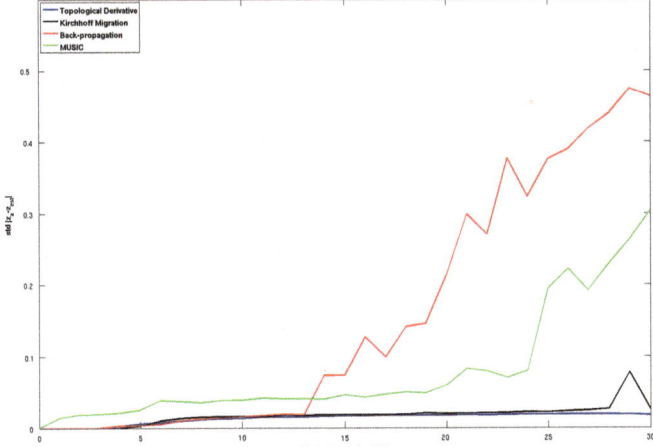

Fig. 18.11 Standard deviations of localization error with respect to measurement noise level for $\mathcal{I}_{MU}, \mathcal{I}_{BP}, \mathcal{I}_{KM}$, and $\mathcal{I}_{TD}$ with 50 uniformly distributed directions of the incident plane waves

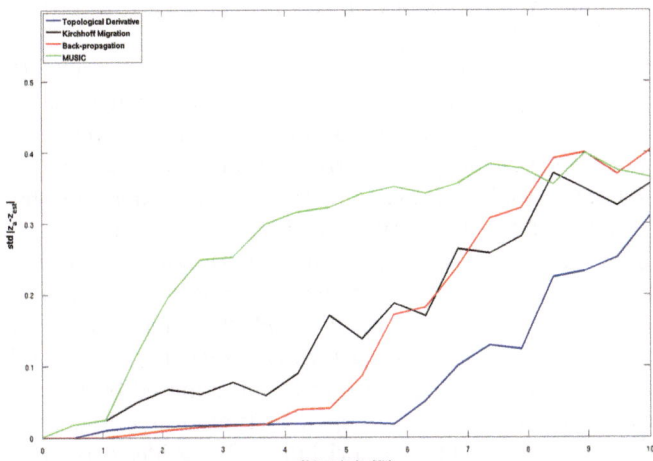

Fig. 18.12 Standard deviations of localization error with respect to clutter noise for $\mathcal{I}_{MU}, \mathcal{I}_{BP}, \mathcal{I}_{KM}$, and $\mathcal{I}_{TD}$ with 50 uniformly distributed directions of the incident plane waves

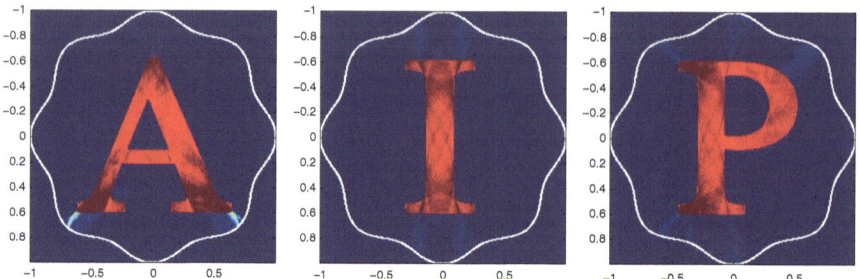

Fig. 18.13 Reconstruction of the letters "A" , "I", and "P" using the modified time-reversal technique

18.5 Target Identification and Tracking

18.5.1 Dictionary Matching Algorithm

We consider a dictionary consisting of 26 Roman capital letters without rotational symmetry. The shapes are defined in such a way that the holes inside the letters are filled.

Let a target be of characteristic size δ and at a distance R from the circular array of transmitters and receivers. We assume that the target is obtained from a certain element B of the dictionary by applying some unknown rotation θ, scaling s and translation z. We set $\delta/R = 0.5$, $s = 2.47, \theta = 6.08, z = (33.35, 73.84)$ and the center of mass of the target at $(33.40, 73.86)$. We first test Algorithm 11.1 on the letter "P". For the noiseless case ($\sigma_{\text{noise}} = 0$), the values of e_n defined in Algorithm 11.1 are plotted in Fig. 18.14a,b. These results suggest that the high-order CGPTs can better distinguish similar shapes such as "P" and "R", since they contain more high frequency information. Nonetheless, the advantage of using high-order CGPTs drops quickly when the data are contaminated by noise. Low-order CGPTs provide more stable results in this situation, see Fig. 18.14c,d.

In the case of noiseless data, Algorithm 11.2 provides correct results with low computational cost. Here we repeat the experiment in Fig. 18.14a,c using Algorithm 11.2, and plot the error e_n defined in Algorithm 11.2 in Fig. 18.15. When data are noisy, Algorithm 11.1 performs significantly better than Algorithm 11.2, as shown by Fig. 18.16 where we compare the two algorithms for identifying letter "A" at various noise levels. Thanks to the debiasing step (11.24), Algorithm 11.1 is much more robust with respect to noise than Algorithm 11.2, in which there is no debiasing and the invariance of the shape descriptors $\mathcal{I}^{(1)}$ and $\mathcal{I}^{(2)}$ may be severely affected by noise (see Fig. 18.16).

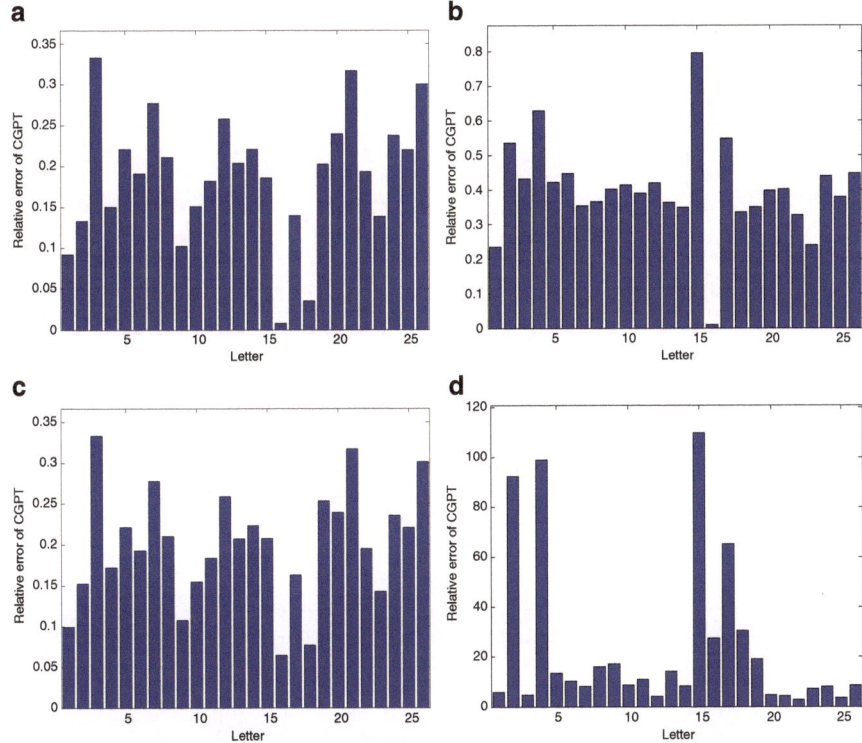

Fig. 18.14 The identification of the letter "P" using the first 2, and 5 orders CGPTs at noise levels $\sigma_{\text{noise}} = 0$ and $\sigma_{\text{noise}} = 0.1$. The bar represents the relative error e_n between the CGPTs of the n-th letter and that of the data, as defined in Algorithm 11.1, and the shortest one in each figure corresponds to the identified letter. For (c) and (d), the experiment has been repeated for 100 times, using independent draws of white noise, and the results are the mean values of all experiments . (a) $\sigma_{\text{noise}} = 0$, order ≤ 2. (b) $\sigma_{\text{noise}} = 0$, order ≤ 5. (c) $\sigma_{\text{noise}} = 0.1$, order ≤ 2. (d) $\sigma_{\text{noise}} = 0.1$, order ≤ 5

18.5.2 Target Location and Orientation Tracking

Here we show the performance of EKF in a full angle of view setting with the shape "A" as target B, which has diameter 10 and is centered at the origin. The path (z_t, θ_t) for $t \geq 1$ is simulated according to the model (11.42) during a period of 10 seconds ($\Delta t = 0.01$), with parameters $\sigma_a = 2, \sigma_\theta = 0.5$, and the initial state $X_0 = (v_0, z_0, \theta_0)^T = (-1, 1, 5, -5, 3\pi/2)^T$. We make sure that the target is always included inside the measurement circle on which $N = 20$ sources/receivers are fixed, see Fig. 18.17. The data stream V_t is generated by first calculating the MSR matrix corresponding to each $D_t, t \geq 1$ then adding a white noise.

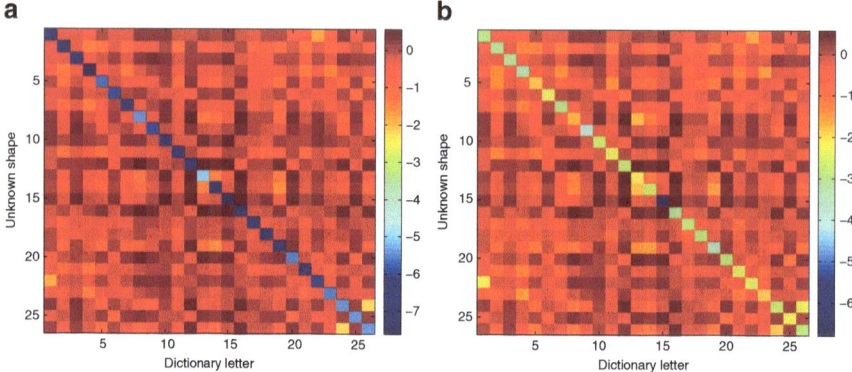

Fig. 18.15 Algorithm 11.2 applied on the all 26 letters using the standard dictionary at noise level $\sigma_{\text{noise}} = 0$. The unknown shapes in (**a**) are exact copies of the standard dictionary, while in (**b**) are perturbed copies. The color indicates the error e_n in logarithmic scale. All letters are correctly identified in both (**a**) and (**b**). (**a**) $\sigma_{\text{noise}} = 0$, order ≤ 5, Standard letters. (**b**) $\sigma_{\text{noise}} = 0$, order ≤ 5, Perturbed letters

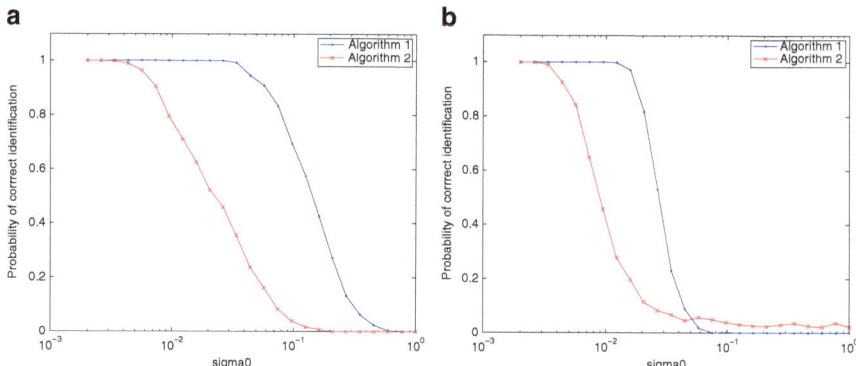

Fig. 18.16 Comparison of Algorithm 11.2 and Algorithm 11.1 on identification of the standard letter "A". At each noise level, the experiment has been repeated 1,000 times, using independent draws of white noise. For each algorithm, the curve represents the percentage of experiments where the letter "A" is correctly identified. (**a**) order ≤ 2. (**b**) order ≤ 3

Suppose that the CGPT of B is correctly determined (for instance, by identifying the target in a dictionary). Then we use the first two orders CGPT $\mathbb{M}$ of B in (11.46b), and take $(0, 0, 10, -0.5, 0)^T$ as an initial guess of X_0 for EKF.

We add 10% and 20% of noise to data, and show the results of tracking in Fig. 18.18a,c and e. We see that EKF can find the true system state, despite of the poor initial guess, and the tracking precision decays as the measurement noise level gets higher. The same experiment with small target (of same shape) of diameter 1 is repeated in Fig. 18.18b,d and f, where the tracking

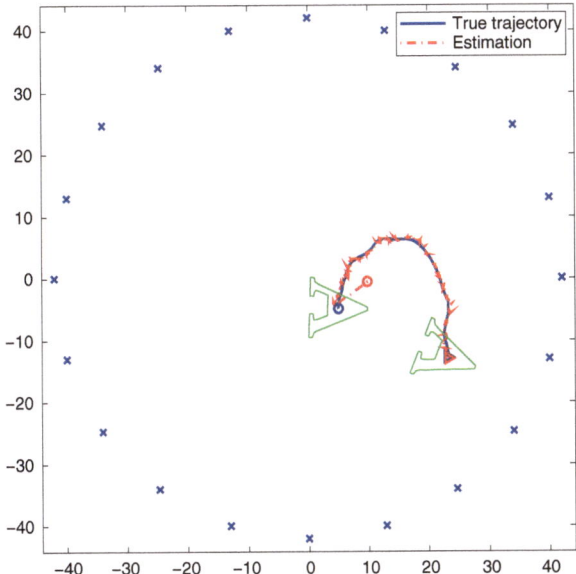

Fig. 18.17 Trajectory of the letter "A" and the estimation by EKF. The initial position is $[5, -5]$ while the initial guess given to EKF is $[10, -0.5]$. The *crosses* indicate the position of sources/receivers, while the *circle* and the *triangle* indicate the starting and the final position of the target, respectively. In *blue* is the true trajectory and in *red* the estimated one

of position remains correct, on the contrary, that of orientation totally fails. Such a result is in accordance with physical intuitions. In fact, the position of a small target can be easily localized in the far field, while its orientation can be correctly determined only in the near field.

18.6 Imaging of Extended Targets

18.6.1 Perturbations of a Disk

In this section we consider an extended target which is a perturbation of the disk D_0 with unit radius $r_0 = 1$. We set $\mu_0 = 1$, $\mu_\star = 5$, $\varepsilon_0 = 1$, and $\varepsilon_\star = 2$. We test the proposed shape reconstruction scheme. On the one hand, in the left picture of Fig. 18.19, we fix the frequency $\omega = 1$ and investigate the performance of the reconstruction method as a function of δ for $\partial D = \partial D_0 + \delta h\nu$ and $h = 1 + 2\cos(3\theta)$. On the other hand, in the right picture of Fig. 18.19, fixing $\omega = 1$ and $\delta = 0.1$, we also test the validity of the reconstruction method as a function of the number of oscillations p

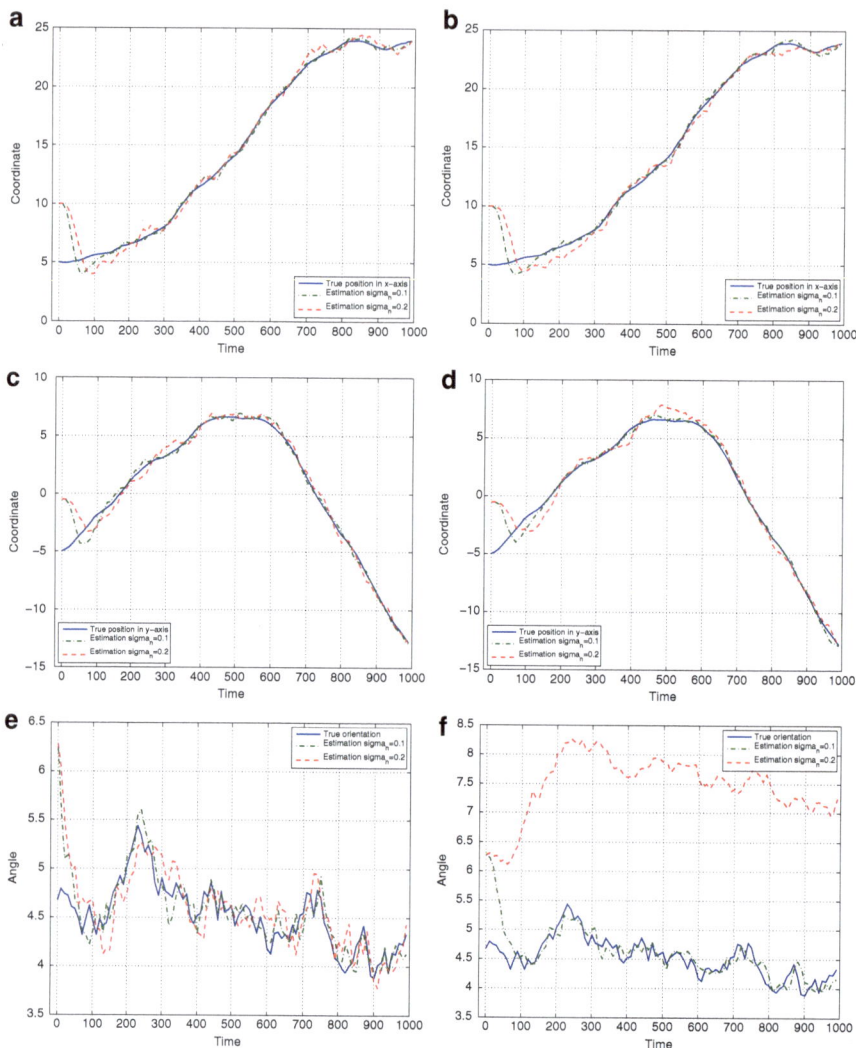

Fig. 18.18 Results of tracking using the configuration of Fig. 18.17 at different noise levels. *First row*: coordinate in x-axis. *Second row*: coordinate in y-axis. *Last row*: orientation. In the first column the target has size 10, while in the second column the target has size 1. The *solid line* always indicates the true system state

of the perturbation $h = 1 + 2\cos(p\theta)$. It turns out that if $p < 2k_0 r_0$, then the numerical scheme works well, as predicted by the resolution theory of Sect. 13.4. In Fig. 18.19, $|D \triangle D_0|$ and $|D \triangle D_6|$ are respectively the symmetric differences between D and D_0 and D and D_6.

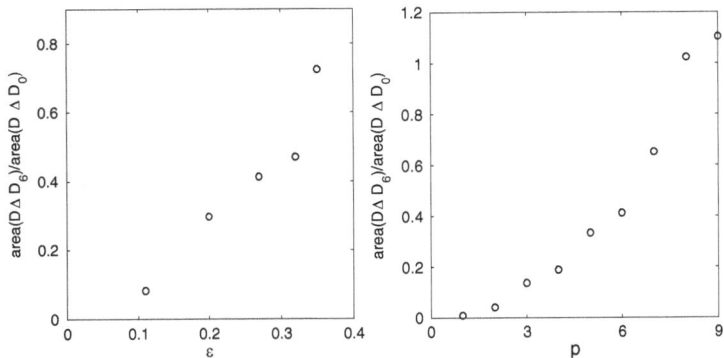

Fig. 18.19 The difference between initial and reconstructed shapes depends on the magnitude of perturbations δ, the radius of the target $r_0 = 1$, and the number p of oscillations of the perturbation. If ε is relatively small and $p < 2k_0 r_0$, then Method 1 works well. D_6 is the reconstructed target after 6 iterations

18.6.2 Nonlinear Optimization Problem

In Fig. 18.20 we reconstruct the shape of extended electromagnetic targets from MSR data. The targets are letters "A", "I", and "P". The algorithm is based on minimizing the discrepancy between the computed and measured MSR data using the optimal control approach described in Chap. 14.

18.7 Invisibility

18.7.1 GPT-Vanishing Structures

Figure 18.21 shows the results of computation of GPT-vanishing structures of order $N = 3$ and 6 with N being the number of layers. We emphasize that the conductivity fluctuates on coatings near the core. When $N = 3$, the maximal conductivity is 5.5158 and the minimal conductivity is 0.4264; When $N = 6$, they are 11.6836 and 0.1706.

Figure 18.22 shows the $\log_{10}$ of $(1, 1)$-entry of the conductivities (matrices) obtained by applying the transform (15.15) to cloaking structures, i.e., $\log_{10}((F_\delta)_*(\sigma \circ \Psi_{\frac{1}{\delta}}))_{11}$ for different N. The structures for different values of N are quite similar. They are obtained by segmenting the structure for $N = 0$ into concentric layers and multiplying the anisotropic conductivity in each layer by the corresponding value in the conductivity profile of the GPT-vanishing structure.

Let λ^k, $j = 1, 2, \ldots$, be the eigenvalues of $\Lambda_{B_2, \emptyset}[1]$ in decreasing order. Let λ^k_{WC}, λ^k_{NC} and λ^k_{EC} be the eigenvalues (in decreasing order) of $\Lambda_{B_2, B_1}[1]$,

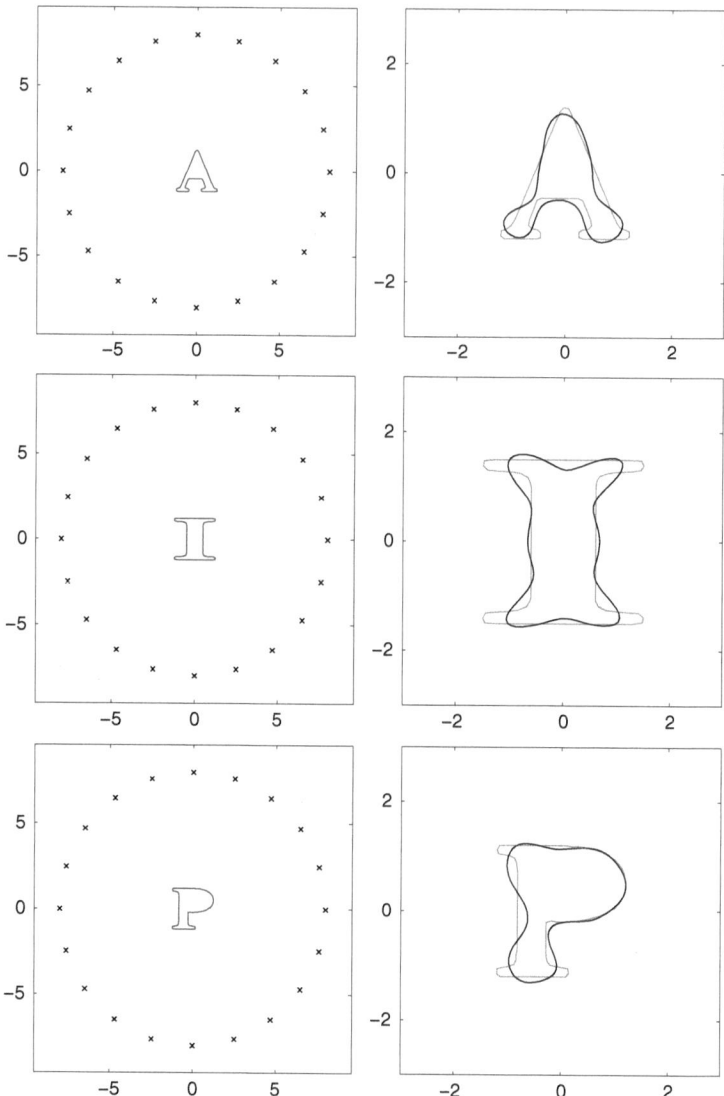

Fig. 18.20 *Left*: true images. *Right*: reconstructed images after 50 iterations. Imaging from the MSR matrix with 20 transmitters and 20 receivers. The frequency $\omega = 2$, $\mu = 1$, $\mu_\star = 1/4$, and $\varepsilon = \varepsilon_\star = 1$

$\Lambda_{B_2,B_1}[(F_\delta)_\ast(1)]$ and $\Lambda_{B_2,B_1}[(F_\delta)_\ast(\sigma \circ \Psi_{\frac{1}{\delta}})]$, respectively. (WC, NC and EC stand for "Without Cloaking", "Near Cloaking" and "Enhanced Cloaking", respectively.) Here $\Lambda_{B_2,B_1}[1]$ is the DtN map where the conductivity of the annulus $B_2 \setminus B_1$ is 1 and the core B_1 is insulated. Figure 18.23 shows the $\log_{10}$ of the discrepancies of the eigenvalues of the DtN maps for

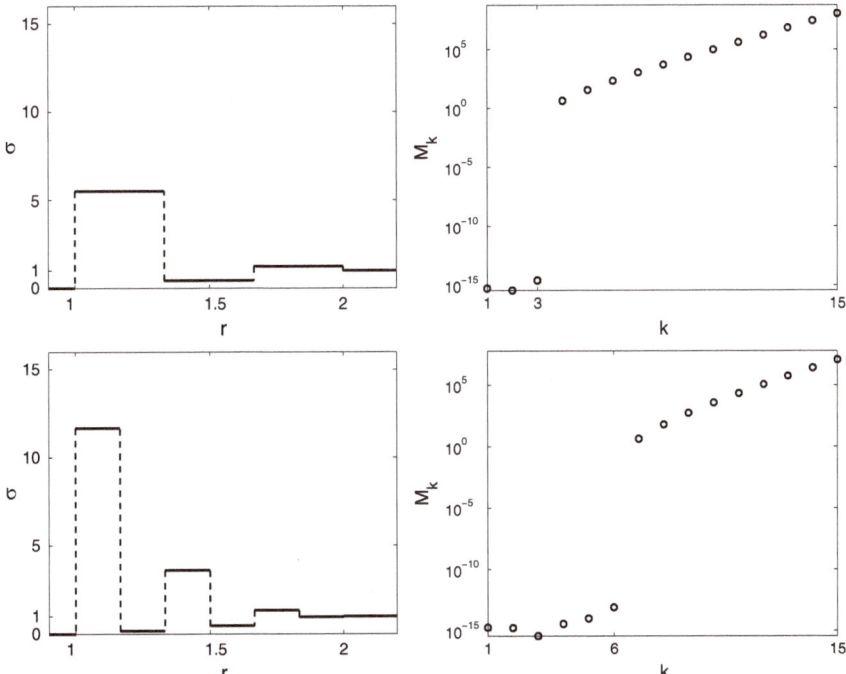

Fig. 18.21 Conductivity profile (*left*) and GPTs (*right*) of the GPT-vanishing structure of order N with the core conductivity being 0. The first row is when $N = 3$ and the second one for $N = 6$

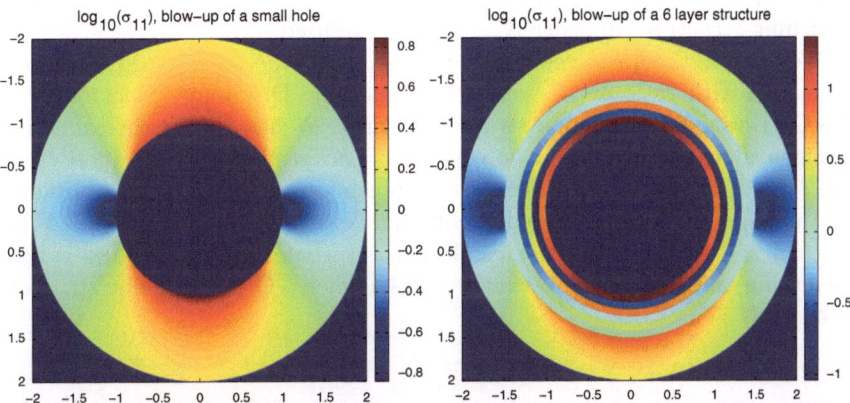

Fig. 18.22 $\log_{10}$ of $(1, 1)$-entry of the conductivities (matrices) obtained by applying the transform (15.15) to the GPT-vanishing structures of different N: $N = 0, 6$, from left to right. $N = 0$ means no coating

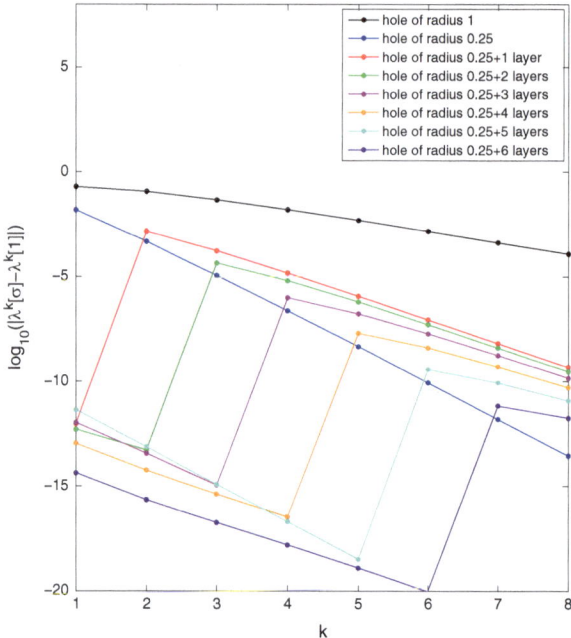

Fig. 18.23 Perturbations of the eigenvalues of the DtN map. The *black line* is for $\log_{10}|\lambda_{\mathrm{WC}}^k - \lambda^k|$, the blue one for $\log_{10}|\lambda_{\mathrm{NC}}^k - \lambda^k|$, and the other colored ones for $\log_{10}|\lambda_{\mathrm{EC}}^k - \lambda^k|$ for $N = 1,\ldots,6$

different structures. The black line represents $\log_{10}|\lambda_{\mathrm{WC}}^k - \lambda^k|$, the blue one $\log_{10}|\lambda_{\mathrm{NC}}^k - \lambda^k|$, and the other colored ones $\log_{10}|\lambda_{\mathrm{EC}}^k - \lambda^k|$ when GPT-vanishing structures of order $N = 1,\ldots,6$ are used.

We observe the quasi-geometric discrepancy of the perturbation triggered by a hole with ratio $\delta^2/4$. We also see that the DtN map associated with the GPT-vanishing cloaking structure of order N has almost the same first N eigenvalues as the one for homogeneous background with conductivity 1. Moreover, GPT-vanishing structures are much less visible than those obtained by the blow-up of an uncoated small hole. Note that the $\sup_k|\lambda_{\mathrm{NC}}^k - \lambda^k|$ is reached at $k = 1$ for the near cloaking (perturbation of eigenvalues is non-increasing) and $\sup_k|\lambda_{\mathrm{EC}}^k - \lambda^k|$ at $k = N+1$ when a GPT-vanishing structure of order N is used.

18.7.2 Scattering Coefficient Vanishing Structures

In this subsection we provide numerical examples of S-vanishing structures of order N at low frequencies, i.e., structures (μ,ε) of the form (15.30) satisfying (15.53). To do so, we use the gradient descent method to minimize over (μ,ε) the quantity

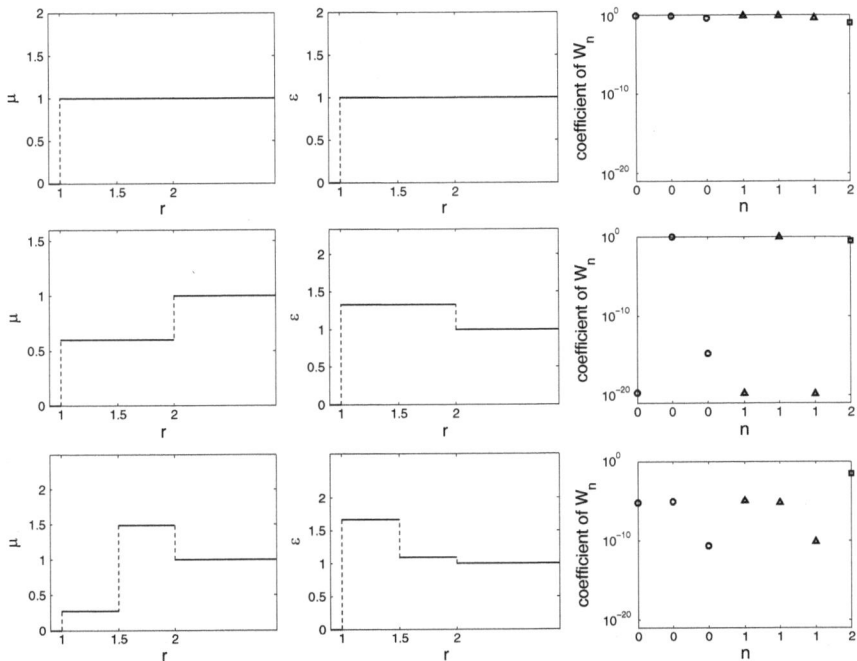

Fig. 18.24 Graphs of the first and second columns show profiles of permeability μ and the permittivity ε. The third column shows the coefficients of $[t^2, t^4, t^4 \log t]$ in the expansion of W_0 (represented by $(0,0,0)$) and W_1 (represented by $(1,1,1)$), and the coefficient of $[t^4]$ in W_2 (represented by 2)

$$|W_n^0[\mu, \varepsilon]|^2 + \sum_{l=1}^{(N-n)} \sum_{j=0}^{M_l} |W_n^{l,j}[\mu, \varepsilon]|^2 \,, \qquad (18.1)$$

where $W_n^0[\mu, \varepsilon]$ and $W_n^{l,j}[\mu, \varepsilon]$ are the coefficients of $W_n[\mu, \varepsilon, t]$ in (15.52). These coefficients are expressed in terms of Bessel functions and their derivatives.

It is quite challenging numerically to minimize the quantity in (18.1) for large N. In numerical examples we take $N = 2$. In this case one can show through tedious computations that the nonzero leading coefficients of $W_n[\mu, \varepsilon, t]$ are as follows:

- $[t^2, t^4, t^4 \log t]$ for $n = 0$, i.e., $W_0^{1,0}$, $W_0^{2,0}$, $W_0^{2,1}$;
- $[t^2, t^4, t^4 \log t]$ for $n = 1$, i.e., W_1^0, $W_1^{1,0}$, $W_1^{1,1}$;
- $[t^4]$ for $n = 2$, i.e., W_2^0.

Results of computations are given in Fig. 18.24 when we use 0, 1, 2 layers ($L = 0, 1, 2$). The computed material parameters are $\mu_1 = 0.6$ and $\varepsilon = 4/3$ when we use one layer, and $\mu = (1.4905, 0.27594)$, $\varepsilon = (1.09271, 1.6702)$ when

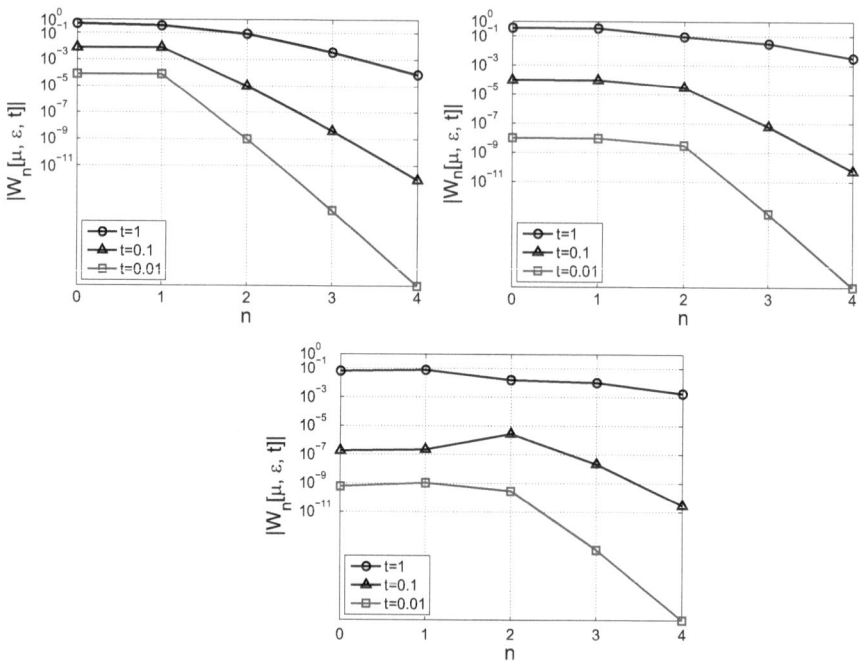

Fig. 18.25 The scattering coefficient, $W_n[\mu, \varepsilon, t]$, for $n = 0, \ldots, 4$ and $t = 1, 0.1, 0.01$ using the permeability and permittivity profiles computed in Fig. 18.24

we use two layers. The first two columns show the distribution of μ and ε for $L = 0, 1, 2$. The last column shows the values of coefficients of $W_n[\mu, \varepsilon, t]$: $W_0^{1,0}, W_0^{2,0}, W_0^{2,1}$ indicated by $(0, 0, 0)$; $W_1^0, W_1^{1,0}, W_1^{1,1}$ indicated by $(1, 1, 1)$; W_2^0 indicated by 2. The figures show that if $L = 0$ then none of the coefficients is zero; if $L = 1$, then the coefficient of t^2 is close to zero; if $L = 2$ all the coefficients are close to zero.

We then compute the scattering coefficients $W_n[\mu, \varepsilon, t]$ using the computed (μ, ε) for $L = 0, 1, 2$. Figure 18.25 shows the results of computations for $n = 0, \ldots, 4$ and $t = 1, 0.1, 0.01$. It clearly shows that $W_n[\mu, \varepsilon, t]$ for $n \leq 2$ gets smaller as the number of layers increases.

Figure 18.26 gives the (real part of the) outer full field u, the outer scattered field u_s (the field inside the cloak is not computed) and the scattering cross-section for the hole of unit radius with Neumann boundary conditions in the three following situations: uncloaked hole, usual near cloaking (with using layers), and S-vanishing structure of order 1. The source wave is a plane wave in the direction $(1\ 0)$ $(\theta = 0)$ at frequency $\omega = \pi$ and $\delta = 0.05$. A few remarks are in order:

- while the change of variables-based cloak shows good performance (it makes the hole with radius 1 looks like a hole with radius δ), using an

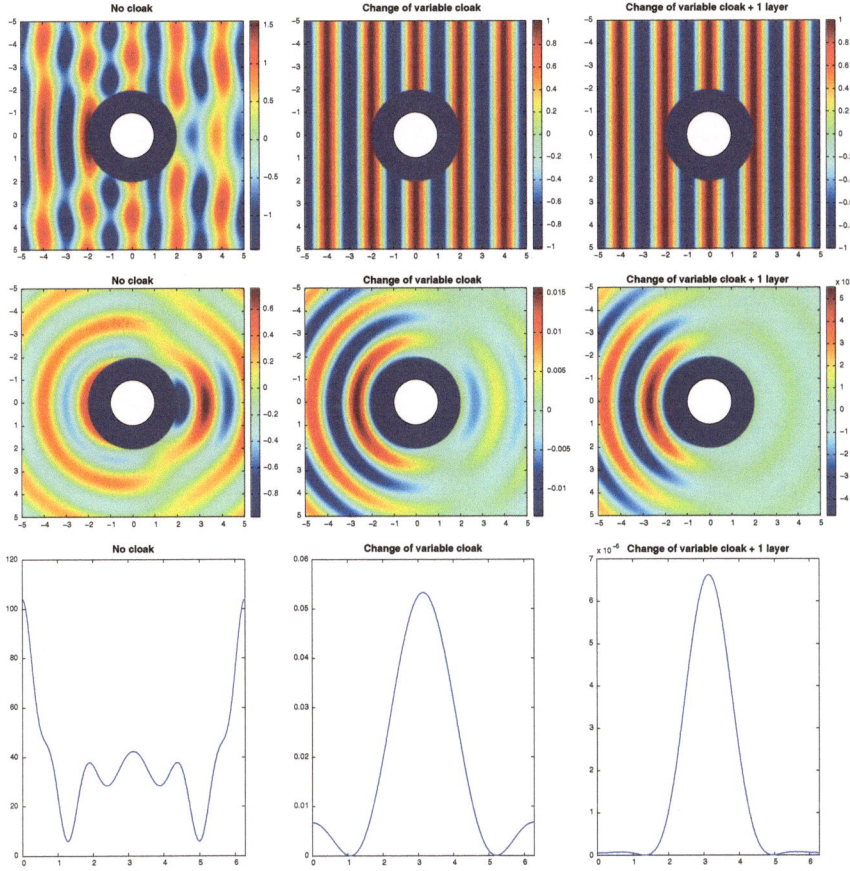

Fig. 18.26 *Top*: outer full field; *middle*: outer scattered field; *bottom*: scattering cross-section; *left*: uncloaked hole; *center*: usual change of variables-based cloak; *right*: S-vanishing cloaking structure of order one

S-vanishing structure greatly improves invisibility. This can be seen by inspecting the second and third rows in Fig. 18.26

- the uncloaked hole has stronger forward scattering (on the right), while the cloaked structures have stronger back-scattering (on the left). Actually, the cloaking structure reduces forward scattering with higher-order than backscattering. This seems to happen because the hole is of the size of the wavelength, thus leaves a shadow. The cloaked holes "appear" of size δ which is much smaller than the wavelength $2\pi/\omega$ and act mostly as weak reflectors.

References

1. M. Abramowitz, I. Stegun (eds.), *Handbook of Mathematical Functions* (National Bureau of Standards, Washington, 1964)
2. A. Alú, N. Engheta, Achieving transparency with plasmonic and metamaterial coatings. Phys. Rev. E **72**, 106623 (2005)
3. A. Alú, N. Engheta, Cloaking and transparency for collections of particles with metamaterial and plasmonic covers. Opt. Express **15**, 7578–7590 (2007)
4. H. Ammari, E. Bretin, J. Garnier, A. Wahab, Time reversal in attenuating acoustic media. Contemp. Math. **548**, 151–163 (2011)
5. H. Ammari, E. Bretin, J. Garnier, V. Jugnon, Coherent interferometry algorithms for photoacoustic imaging. SIAM J. Numer. Anal. **50**, 2259–2280 (2012)
6. H. Ammari, E. Bretin, V. Jugnon, A. Wahab, Photo-acoustic imaging for attenuating acoustic media, in *Mathematical Modeling in Biomedical Imaging II*. Lecture Notes in Mathematics, vol. 2035 (Springer, Berlin, 2011), pp. 57–84
7. H. Ammari, T. Boulier, J. Garnier, Modeling active electrolocation in weakly electric fish. SIAM J. Imaging Sci. **5**, 285–321 (2013)
8. H. Ammari, T. Boulier, J. Garnier, W. Jing, H. Kang, H. Wang, Target identification using dictionary matching of generalized polarization tensors. Found. Comput. Math. doi:10.1007/s10208-013-9168-6
9. H. Ammari, T. Boulier, J. Garnier, H. Kang, H. Wang, Tracking of a mobile target using generalized polarization tensors. SIAM J. Imaging Sci. **6**, 1477–1498 (2013)
10. H. Ammari, T. Boulier, J. Garnier, H. Wang, Shape identification and classification in electrolocation (submitted)
11. H. Ammari, D. Chung, H. Kang, H. Wang, Invariance properties of generalized polarization tensors and design of shape descriptors in three dimensions (submitted)
12. H. Ammari, G. Ciraolo, H. Kang, H. Lee, G. Milton, Spectral theory of a Neumann-Poincaré-type operator and analysis of cloaking due to anomalous localized resonance. Arch. Ration. Mech. Anal. **208**, 667–692 (2013)
13. H. Ammari, G. Ciraolo, H. Kang, H. Lee, G. Milton, Spectral theory of a Neumann-Poincaré-type operator and analysis of anomalous localized resonance II. Contemp. Math. (to appear)
14. H. Ammari, G. Ciraolo, H. Kang, H. Lee, G. Milton, Anomalous localized resonance using a folded geometry in three dimensions. Proc. R. Soc. A **469**, 20130048 (2013)
15. H. Ammari, G. Ciraolo, H. Kang, H. Lee, K. Yun, Spectral analysis of the Neumann-Poincaré operator and characterization of the stress blow-up in anti-plane elasticity. Arch. Ration. Mech. Anal. **208**, 275–304 (2013)

16. H. Ammari, Y. Deng, H. Kang, H. Lee, Reconstruction of inhomogeneous conductivities via generalized polarization tensors. Ann. IHP Anal. Non Lin. doi:10.1016/j-anihpc.2013.07.008

17. H. Ammari, P. Garapon, L. Guadarrama Bustos, H. Kang, Transient anomaly imaging by the acoustic radiation force. J. Differ. Equ. **249**, 1579–1595 (2010)

18. H. Ammari, J. Garnier, V. Jugnon, Detection, reconstruction, and characterization algorithms from noisy data in multistatic wave imaging (submitted)

19. H. Ammari, J. Garnier, V. Jugnon, H. Kang, Direct reconstruction methods in ultrasound imaging of small anomalies, in *Mathematical Modeling in Biomedical Imaging II*. Lecture Notes in Mathematics, vol. 2035 (Springer, Berlin, 2011), pp. 31–55

20. H. Ammari, J. Garnier, V. Jugnon, H. Kang, Stability and resolution analysis for a topological derivative based imaging functional. SIAM J. Control Optim. **50**, 48–76 (2012)

21. H. Ammari, J. Garnier, H. Kang, M. Lim, K. Sølna, Multistatic imaging of extended targets. SIAM J. Imaging Sci. **5** (2012), 564–600

22. H. Ammari, J. Garnier, H. Kang, M. Lim, S. Yu, Generalized polarization tensors for shape description. Numer. Math. doi:10.1007/s00211-013-0561-5 (to appear)

23. H. Ammari, J. Garnier, H. Kang, W.K. Park, K. Sølna, Imaging schemes for perfectly conducting cracks. SIAM J. Appl. Math. **71**, 68–91 (2011)

24. H. Ammari, J. Garnier, K. Sølna, A statistical approach to target detection and localization in the presence of noise. Waves Random Complex Media **22**, 40–65 (2012)

25. H. Ammari, J. Garnier, K. Sølna, Resolution and stability analysis in full-aperture, linearized conductivity and wave imaging. Proc. Am. Math. Soc. **141**, 3431–3446 (2013)

26. H. Ammari, E. Iakovleva, D. Lesselier, Two numerical methods for recovering small electromagnetic inclusions from scattering amplitude at a fixed frequency. SIAM J. Sci. Comput. **27**, 130–158 (2005)

27. H. Ammari, H. Kang, High-order terms in the asymptotic expansions of the steady-state voltage potentials in the presence of conductivity inhomogeneities of small diameter. SIAM J. Math. Anal. **34**, 1152–1166 (2003)

28. H. Ammari, H. Kang, Properties of generalized polarization tensors. SIAM Multiscale Model. Simul. **1**, 335–348 (2003)

29. H. Ammari, H. Kang, *Reconstruction of Small Inhomogeneities from Boundary Measurements*. Lecture Notes in Mathematics, vol. 1846 (Springer, Berlin, 2004)

30. H. Ammari, H. Kang, Boundary layer techniques for solving the Helmholtz equation in the presence of small inhomogeneities. J. Math. Anal. Appl. **296**, 190–208 (2004)

31. H. Ammari, H. Kang, *Polarization and Moment Tensors: With Applications to Inverse Problems and Effective Medium Theory*. Applied Mathematical Sciences, vol. 162 (Springer, New York, 2007)

32. H. Ammari, H. Kang, E. Kim, M. Lim, Reconstruction of closely spaced small inclusions. SIAM J. Numer. Anal. **42**, 2408–2428 (2005)

33. H. Ammari, H. Kang, M. Lim, Polarization tensors and their applications, Proceedings of the second International Conference on Inverse Problems: recent developments and numerical approaches, Shanghai, 2004. J. Phys. Conf. Ser. **12**, 13–22 (2005)

34. H. Ammari, H. Kang, M. Lim, H. Lee, Enhancement of near-cloaking using generalized polarization tensors vanishing structures. Part I: the conductivity problem. Commun. Math. Phys. **317**, 253–266 (2013)

35. H. Ammari, H. Kang, M. Lim, H. Lee, Enhancement of near-cloaking. Part II: the Helmholtz equation. Commun. Math. Phys. **317**, 485–502 (2013)

36. H. Ammari, H. Kang, M. Lim, H. Lee, Enhancement of near cloaking for the full Maxwell equations. SIAM J. Appl. Math. (to appear)

37. H. Ammari, H. Kang, M. Lim, H. Zribi, The generalized polarization tensors for resolved imaging. Part I: shape reconstruction of a conductivity inclusion. Math. Comput. **81**, 367–386 (2012)

38. H. Ammari, H. Kang, E. Kim, J.-Y. Lee, The generalized polarization tensors for resolved imaging. Part II: shape and electromagnetic parameters reconstruction of an electromagnetic inclusion from multistatic measurements. Math. Comput. **81**, 839–860 (2012)

39. H. Ammari, A. Khelifi, Electromagnetic scattering by small dielectric inhomogeneities. J. Math. Pures Appl. **82**, 749–842 (2003)

40. H. Ammari, J.K. Seo, An accurate formula for the reconstruction of conductivity inhomogeneities. Adv. Appl. Math. **30**, 679–705 (2003)

41. H. Ammari, M.P. Tran, H. Wang, Shape identification and classification in echolocation, preprint, 2013

42. J. Baik, J.W. Silverstein, Eigenvalues of large sample covariance matrices of spiked population models. J. Multivar. Anal. **97**, 1382–1408 (2006)

43. G. Bao, H. Liu, Nearly cloaking the full Maxwell equations, arXiv:1210.2447v1

44. C. Bardos, A mathematical and deterministic analysis of the time-reversal mirror, in *Inside Out: Inverse Problems and Applications*. Mathematical Sciences Research Institute Publications, vol. 47 (Cambridge University Press, Cambridge, 2003), pp. 381–400

45. C. Bardos, M. Fink, Mathematical foundations of the time reversal mirror. Asymp. Anal. **29**, 157Ũ182 (2002)

46. F. Benaych-Georges, R.R. Nadakuditi, The eigenvalues and eigenvectors of finite, low rank perturbations of large random matrices. Adv. Math. **227**, 494–521 (2011)

47. A.P. Berens, NDE reliability data analysis, in *ASM Handbook*, vol. 17 (ASM International, Materials Park, 1989), pp. 689–701

48. E. Beretta, E. Francini, Asymptotic formulas for perturbations in the electromagnetic fields due to the presence of thin inhomogeneities. Contemp. Math. **333**, 49–62 (2003)

49. J. Bergh, J. Löfström, *Interpolation Spaces. An Introduction*. Grundlehren der Mathematischen Wissenschaften, vol. 223 (Springer, Berlin, 1976)

50. N. Bleistein, J.K. Cohen, J.W. Stockwell Jr., *Mathematics of Multidimensional Seismic Imaging, Migration, and Inversion* (Springer, New York, 2001)

51. L. Borcea, G. Papanicolaou, C. Tsogka, Theory and applications of time reversal and interferometric imaging. Inverse Probl. **19**, 134–164 (2003)

52. L. Borcea, G. Papanicolaou, C. Tsogka, Interferometric array imaging in clutter. Inverse Probl. **21**, 1419–1460 (2005)

53. L. Borcea, G.C. Papanicolaou, C. Tsogka, J.G. Berrymann, Imaging and time reversal in random media. Inverse Probl. **18**, 1247–1279 (2002)

54. M. Born, E. Wolf, *Principles of Optics: Electromagnetic Theory of Propagation, Interference and Diffraction of Light*, 6th edn. (Cambridge University Press, Cambridge, 1997)

55. L. Breiman, *Probability* (Addison-Wesley, Reading, 1968); reprinted by Society for Industrial and Applied Mathematics, Philadelphia, 1992

56. W.L. Briggs, V.E. Henson, *The DFT: An Owner Manual for the Discrete Fourier Transform* (Society for Industrial and Applied Mathematics, Philadelphia, 1995)

57. M. Brühl, M. Hanke, M.S. Vogelius, A direct impedance tomography algorithm for locating small inhomogeneities. Numer. Math. **93**, 635–654 (2003)

58. K. Bryan, T. Leise, Impedance imaging, inverse problems, and Harry Potter's Cloak. SIAM Rev. **52**, 359–377 (2010)

59. M. Burger, S.J. Osher, A survey on level set methods for inverse problems and optimal design. Eur. J. Appl. Math. **16**, 263–301 (2005)

60. Y. Capdeboscq, A.B. Karrman, J.C. Nédélec, Numerical computation of approximate generalized polarization tensors. Appl. Anal. **91**, 1189–1203 (2012)

61. Y. Capdeboscq, M.S. Vogelius, Optimal asymptotic estimates for the volume of internal inhomogeneities in terms of multiple boundary measurements. Math. Model. Numer. Anal. **37**, 227–240 (2003)

62. D.J. Cedio-Fengya, S. Moskow, M.S. Vogelius, Identification of conductivity imperfections of small diameter by boundary measurements: continuous dependence and computational reconstruction. Inverse Probl. **14**, 553–595 (1998)

63. A. Chai, M. Moscoso, G. Papanicolaou, Array imaging using intensity-only measurements. Inverse Probl. **27**, 015005 (2011)

64. A. Chai, M. Moscoso, G. Papanicolaou, Robust imaging of localized scatterers using the singular value decomposition and l_1 minimization. Inverse Probl. **29**, 025016 (2013)

65. D.H. Chambers, J.G. Berryman, Analysis of the time-reversal operator for a small spherical scatterer in an electromagnetic field. IEEE Trans. Antennas Propag. **52**, 1729–1738 (2004)

66. D.H. Chambers, J.G. Berryman, Time-reversal analysis for scatterer characterization. Phys. Rev. Lett. **92**, 023902-1 (2004)

67. H. Cheng, W. Crutchfield, Z. Gimbutas, L. Greengard, J. Huang, V. Rokhlin, N. Yarvin, J. Zhao, Remarks on the implementation of the wideband FMM for the Helmholtz equation in two dimensions. Contemp. Math. **408**, 99–110 (2006)

68. D. Colton, R. Kress, *Inverse Acoustic and Electromagnetic Scattering Theory*. Applied Mathematical Sciences, vol. 93 (Springer, New York, 1992)

69. J.L. Crassidis, J.L. Junkins, *Optimal Estimation of Dynamic Systems* (CRC Press, Boca Raton, 2004)

70. A.C. Davison, *Statistical Models* (Cambridge University Press, Cambridge, 2003)

71. J.A. Decker Jr., Hadamard-transform image scanning. Appl. Opt. **9**, 1392–1395 (1970)

72. A.J. Devaney, A filtered backpropagation algorithm for diffraction tomography. Ultrason. Imaging **4**, 336–350 (1982)

73. A.J. Devaney, Time reversal imaging of obscured targets from multistatic data. IEEE Trans. Antennas Propag. **523**, 1600–1610 (2005)

74. H.W. Engl, M. Hanke, A. Neubauer, *Regularization of Inverse Problems* (Kluwer, Dordrecht, 1996)

75. A. Erdélyi, W. Magnus, F. Oberhettinger, F.G. Tricomi (eds.), *Higher Transcendental Functions*, vol. II (McGraw-Hill, New York, 1953)

76. M. Fink, Time-reversal acoustics. Contemp. Math. **408**, 151–179 (2006)

77. G.B. Folland, *Introduction to Partial Differential Equations* (Princeton University Press, Princeton, 1976)

78. G.B. Folland, *Real Analysis: Modern Techniques and Their Applications*, 1st edn. (Wiley, New York, 1984)

79. J.P. Fouque, J. Garnier, G. Papanicolaou, K. Sølna, *Wave Propagation and Time Reversal in Randomly Layered Media* (Springer, New York, 2007)

80. A. Friedman, M.S. Vogelius, Identification of small inhomogeneities of extreme conductivity by boundary measurements: a theorem on continuous dependence. Arch. Ration. Mech. Anal. **105**, 299–326 (1989)

81. J. Garnier, Use of random matrix theory for target detection, localization, and reconstruction. Contemp. Math. **548**, 1–19 (2011)

82. J. Garnier, G. Papanicolaou, Resolution analysis for imaging with noise. Inverse Probl. **26**, 074001 (2010)

83. A. Greenleaf, Y. Kurylev, M. Lassas, G. Uhlmann, Approximate quantum cloaking and almost trapped states. Phys. Rev. Lett. **101**, 220404 (2008)

84. A. Greenleaf, Y. Kurylev, M. Lassas, G. Uhlmann, Full-wave invisibility of active devices at all frequencies. Commun. Math. Phys. **275**, 749–789 (2007)

85. A. Greenleaf, Y. Kurylev, M. Lassas, G. Uhlmann, Cloaking devices, electromagnetic wormholes, and transformation optics. SIAM Rev. **51**, 3–33 (2009)

86. A. Greenleaf, Y. Kurylev, M. Lassas, G. Uhlmann, Cloaking a sensor via transformation optics. Phys. Rev. E **83**, 016603 (2011)

87. A. Greenleaf, Y. Kurylev, M. Lassas, G. Uhlmann, Approximate quantum and acoustic cloaking. J. Spectral Theory **1**, 27–80 (2011)

88. D. Grieser, The plasmonic eigenvalue problem, arXiv:1208:3120

89. F. Guevara Vasquez, G.W. Milton, D. Onofrei, Active exterior cloaking for the 2D Laplace and Helmholtz Equations. Phys. Rev. Lett. **103**, 073901 (2009)
90. F. Guevara Vasquez, G.W. Milton, D. Onofrei, Broadband exterior cloaking. Opt. Express **17**, 14800–14805 (2009)
91. J. Hadamard, Résolution d'une question relative aux déterminants. Bull. Sci. Math. **17**, 30–31 (1893)
92. M. Hanke, A. Neubauer, O. Scherzer, A convergence analysis of the Landweber iteration for nonlinear ill-posed problems. Numer. Math. **72**, 21–37 (1995)
93. D.J. Hansen, M.S. Vogelius, High frequency perturbation formulas for the effect of small inhomogeneities. J. Phys. Conf. Ser. **135**, 012106 (2008)
94. H. Hashemi, A. Oskooi, J.D. Joannopoulos, S.G. Johnson, General scaling limitations of ground-plane and isolated-object cloaks. Phys. Rev. A **84**, 023815 (2011)
95. J. Helsing, K.-M. Perfekt, On the polarizability and capacitance of the cube. Appl. Comput. Harmon. Anal. **34**, 445–468 (2013)
96. M.V. de Hoop, L. Qiu, O. Scherzer, Local analysis of inverse problems: Hölder stability and iterative reconstruction. Inverse Probl. **28**, 045001 (2012)
97. R.A. Horn, C.R. Johnson, *Matrix Analysis* (Cambridge University Press, Cambridge, 1985)
98. S. Hou, K. Sølna, H. Zhao, A direct imaging algorithm for extended targets. Inverse Probl. **22**, 1151–1178 (2006)
99. S. Hou, K. Sølna, H. Zhao, Imaging of location and geometry for extended targets using the response matrix. J. Comput. Phys. **199**, 317–338 (2004)
100. V. Isakov, *Inverse Source Problems* (American Mathematical Society, Providence, 1990)
101. D.J. Jobson, S.J. Katzberg, R.B. Spiers Jr., Signal-to-noise analysis and evaluation of the Hadamard imaging technique, NASA Technical Note, D-8377, 1977
102. I.M. Johnstone, On the distribution of the largest eigenvalue in principal components analysis. Ann. Stat. **29**, 295–327 (2001)
103. J. Kaipio, E. Somersalo, *Statistical and Computational Inverse Problems*. Applied Mathematical Sciences, vol. 160 (Springer, New York, 2005)
104. K. Kalimeris, O. Scherzer, Photoacoustic imaging in attenuating acoustic media based on strongly causal models. Math. Methods Appl. Sci. **36**, 2254–2264 (2013)
105. R.E. Kalman, A new approach to linear filtering and prediction problems. Trans. ASME—J. Basic Eng. **82**, 35–45 (1960)
106. H. Kang, G.W. Milton, On conjectures of Pólya-Szegö and Eshelby, in *Inverse Problems, Multi-scale Analysis and Effective Medium Theory*. Contemporary Mathematics, vol. 408 (Amer. Math. Soc., Rhode Island, 2006), pp. 75–80
107. H. Kang, G.W. Milton, Solutions to the Pólya-Szegö conjecture and the weak Eshelby conjecture. Arch. Ration. Mech. Anal. **188**, 93–116 (2008)
108. H. Kang, J.K. Seo, Layer potential technique for the inverse conductivity problem. Inverse Probl. **12**, 267–278 (1996)
109. H. Kang, J.K. Seo, Identification of domains with near-extreme conductivity: global stability and error estimates. Inverse Probl. **15**, 851–867 (1999)
110. H. Kang, J.K. Seo, Inverse conductivity problem with one measurement: uniqueness of balls in R^3. SIAM J. Appl. Math. **59**, 1533–1539 (1999)
111. H. Kang, J.K. Seo, Recent progress in the inverse conductivity problem with single measurement, in *Inverse Problems and Related Fields* (CRC Press, Boca Raton, 2000), pp. 69–80
112. S.M. Kay, *Fundamentals of Statistical Signal Processing, Detection Theory* (Prentice-Hall, Englewood Cliffs, 1998)
113. J.B. Keller, R.M. Lewis, Asymptotic methods for partial differential equations: the reduced wave equation and Maxwell's equations, in *Surveys in Applied Mathematics*, vol. 1, ed. by J.B. Keller, D.W. McLaughlin, G.C. Papanicolaou (Plenum Press, New York, 1995), pp. 1–82

114. O.D. Kellogg, *Foundations of Potential Theory* (Dover, New York, 1953)
115. D. Khavinson, M. Putinar, H.S. Shapiro, Poincaré's variational problem in potential theory. Arch. Ration. Mech. Anal. **185**, 143–184 (2007)
116. L. Klimes, Correlation functions of random media. Pure Appl. Geophys. **159**, 1811–1831 (2002)
117. R.V. Kohn, J. Lu, B. Schweizer, M.I. Weinstein, A variational perspective on cloaking by anomalous localized resonance, preprint, arXiv:1210.4823
118. R.V. Kohn, D. Onofrei, M.S. Vogelius, M.I. Weinstein, Cloaking via change of variables for the Helmholtz equation. Commun. Pure Appl. Math. **63**, 973–1016 (2010)
119. R.V. Kohn, H. Shen, M.S. Vogelius, M.I. Weinstein, Cloaking via change of variables in electric impedance tomography. Inverse Probl. **24**, 015016 (2008)
120. O. Kwon, J.K. Seo, J.R. Yoon, A real-time algorithm for the location search of discontinuous conductivities with one measurement. Commun. Pure Appl. Math. **55**, 1–29 (2002)
121. L. Landweber, An iteration formula for Fredholm integral equations of the first kind. Am. J. Math. **73**, 615–624 (1951)
122. M. Lassas, T. Zhou, Two dimensional invisibility cloaking for Helmholtz equation and non-local boundary conditions. Math. Res. Lett. **18**, 473–488 (2011)
123. C.L. Lawson, R.J. Hanson, *Solving Least Squares Problems*. Classics in Applied Mathematics, vol. 15 (Society for Industrial and Applied Mathematics, Philadelphia, 1995). Revised reprint of the 1974 original
124. R. Lipton, Inequalities for electric and elastic polarization tensors with applications to random composites. J. Mech. Phys. Solids **41**, 809–833 (1993)
125. H. Liu, Virtual reshaping and invisibility in obstacle scattering. Inverse Probl. **25**, 044006 (2009)
126. S. Mallat, *A Wavelet Tour of Signal Processing* (Academic, San Diego, 1998)
127. V.A. Marcenko, L.A. Pastur, Distributions of eigenvalues of some sets of random matrices. Math. USSR-Sb. **1**, 507–536 (1967)
128. T.D. Mast, A. Nachman, R.C. Waag, Focusing and imagining using eigenfunctions of the scattering operator. J. Acoust. Soc. Am. **102**, 715–725 (1997)
129. W. McLean, *Strongly Elliptic Systems and Boundary Integral Equations* (Cambridge University Press, Cambridge, 2000)
130. M.L. Mehta, *Random Matrices* (Academic, San Diego, 1991)
131. G.W. Milton, *The Theory of Composites*. Cambridge Monographs on Applied and Computational Mathematics (Cambridge University Press, Cambridge, 2001)
132. G. Milton, N.A. Nicorovici, On the cloaking effects associated with anomalous localized resonance. Proc. R. Soc. A **462**, 3027–3059 (2006)
133. G. Milton, N.A. Nicorovici, R.C. McPhedran, V.A. Podolskiy, A proof of superlensing in the quasistatic regime, and limitations of superlenses in this regime due to anomalous localized resonance. Proc. R. Soc. A **461**, 3999–4034 (2005)
134. F. Natterer, *The Mathematics of Computerized Tomography*. Classics in Applied Mathematics (Society for Industrial and Applied Mathematics, Philadelphia, 2001)
135. J.C. Nédélec, *Acoustic and Electromagnetic Equations. Integral Representations for Harmonic Problems*. Applied Mathematical Sciences, vol. 144 (Springer, New York, 2001)
136. II.M. Nguyen, Cloaking via change of variables for the Helmholtz equation in the whole space. Commun. Pure Appl. Math. **63**, 1505–1524 (2010)
137. H. Nguyen, M. Vogelius, A representation formula for the voltage perturbations caused by diametrically small conductivity inhomogeneities. Proof of uniform validity. Ann. I. H. Poincaré-AN **26**, 2283–2315 (2009)
138. N.-A.P. Nicorovici, R.C. McPhedran, G.W. Milton, Optical and dielectric properties of partially resonant composites. Phys. Rev. B **49**, 8479–8482 (1994)
139. N.-A.P. Nicorovici, G.W. Milton, R.C. McPhedran, L.C. Botten, Quasistatic cloaking of two-dimensional polarizable discrete systems by anomalous resonance. Opt. Express **15**, 6314–6323 (2007)

140. J. Nocedal, S.J. Wright, *Numerical Optimization* (Springer, New York, 1999)
141. H.J. Nussbaumer, *Fast Fourier Transform and Convolution Algorithms* (Springer, New York, 1982)
142. S. Osher, J.A. Sethian, Fronts propagating with curvature-dependent speed: algorithms based on Hamilton-Jacobi formulations. J. Comput. Phys. **79**, 12–49 (1988)
143. D. Paul, Asymptotics of sample eigenstructure for a large dimensional spiked covariance model. Stat. Sinica **17**, 1617–1642 (2007)
144. J.B. Pendry, D. Schurig, D.R. Smith, Controlling electromagnetic fields. Science **312**, 1780–1782 (2006)
145. I.G. Petrovsky, *Lectures on Partial Differential Equations* (Dover, New York, 1954)
146. C.W. Qiu, L. Hu, B. Zhang, B.I. Wu, S.G. Johnson, J.D. Joannopoulos, Spherical cloaking using nonlinear transformations for improved segmentation into concentric isotropic coatings. Opt. Express **17**, 13467–13478 (2009)
147. F. Santosa, A level-set approach for inverse problems involving obstacles. ESAIM: COCV **1**, 17–33 (1996)
148. O. Scherzer, M. Grasmair, H. Grossauer, M. Haltmeier, F. Lenzen, *Variational Methods in Imaging*. Applied Mathematical Sciences, vol. 167 (Springer, New York, 2009)
149. A. Shabalin, A. Nobel, Reconstruction of a low-rank matrix in the presence of Gaussian noise, preprint, arXiv:1007.4148v1
150. D. Slepian, Some comments on Fourier analysis, uncertainty and modeling. SIAM Rev. **25**, 379–393 (1983)
151. N.J.A. Sloane, T. Fine, P.G. Phillips, M. Harwit, Codes for multiplex spectrometry. Appl. Opt. **8**, 2103–2106 (1969)
152. G.W. Stewart, Perturbation theory for the singular value decomposition, in *SVD and Signal Processing, II: Algorithms, Analysis and Applications* (Elsevier, Amsterdam, 1990), pp. 99–109
153. M.E. Taylor, *Partial Differential Equations I. Basic Theory*. Applied Mathematical Sciences, vol. 115 (Springer, New York, 1996)
154. C.W. Therrien, *Discrete Random Signals and Statistical Signal Processing* (Prentice-Hall, Englewood Cliffs, 1992)
155. S. Tretyakov, P. Alitalo, O. Luukkonen, C. Simovski, Broadband electromagnetic cloaking of long cylindrical objects. Phys. Rev. Lett. **103**, 103905 (2009)
156. Y.A. Urzhumov, N.B. Kundtz, D.R. Smith, J.B. Pendry, Cross-section comparisons of cloaks designed by transformation optical and optical conformal mapping approaches. J. Opt. **13**, 024002 (2011)
157. M.S. Vogelius, D. Volkov, Asymptotic formulas for perturbations in the electromagnetic fields due to the presence of inhomogeneities. Math. Model. Numer. Anal. **34**, 723–748 (2000)
158. G.N. Watson, *Theory of Bessel Functions*, 2nd edn. (Cambridge University Press, Cambridge, 1944)
159. P.-A. Wedin, Perturbation bounds in connection with singular value decomposition. BIT Numer. Math. **12**, 99–111 (1972)
160. G. Welch, G. Bishop, An introduction to the Kalman filter (Technical Report 95-041), University of North Carolina at Chapel Hill, 2001 & SIGGRAPH 2001, Los Angeles, CA, 12–17 August (ACM, New York, 2001)
161. R. Wong, Asymptotic expansion of $\int_0^{\pi/2} J_\nu^2(\lambda \cos\theta) d\theta$. Math. Comput. **50**, 229–234 (1988)
162. H. Zhao, Analysis of the response matrix for an extended target. SIAM J. Appl. Math. **64**, 725–745 (2004)

Index

H. Ammari et al., *Mathematical and Statistical Methods for Multistatic Imaging*,   359
Lecture Notes in Mathematics 2098, DOI 10.1007/978-3-319-02585-8,
© Springer International Publishing Switzerland 2013

LECTURE NOTES IN MATHEMATICS   🐴 Springer

Edited by J.-M. Morel, B. Teissier; P.K. Maini

Editorial Policy (for the publication of monographs)

1. Lecture Notes aim to report new developments in all areas of mathematics and their applications - quickly, informally and at a high level. Mathematical texts analysing new developments in modelling and numerical simulation are welcome.

   Monograph manuscripts should be reasonably self-contained and rounded off. Thus they may, and often will, present not only results of the author but also related work by other people. They may be based on specialised lecture courses. Furthermore, the manuscripts should provide sufficient motivation, examples and applications. This clearly distinguishes Lecture Notes from journal articles or technical reports which normally are very concise. Articles intended for a journal but too long to be accepted by most journals, usually do not have this "lecture notes" character. For similar reasons it is unusual for doctoral theses to be accepted for the Lecture Notes series, though habilitation theses may be appropriate.

2. Manuscripts should be submitted either online at www.editorialmanager.com/lnm to Springer's mathematics editorial in Heidelberg, or to one of the series editors. In general, manuscripts will be sent out to 2 external referees for evaluation. If a decision cannot yet be reached on the basis of the first 2 reports, further referees may be contacted: The author will be informed of this. A final decision to publish can be made only on the basis of the complete manuscript, however a refereeing process leading to a preliminary decision can be based on a pre-final or incomplete manuscript. The strict minimum amount of material that will be considered should include a detailed outline describing the planned contents of each chapter, a bibliography and several sample chapters.

   Authors should be aware that incomplete or insufficiently close to final manuscripts almost always result in longer refereeing times and nevertheless unclear referees' recommendations, making further refereeing of a final draft necessary.

   Authors should also be aware that parallel submission of their manuscript to another publisher while under consideration for LNM will in general lead to immediate rejection.

3. Manuscripts should in general be submitted in English. Final manuscripts should contain at least 100 pages of mathematical text and should always include

   – a table of contents;
   – an informative introduction, with adequate motivation and perhaps some historical remarks: it should be accessible to a reader not intimately familiar with the topic treated;
   – a subject index: as a rule this is genuinely helpful for the reader.

   For evaluation purposes, manuscripts may be submitted in print or electronic form (print form is still preferred by most referees), in the latter case preferably as pdf- or zipped ps-files. Lecture Notes volumes are, as a rule, printed digitally from the authors' files. To ensure best results, authors are asked to use the LaTeX2e style files available from Springer's web-server at:

   ftp://ftp.springer.de/pub/tex/latex/svmonot1/ (for monographs) and
   ftp://ftp.springer.de/pub/tex/latex/svmultt1/ (for summer schools/tutorials).

Additional technical instructions, if necessary, are available on request from lnm@springer.com.

4. Careful preparation of the manuscripts will help keep production time short besides ensuring satisfactory appearance of the finished book in print and online. After acceptance of the manuscript authors will be asked to prepare the final LaTeX source files and also the corresponding dvi-, pdf- or zipped ps-file. The LaTeX source files are essential for producing the full-text online version of the book (see http://www.springerlink.com/openurl.asp?genre=journal&issn=0075-8434 for the existing online volumes of LNM). The actual production of a Lecture Notes volume takes approximately 12 weeks.

5. Authors receive a total of 50 free copies of their volume, but no royalties. They are entitled to a discount of 33.3 % on the price of Springer books purchased for their personal use, if ordering directly from Springer.

6. Commitment to publish is made by letter of intent rather than by signing a formal contract. Springer-Verlag secures the copyright for each volume. Authors are free to reuse material contained in their LNM volumes in later publications: a brief written (or e-mail) request for formal permission is sufficient.

Addresses:

Professor J.-M. Morel, CMLA,
École Normale Supérieure de Cachan,
61 Avenue du Président Wilson, 94235 Cachan Cedex, France
E-mail: morel@cmla.ens-cachan.fr

Professor B. Teissier, Institut Mathématique de Jussieu,
UMR 7586 du CNRS, Équipe "Géométrie et Dynamique",
175 rue du Chevaleret
75013 Paris, France
E-mail: teissier@math.jussieu.fr

For the "Mathematical Biosciences Subseries" of LNM:

Professor P. K. Maini, Center for Mathematical Biology,
Mathematical Institute, 24-29 St Giles,
Oxford OX1 3LP, UK
E-mail: maini@maths.ox.ac.uk

Springer, Mathematics Editorial, Tiergartenstr. 17,
69121 Heidelberg, Germany,
Tel.: +49 (6221) 4876-8259

Fax: +49 (6221) 4876-8259
E-mail: lnm@springer.com